INTERNATIONAL

# Longman Mathematics for IGCSE Book 1

D A Turner, I A Potts,
W R J Waite, B V Hony

PEARSON
Longman

Pearson Education Limited
Edinburgh Gate
Harlow
Essex
CM20 2JE
England

Fifth impression 2007

ISBN: 978-1-4058-0211-6

Prepared for publication by Peter and Jan Simmonett.

Printed in China
CTPSC/05

## Acknowledgements

We are grateful to Edexcel and OCR for permission to reproduce copyright examination questions. Edexcel and OCR can accept no responsibility whatsoever for the accuracy or method of working in the answers, where given.

We are grateful to the following for permission to reproduce photographs:

**Corbis:** p.203 (Tibor Bognár), p.305 (Michael Prince); **Kobal Collection:** p.151; **Lonely Planet Images:** p.198 (Donald C & Priscilla Alexander Eastman)

Every effort has been made to trace the copyright holders and we apologise in advance for an unintentional omissions. We would be pleased to insert the appropriate acknowledgment in any subsequent edition of this publication.

Picture research by Sandra Hilsdon

# Contents

# Course Structure

## Unit 4

### Number 4
Inverse percentages
Rounding
Upper and lower bounds
Estimating
Estimating using standard
  form

### Algebra 4
Change of subject
Using formulae

### Graphs 4
Quadratic graphs
  $y = ax^2 + bx + c$
Solving $ax^2 + bx + c = 0$
  using graphs

### Shape and space 4
Circles
Similar triangles
Pythagoras' theorem

### Handling data 4
Probability

### Summary 4

### Examination practice 4

## Unit 5

### Number 5
Proportion
Positive integer powers of
  numbers
Simple recurring decimals

### Algebra 5
Multiplying brackets
Factorising quadratic
  expressions with two terms
Factorising quadratic
  expressions with three terms
Solving quadratic equations
  by factorisation
Problems leading to quadratic
  equations

### Sequences 5
Continuing sequences
Formula for sequences
The difference method
Finding a formula for a
  sequence

### Shape and space 5
Basic transformations
Combining transformations
Enlargements

### Handling data 5
Distributions
Quartiles
Measures of spread
Cumulative frequency

### Summary 5

### Examination practice 5

### Numeracy practice
Skills practice 1–5

### Challenges

### Fact finders
Great white shark
This troubled planet
The Channel Tunnel
Recycling
The Solar System

### How to tackle investigations

### Index

# Preface

This two-book series is written for students following the IGCSE Higher tier specification for the Edexcel examination board. It comprises a Students' Book for each year of the course.

The course has been structured to enable these two books to be used in a sequential manner, both in the classroom and by students working on their own.

Each book contains five units of work. Each unit contains five sections in the topic areas: Number, Algebra, Graphs & Sequences, Shape & Space and Handling Data.

In each unit, there are concise explanations and worked examples, plus numerous exercises that will help students to build up confidence.

**Paired questions**, with answers to the odd-numbered questions at the end of the Students' Book, allow students to check their answers and monitor their own progress. More difficult questions, to stretch the more able student, appear at the end of some exercises, and are identified by blue question numbers.

**Parallel exercises** are provided allowing students to consolidate basic principles before being challenged with more difficult questions.

♦ **Non-starred exercises** are designed for students working towards IGCSE grades B/C.
♦ **Starred exercises** are designed to challenge students working towards IGCSE grades A/A*.

Real data has been used where possible and both within the sections and at the end of each unit, challenges and investigations encourage students to think for themselves.

♦ **Challenges** provide questions applying the basic principles in unusual situations.
♦ **Investigations** prepare students for independent thought.

Consolidation is a recurring theme throughout the course and general skills are reinforced at the end of each unit.

♦ **Pairs of parallel revision exercises** appear at the end of each of the five sections within each unit.
♦ **Numeracy practice exercises** provide opportunities for students to flex their basic arithmetic and algebraic skills.
♦ **Fact finders** test numerical comprehension of real data. Students use the information supplied to answer thought-provoking questions. The shaded questions are more challenging.
♦ **Summaries** précis the major points of each unit.
♦ **Examination practice papers** test students' understanding of material and terminate each of the units in Book 1 and Book 2.

**Revision** is vital to the success of every student and has been covered fully in Book 2:

♦ Summaries of all topics in Book 1 plus Revision exercises.
♦ Four examination exercises covering the entire syllabus with cross-referencing to the text for students who might require assistance.

# Number 1

## Simplifying fractions

> **Remember**
>
> A fraction has been simplified when the numerator (the top) and the denominator (the bottom) are expressed as whole numbers cancelled down as far as possible.
>
> ♦   $\frac{28}{42} \to \frac{14}{21} \to \frac{2}{3}$          ♦   $\frac{0.7}{1.4} \to \frac{7}{14} \to \frac{1}{2}$

Fractions are important when working with probabilities and ratios. They are also used in many other calculations in everyday life.

> **Remember**
>
> ♦   To change a fraction or a decimal into a percentage, multiply it by 100.
>
> ♦   To change a percentage into a fraction, rewrite % as division by 100.
>
> ♦   To write a fraction as a decimal, divide the top number by the bottom number.

## Exercise 1

Copy and complete this table, giving the fractions in their lowest terms.

| | Fraction | Decimal | Percentage |
|---|---|---|---|
| **1** | | 0.75 | |
| **2** | | 0.2 | |
| **3** | | | 25% |
| **4** | | | 12.5% |
| **5** | $\frac{3}{20}$ | | |
| **6** | $\frac{3}{40}$ | | |
| **7** | | 0.35 | |
| **8** | | 0.375 | |

Change each of these to a mixed number.

**9** $\frac{8}{3}$             **10** $\frac{13}{4}$             **11** $\frac{17}{5}$             **12** $\frac{19}{7}$

Change each of these to an improper fraction.

**13** $2\frac{1}{3}$             **14** $3\frac{3}{5}$             **15** $1\frac{5}{6}$             **16** $5\frac{6}{7}$

Simplify and write each of these as a single fraction.

**17** $\frac{6}{21}$      **18** $\frac{14}{21}$      **19** $\frac{15}{90}$      **20** $\frac{105}{165}$

**21** $\frac{0.7}{1.4}$      **22** $\frac{1.2}{3.2}$      **23** $\frac{0.9}{12}$      **24** $\frac{0.8}{12}$

**25** $5 \times \frac{3}{18}$      **26** $\frac{5}{18} \times 3$      **27** $4 \times \frac{7}{42}$      **28** $\frac{2}{35} \times 14$

**29** $\frac{15}{27} \times 0.8$      **30** $\frac{21}{28} \times 0.5$      **31** $0.3 \times \frac{7}{12}$      **32** $0.2 \times \frac{3}{8}$

Simplify and write each of these as an ordinary number.

**33** $68 \div 0.1$      **34** $9.1 \div 0.01$      **35** $765 \times 0.001$      **36** $9.5 \times 0.01$

**37** $\frac{7.8}{0.2}$      **38** $\frac{0.62}{0.2}$      **39** $\frac{36}{1.5}$      **40** $\frac{27}{1.5}$

**41** $25 \times \frac{105}{100}$      **42** $34 \times \frac{126}{100}$      **43** $46 \times \frac{91}{100}$      **44** $58 \times \frac{86}{100}$

# Directed numbers and order of operations

This section will remind you of how to work with negative numbers, and why it is important to do calculations in the correct order.

### Remember

◆ Directed numbers

$3 + (-4) = 3 - 4 = -1$
$3 - (-4) = 3 + 4 = 7$
$(-3) + (-4) = -3 - 4 = -7$
$(-3) - (-4) = -3 + 4 = 1$
$6 \times (-2) = -12$
$6 \div (-2) = -3$
$(-6) \div (-2) = 3$
$(-6) \times (-2) = 12$

◆ Order of operations

The mnemonic **BIDMAS** may help you to remember the correct priority of operations when doing calculations working from left to right:

**B**rackets
**I**ndices
**D**ivision and/or **M**ultiplication
**A**ddition and/or **S**ubtraction

# Exercise 2

Calculate these.

**1** $(-5) + 10$      **2** $6 - (14)$      **3** $10 - (-3)$

**4** $(-12) + (-5)$      **5** $17 + (-4)$      **6** $(-16) - 4$

**7** $(-7) - (-4)$      **8** $(-13) + 3$      **9** $(-4) \times 3$

**10** $(-7) \times (-4)$      **11** $12 \div (-2)$      **12** $(-12) \div 2$

**13** $\frac{(-16)}{8}$      **14** $\frac{20}{(-4)}$      **15** $\frac{(-24)}{(-8)}$

**16** $\frac{(-30) - 6}{(-12)}$      **17** $4 \times 3 - 2$      **18** $4 \times (3 - 2)$

**19** $4 - 2 \times 3$

**20** $(4 - 2) \times 3$

**21** $20 \div 1 + 3$

**22** $20 \div (1 + 3)$

**23** $16 + 4 \div 2$

**24** $(16 + 4) \div 2$

**25** $16 + \frac{4}{2}$

**26** $\frac{16 + 4}{2}$

**27** $16 \div 2 + 4$

**28** $16 \div (2 + 4)$

**29** $\frac{16}{2} + 4$

**30** $2 \times 3^2$

**31** $(2 \times 3)^2$

**32** $4 + 3^2$

**33** $(4 + 3)^2$

**34** $2 + 2 \times 6^2$

**35** $(2 + 2) \times 6^2$

**36** $(2 + 2 \times 6)^2$

# Percentages

Percentages are used to compare quantities. The unit of comparison is 100, and this is why the term 'per cent' is used ('per' means divide, and 'cent' means 100).

---

### Example 1

Kathay Pacific increased ticket prices by 8%.  Calculate the new price of a $2450 ticket.

*Using the ratio line*

Let $x$ be the new price in dollars ($).

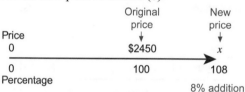

$\frac{x}{108} = \frac{2450}{100}$  (Multiply both sides by 108)

$x = \frac{2450}{100} \times 108$

The new price is $2646.

*Using the multiplying factor*

$100\% + 8\% = 108\%$.

$\therefore$ multiplying factor $= \frac{108}{100} = 1.08$

New price

$= \frac{108}{100} \times 2450$

$= \$2646$

### Example 2

In 1348–49, the population of England was 4 million. The Black Death reduced the population by 37.5%. Find the new population.

*Using the ratio line*

Let $x$ be the new population in millions.

Population

| 0 | New $x$ | Original 4 million |

| 0 | 62.5 | 100 |

Percentage

37.5% reduction

$\frac{x}{62.5} = \frac{4}{100}$

$x = \frac{4 \times 62.5}{100}$

$x = 2.5$ million

*Using the multiplying factor*

$100\% - 37.5\% = 62.5\%$

$\therefore$ multiplying factor $= \frac{62.5}{100} = 0.625$

New population

$= 0.625 \times 4$ million

$= 2.5$ million

---

The multiplying factor method is easier to use in advanced problems on percentage increases and decreases.

# Exercise 3

For Questions 1–6, do the calculation.

1  5% of 36

2  8% of 3.4 km

3  12% of 46 m

4  6% of 23 cm

5  9% of 7.6

6  15% of 84 litres

7  Increase £30 by 6%.

8  Increase £60 by 8%.

9  Reduce £50 by 9%.

10  Reduce £80 by 11%.

11  Friedrich buys a car for $15 000, and sells it for $12 750. What is his percentage loss?

12  Kurt buys a boat for $9000, and sells it for $8460. What is the percentage loss?

13  A bicycle is bought for $250, and sold for $275. What is the percentage profit?

14  Liesl buys a bottle of perfume for €5.60, and sells it for €6.44. What is the percentage profit?

15  Louisa throws the javelin 34 m. Then she improves by 1.7 m. What is her percentage improvement?

16  Hans jumps 5.50 m in the long jump. His next jump improves this by 8%.
    How far did he jump that time?

17  A rare stamp is bought for €7800 and increases in value by €468.
    What is the percentage increase in value?

18  A caravan was bought for $12 000. It then decreases in value by 24%. What is its new value?

19  A washing machine costs €640 plus an installation charge of 7.5%. What is the total cost?

20  Using percentages, comment on these figures.

|  | 1959 | 1975 | 1991 |
|---|---|---|---|
| **Life expectancy of men in the UK** | 74 years | 76 years | 79 years |

# Exercise 3★

For Questions 1–6, do the calculation.

1  1.5% of 50

2  4.2% of 500 kg

3  5.7% of $3000

4  1.8% of 650 km

5  7.5% of 700

6  12.5% of 850

7  Increase 67 km by 1.5%.

8  Increase 84 kg by 2.5%.

9  Decrease $87 by 8%.

10  Decrease $98 by 9%.

11  Marta buys a bicycle for €350, reduced from €402.50. What is her percentage profit?

12  Gretl buys a computer for $950, and sells it at a loss of 12%. What is the selling price?

13  A rare stamp is bought for €34 and sold for €38.25. What is the percentage profit?

14  Brigitta buys a table for $960. It is sold at auction at a loss of $374.40. What is the percentage loss?

15  What is the percentage error if I use a value of $3\frac{1}{7}$ for $\pi$?

16  What will my percentage saving be if I buy something in a sale that offers 3 for the price of 2?

17  A is 40% of B. What percentage is B of A?

18  A transatlantic airline ticket costs $320 in the US, and $360 in Europe. As a percentage, how much cheaper is the ticket in the US? As a percentage, how much more expensive is the ticket in Europe?

**19** A survey showed that the ratio of telephones to people in the world was $1:8$. Of the world's telephones, 0.012% were in Ethiopia. In Ethiopia, 0.2% of the population had telephones. At the time of the survey, there were 4000 million people in the world. What was the population of Ethiopia?

**20** For both sizes of tin for each year, calculate the number of grams per penny correct to 4 significant figures. Analyse your answers using percentages. Comment on your results. This table shows contents (in g) and price (in p) for two sizes of tins of baked beans in the UK.

| Baked beans | 1976 mass (g) | 1976 cost (p) | 1986 mass (g) | 1986 cost (p) | 1996 mass (g) | 1996 cost (p) |
|---|---|---|---|---|---|---|
| Large tin | 794 | 28.5 | 840 | 51 | 840 | 62 |
| Small tin | 142 | 7 | 150 | 15 | 150 | 24 |

# Standard form with positive indices

You can write the very large number 100 000 000 more simply as $1 \times 10^8$ using **standard form**. All numbers can be written in standard form, for example:

$$2904 = 2.904 \times 1000 = 2.904 \times 10^3$$

A standard form number can be converted back to an ordinary number:

$$5.6 \times 10^5 = 5.6 \times 100\,000 = 560\,000$$

## Key Points

♦ In **standard form**, a million is written as $1 \times 10^6$.

On a calculator this is 1 EXP 6 , and is displayed as 1.06 .

♦ **Standard form** is always written as $a \times 10^b$, where $a$ is between 1 and 10, but never equal to 10, and $b$ is an integer (a whole number).

**Example 3**

Convert 549 into standard form.

549_____ *divide by 100* $\longrightarrow$ 5.49
So *multiply by 100* to compensate.
$549 = 5.49 \times 100 = 5.49 \times 10^2$

**Example 4**

Convert 7 670 000 into standard form.

7 6 7 0 0 0 0. *move the decimal point*
↑ ↑ ↑ ↑ ↑ ↑ *six places to make* 7.6 7
So, $7\,670\,000 = 7.67 \times 10^6$

N.B. 'Moving the decimal point one place to the left' divides the number by 10.

## Activity 1

In the human brain, there are about 100 000 000 000 neurons, and over the human lifespan 1 000 000 000 000 000 neural connections are made.

◆ Write these numbers in standard form.
◆ Calculate the approximate number of neural connections made per second in an average human lifespan of 75 years.

## Exercise 4

Calculate these, and write each answer in standard form.

| | | | |
|---|---|---|---|
| **1** $10^2 \times 10^3$ | **2** $10^4 \times 10^1$ | **3** $10^7 \times 10^3$ | **4** $10^4 \times 1$ million |
| **5** $10^1 \times 10^2$ | **6** $10^2 \times 10^5$ | **7** $10^5 \times 10^8$ | **8** 1 million $\times 10^5$ |
| **9** $10^5 \div 10^3$ | **10** $10^6 \div 10$ | **11** $\dfrac{10^7}{10^4}$ | **12** 10 million $\div 10^5$ |
| **13** $10^6 \div 10^3$ | **14** $\dfrac{10^{12}}{10^4}$ | **15** $10^{10} \div 10^9$ | **16** $10^8 \div 1$ million |

Write each of these in standard form.

| | | | |
|---|---|---|---|
| **17** 456 | **18** 67.8 | **19** 123.45 | **20** 67 million |
| **21** 568 | **22** 38.4 | **23** 706.05 | **24** 123 million |

Write each of these as an ordinary number.

| | | | |
|---|---|---|---|
| **25** $4 \times 10^3$ | **26** $5.6 \times 10^4$ | **27** $4.09 \times 10^6$ | **28** $6.789 \times 10^5$ |
| **29** $5.6 \times 10^2$ | **30** $6.5 \times 10^4$ | **31** $7.97 \times 10^6$ | **32** $9.876 \times 10^5$ |

**33** The approximate area of all the land on Earth is $10^8$ square miles. The area of the British Isles is $10^5$ square miles. How many times greater is the Earth's area?

**34** The area of the surface of the largest known star is about $10^{15}$ square miles. The area of the surface of the Earth is about $10^{11}$ square miles. How many times greater is the star's area?

Calculate these, and write each answer in standard form.

**35** $(2 \times 10^4) \times (4.2 \times 10^5)$     **36** $(6.02 \times 10^5) \div (4.3 \times 10^3)$     **37** $(4.5 \times 10^{12}) \div (9 \times 10^{10})$

## Exercise 4★

Write each of these in standard form.

| | | | |
|---|---|---|---|
| **1** 45 089 | **2** 87 050 | **3** 29.83 million | **4** 0.076 54 billion |

Calculate these, and write each answer in standard form.

| | | | |
|---|---|---|---|
| **5** $10 \times 10^2$ | **6** $(10^3)^2$ | **7** $\dfrac{10^9}{10^4}$ | **8** 10 million $\div 10^6$ |
| **9** $10^{12} \times 10^9$ | **10** $(10^2)^4$ | **11** $10^7 \div 10^7$ | **12** $\dfrac{10^{12}}{1 \text{ million}}$ |

Calculate these, and write each answer in standard form.

**13** $(5.6 \times 10^5) + (5.6 \times 10^6)$        **14** $(4.5 \times 10^4) \times (6 \times 10^3)$

**15** $(3.6 \times 10^4) \div (9 \times 10^2)$        **16** $(7.87 \times 10^4) - (7.87 \times 10^3)$

Calculate these, and write each answer in standard form.

**17** $(4.5 \times 10^5)^3$    **18** $(3 \times 10^8)^5$    **19** $10^{12} \div (4 \times 10^7)$

**20** $(3.45 \times 10^8) + 10^6$    **21** $10^9 - (3.47 \times 10^7)$    **22** $10^{16} \div (2.5 \times 10^{12})$

You will need the information in this table to answer Questions 23, 24 and 25.

| Celestial body | Approximate distance from Earth (miles) |
| --- | --- |
| Sun | $10^8$ |
| Saturn | $10^9$ |
| Andromeda Galaxy (nearest major galaxy) | $10^{19}$ |
| Quasar OQ172 (one of the remotest objects known) | $10^{22}$ |

Copy and complete these sentences.

**23** The Andromeda Galaxy is … times further away from the Earth than Saturn.

**24** The quasar OQ172 is … times further away from the Earth than the Andromeda Galaxy.

**25** To make a scale model showing the distances of the four bodies from the Earth, a student marks the Sun 1 cm from the Earth.

How far along the line should the other three celestial bodies be placed?

# Significant figures and decimal places

It is often useful to simplify numbers by writing them either correct to so many **significant figures** (s.f.) or correct to so many **decimal places** (d.p.).

---

**Example 5**

Write 672 900 correct to 3 significant figures.
672 900 → 673 000 (to 3 s.f.)

4th s.f. = 9; 9 > 5. ∴ 2 rounds up to 3.
(672 900 is closer in value to 673 000 than to 672 000.)

**Example 7**

Write 6.4873 correct to 2 decimal places.
6.4873 → 6.49 (to 2 d.p.)

3rd d.p. = 7; 7 > 5. ∴ 8 rounds up to 9.
(6.4873 is closer in value to 6.49 than to 6.48.)

**Example 6**

Write 0.007 645 correct to 2 significant figures.
0.007 645 → 0.0076 (to 2 s.f.)

3rd s.f. = 4; 4 < 5. ∴ 6 is not rounded up to 7.
(0.007 645 is closer in value to 0.0076 than to 0.0077.)

**Example 8**

Write 23.428 correct to 1 decimal place.
23.428 → 23.4 (to 1 d.p.)

2nd d.p. = 2; 2 < 5. ∴ 4 is not rounded up to 5.
(23.428 is closer in value to 23.4 than to 23.5.)

---

# Exercise 5

Write each of these correct to 1 significant figure.

**1** 783          **2** 87 602          **3** 10.49          **4** 5049

Write each of these correct to 3 significant figures.

**5** 3738          **6** 80 290          **7** 45.703          **8** 89 508

Correct each of these to 2 significant figures.

**9** 0.439          **10** 0.555          **11** 0.0688          **12** 0.006 78

Correct each of these to 3 significant figures.

**13** 0.5057          **14** 0.1045          **15** 0.049 549          **16** 0.000 5679

Write each of these to 2 decimal places.

**17** 34.777          **18** 0.654          **19** 8.997          **20** 2.0765

Write each of these to 1 decimal place.

**21** 3.009          **22** 9.09          **23** 6.96          **24** 78.1818

Write 105 678 in standard form correct to

**25** 1 s.f.          **26** 2 s.f.          **27** 3 s.f.          **28** 4 s.f.

Write 98 765 in standard form correct to

**29** 1 s.f.          **30** 2 s.f.          **31** 3 s.f.          **32** 4 s.f.

# Exercise 6 (Revision)

For Questions 1 and 2, do the calculation, and write each answer in standard form.

**1** $10^3 \times 10^2$          **2** $10^4 \div 10^2$

**3** Write 4566 correct to 3 significant figures.

**4** Write 4566 in standard form correct to 3 significant figures.

**5** Write $3.7 \times 10^3$ as an ordinary number.

**6** Write 48% as a fraction in its lowest terms.          **7** Write 48% as a decimal.

**8** Write $2\frac{5}{8}$ as an improper fraction.          **9** Write $2\frac{5}{8}$ as an ordinary number.

Simplify and write each of these as a fraction.

**10** $\frac{6}{18}$          **11** $\frac{0.6}{18}$          **12** $4 \times \frac{7}{36}$          **13** $\frac{5}{24} \times 0.3$

Simplify and write each of these as an ordinary number.

**14** $42 \times 0.1$          **15** $42 \div 0.1$          **16** $\frac{7.2}{1.8}$          **17** $36 \times \frac{40}{100}$

Write 0.045 67 correct to

**18** 2 decimal places          **19** 3 decimal places          **20** 2 significant figures

Calculate these, and write each answer in standard form.

**21** $(2.3 \times 10^3) \times (4 \times 10^5)$      **22** $(4.5 \times 10^7) \div (9 \times 10^2)$      **23** Find 3% of 68.

**24** Decrease 48 km by 6%.

**25** Berthe buys a book for $12.00, and sells it for $12.60. What is her percentage profit?

**26** An audio system was bought for €480, and then sold at a loss of €72. What is the loss as a percentage of the buying price?

## Exercise 6★ (Revision)

Simplify these, and write each answer as a fraction in its lowest terms.

**1** $\frac{0.6}{2.4}$

**2** $2.4 \times \frac{5}{28}$

**3** $\frac{0.5}{18} \times 15$

Write 8095.0501, in standard form, correct to

**4** 2 significant figures      **5** 3 significant figures

Calculate these.

**6** $45.6 \times 0.001$      **7** $45.6 \div 0.001$

Calculate these, and write each answer in standard form.

**8** $(10^6)^2$

**9** $10^4 \div 10^2$

**10** $(1 \text{ million})^2 \div 10^4$

Calculate these, and write each answer in standard form.

**11** $(2.6 \times 10^3) \times (6.1 \times 10^4)$      **12** $(4.8 \times 10^7) + (4.8 \times 10^6)$      **13** $(9.6 \times 10^7) \div (1.2 \times 10^5)$

**14** Which of these gives the larger result, and by how much: increasing 80 by 10%, or reducing 120 by 30%?

**15** Rolfe buys a house for €900 000, and sells it for €1 062 000. What is his percentage profit?

**16** Margaretta runs the 100 m in 13.50 s. She then runs it 4% faster. What is her new time?

**17** Franz has a salary of $34 000. He is given a rise of 18%. Then he is given another rise of 7%. What is his final salary?

**18** A shopkeeper sells one type of pen for 99 cents and makes 10% profit. He sells another type of pen for 99 cents and makes a 10% loss. On the two sales together, did the shopkeeper gain, lose or break even?

# Algebra 1

Algebraic expressions contain letters which stand for numbers but can be treated in the same way as expressions containing numbers.

## Activity 2

Think of a number. Add 7 and then double the answer. Subtract 10, halve the result, and then subtract the number you originally thought of. Algebra can show you why the answer is always 2.

| Think of a number: | $x$ |
| --- | --- |
| Add 7: | $x + 7$ |
| Double the result: | $2x + 14$ |
| Subtract 10: | $2x + 4$ |
| Halve the result: | $x + 2$ |
| Subtract the original number: | 2 |

♦ Make up two magic number tricks of your own, one like the one above and another that is longer. Check that they work using algebra. Then test them on a friend.

♦ Think of a number. Double it, add 12, halve the result, and then subtract the original number.
   ▶ Use algebra to find the answer.
   ▶ If you add a number other than 12, the answer will change. Work out the connection between the number you add and the answer.

# Simplifying algebraic expressions

You will often find it useful to simplify algebraic expressions before using them.

## Investigate

Investigate what happens when you substitute various values (positive or negative) for $x$ in these expressions.

$$x + 1 \quad \text{and} \quad \frac{x^3 + x^2 + x + 1}{x^2 + 1}$$

What is your conclusion? Which expression would you rather use?

> **Remember**
>
> You can only add or subtract **like** terms.
>
> ◆ $3ab + 2ab = 5ab$, but the terms in $3ab + b$ cannot be added together.
>
> ◆ $3a^2 + 2a^2 = 5a^2$, but the terms in $3a^2 + 2a$ cannot be added together.
>
> You can **check** your simplifications by substituting numbers.

# Exercise 7

Simplify these as much as possible.

1 $9ab - 5ab$

2 $11uv - 8uv$

3 $5xy + 2yx$

4 $3ab + 5ba$

5 $4pq - 7qp$

6 $3yx - 6xy$

7 $2xy + y - 3xy$

8 $a + 3ab - 4ab$

9 $x - 3x + 2 - 4x$

10 $y + 2 - 4y - 6$

11 $7cd - 8dc + 3cd$

12 $9ab + 5ba - 13ab$

13 $6xy - 12xy + 2xy$

14 $3pq - 11pq + 5pq$

15 $4ab + 10bc - 2ab - 5cb$

16 $5ab - 2ab + 3bc - ba$

17 $3ba - ab + 3ab - 5ab$

18 $7cd - 8dc + 3cd - 2cd$

19 $4gh - 5jk - 2gh + 7$

20 $2bc + 3ab + 2 - 5bc$

21 $2p^2 - 5p^2 + 2p - 4p$

22 $3p^3 + 2p^2 - 2p^3 + 5p^2$

23 $2x^2y - xy^2 + 3yx^2 - 2y^2x$

24 $5ab^3 - 4ab^2 + 2b^2a - 2b^3a$

# Exercise 7★

Simplify these as much as possible.

1 $7xy + 5xy - 13xy$

2 $3xy - 10xy + 5xy$

3 $7ab - b - 3ab$

4 $9xy + 2xy - x$

5 $2ab - 3ba + 7ab$

6 $6xy + 5yx - 7yx$

7 $12ab - 6ba + ba - 7ab$

8 $2ab - 5ab + 6ab - 3ba$

9 $4ab + 10bc - ba - 7cb$

10 $9fg + 8gh - 7gf + 3hg$

11 $q^2 + q^3 + 2q^2 - q^3$

12 $3r^4 - r^3 - r^4 + r^3$

13 $x^2 - 5x + 4 - x^2 + 6x - 3$

14 $2 + 3x^2 + x^4 - 3x^2 + 1$

15 $5a^2 + a^3 - 3a^2 + a$

16 $x^5 - 5x^3 + 2 - 2x^3$

17 $h^3 + 5h - 3 - 4h^2 - 2h + 7 + 5h^2$

18 $4h - 5h^2 + 2 - h^3 + 3h^2 - 4h + 3h^3$

19 $3a^2b - 2ab + 4ba^2 - ba$

20 $4fg^2 - 5g^2f + 3fg - 7fg^2$

21 $0.7a^2b^3c - 0.4b^2a^3c + 0.3cb^3a^2 - 0.2a^3cb^2 + 0.3$

22 $0.8x^5y^2z^3 - 0.3x^5y^2z^2 + 0.2z^3y^2x^5 + 0.3z^2y^2x^5 - y^2x^5z^3$

23 $2pq^2r^5 - pq^2r^4 - (r^4pq^2 - 2q^2r^5p)$

24 $[5g^2f^2h^3 + 2f^2h^2g^2 - 2h^2g^3f^2] - [2g^2f^2h^2 - 4g^2h^3f^2 - (2hgf)^2]$

## Simplifying algebraic products

The multiplication sign is often left out.

> **Remember**
>
> $3ab$ means $3 \times a \times b$.

## Exercise 8

Simplify these.

**1** $3 \times 2a$

**2** $4 \times 7x$

**3** $2x \times x$

**4** $y \times 3y$

**5** $3x \times x^2$

**6** $a^3 \times 2a$

**7** $5a^3 \times 3a^2$

**8** $4b^2 \times 2b^4$

**9** $2t \times 3s$

**10** $4r \times 5t$

**11** $4r \times s^2$

**12** $3d \times e^3$

**13** $2a^2 \times b^2$

**14** $3v^2 \times 3u^2$

**15** $2y \times 2y \times y$

**16** $4r \times 3r \times 2r$

**17** $2x^2 \times 3 \times 2x$

**18** $3y \times y^3 \times y$

**19** $(2a)^2 \times 5a$

**20** $(3b)^2 \times 3b$

## Exercise 8★

Simplify these.

**1** $8a \times a^2$

**2** $x^3 \times 3x$

**3** $5x^3 \times 3y^2 \times x$

**4** $4y^5 \times x \times 2y$

**5** $a^2 \times 2a^4 \times 3a$

**6** $b^3 \times 6b \times 5b^3$

**7** $(3y)^2 \times 2y$

**8** $(2a)^3 \times 4a$

**9** $6xy^2 \times 2x^3 \times 3xy$

**10** $4x^3y^4 \times 2y \times 3xy^2$

**11** $5abc \times 2ab^2c^3 \times 3ac$

**12** $2bc^2 \times 3ab^3 \times 4abc^4$

**13** $7x \times 2y^2 \times (2y)^2$

**14** $(3x)^2 \times 3x^2 \times 5y$

**15** $2xy^2 \times 3x^2y + 4x^3y^3$

**16** $4xy^4 \times 2xy + x^2y^5$

**17** $x^2y^3 \times 3xy - 2x^3y^2$

**18** $5x^4y \times x^2y + 3x^5y^2$

**19** $(2ab)^2 \times 5a^2b^4 - 2a^2b^5 \times 3a^2b$

**20** $7xy^4 \times 2x^5y - 3x^2y^2 \times (2x^2)^2 \times y^3$

## Simplifying algebraic expressions with brackets

To simplify an expression with brackets, multiply each term inside the bracket by the term outside the bracket.

> **Example 1**
>
> Simplify $2(3 + x)$.
>
> $2(3 + x) = 2 \times 3 + 2 \times x = 6 + 2x$
>
> The diagram helps show that $2(3 + x) = 6 + 2x$.
>
> The area of the whole rectangle is $2(3 + x)$.
>
> The area of rectangle A is 6.
>
> The area of rectangle B is $2x$.

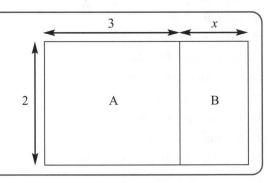

**Remember**

$3(x + y)$  means  $3 \times (x + y) = 3 \times x + 3 \times y = 3x + 3y$

Be very careful with negative signs outside a bracket.

$-2(a - 3)$  means  $-2 \times (a - 3) = (-2) \times (a) + (-2) \times (-3) = -2a + 6$

When multiplying, the number 1 is usually left out.

$-(2x + 3)$  means  $-1 \times (2x + 3) = (-1) \times (2x) + (-1) \times (3) = -2x - 3$

## Exercise 9

Remove the brackets and simplify these if possible.

| | | |
|---|---|---|
| 1  $5(2 + 3a)$ | 2  $3(4 + 7x)$ | 3  $2(b - 4c)$ |
| 4  $6(v - 5w)$ | 5  $-3(2a + 8)$ | 6  $-9(3g + 5)$ |
| 7  $-4(3 - x)$ | 8  $-7(4 - y)$ | 9  $-(a - 2b)$ |
| 10  $-(x - 5y)$ | 11  $3a + 2(a + 2b)$ | 12  $2(x + y) + 5x$ |
| 13  $3(t - 4) - 6$ | 14  $5(v - 2) - 7$ | 15  $7x - (x - y)$ |
| 16  $9a - (2b - a)$ | 17  $0.4(x - 3y) + 0.5(2x + 3y)$ | 18  $0.2(b + 3c) + 0.3(b - 2c)$ |
| 19  $1.1(a + 3) - 5(3 - 0.2a)$ | 20  $5(2 + 0.5t) - 0.9(1 - 2t)$ | |

## Exercise 9★

Remove the brackets and simplify these if possible.

| | | |
|---|---|---|
| 1  $4(3m - 2)$ | 2  $5(2n - 6)$ | 3  $2(x - y + z)$ |
| 4  $3(a + b - c)$ | 5  $5(3a + b - 4c)$ | 6  $3(5a - 2b - 3c)$ |
| 7  $\frac{1}{2}\left(4x - 6y + 8\right)$ | 8  $\frac{1}{3}\left(15x + 9y - 12\right)$ | 9  $5x - 3(2x - y)$ |
| 10  $7x - 6(x - 2y)$ | 11  $0.4(2 - x) - (x + 3)$ | 12  $0.3(3 - 2z) - (2 + z)$ |
| 13  $\frac{3}{4}\left(4x - 8y\right) - \frac{3}{5}\left(15x - 5y\right)$ | 14  $\frac{2}{7}\left(14p - 21q\right) - \frac{2}{3}\left(6p + 3q\right)$ | 15  $5x - 7y - 0.4(x - 2y + z)$ |
| 16  $2x - 0.6(2x - 3y - z) - z$ | 17  $0.3(2a - 6b + 1) - 0.4(3a + 6b - 1)$ | |
| 18  $0.5(2a - 3b + 4c) - 0.7(a - 2b + 3c)$ | 19  $0.3x(0.2x - y) - 4y(x + 0.3y) + 0.5x(y - x)$ | |
| 20  $0.5a(2a + 0.3b) - 0.7b(a + 2b) + 0.3b(b - 0.2a)$ | | |

## Solving equations

If is often easier to solve mathematical problems using algebra. Let the unknown quantity be $x$ and then write down the facts in the form of an equation.

There are six basic types of equation:

$$x + 3 = 12 \qquad x - 3 = 12 \qquad 3 - x = 12$$

$$3x = 12 \qquad \frac{x}{3} = 12 \qquad \frac{3}{x} = 12$$

Solving an equation means getting $x$ on its own, on one side of the equation.

**Remember**

To solve equations, do the **same** thing to both sides.

Always check your answer.

**Example 2**

$x + 3 = 12$ (Subtract 3 from both sides)

$x = 9$ (Check: $9 + 3 = 12$)

**Example 3**

$x - 3 = 12$ (Add 3 to both sides)

$x = 15$ (Check: $15 - 3 = 12$)

**Example 4**

$3 - x = 12$ (Add $x$ to both sides)

$3 = 12 + x$ (Subtract 12 from both sides)

$-12 + 3 = x$

$x = -9$ (Check: $3 - (-9) = 12$)

**Example 5**

$3x = 12$ (Divide both sides by 3)

$x = 4$ (Check: $3 \times 4 = 12$)

**Example 6**

$\dfrac{x}{3} = 12$ (Multiply both sides by 3)

$x = 36$ (Check: $36 \div 3 = 12$)

**Example 7**

$\dfrac{3}{x} = 12$ (Multiply both sides by $x$)

$3 = 12x$ (Divide both sides by 12)

$\dfrac{1}{4} = x$ (Check: $3 \div \dfrac{1}{4} = 12$)

# Exercise 10

Solve these for $x$.

1 $5x = 20$

2 $36 = 3x$

3 $x + 5 = 20$

4 $3 = 36 + x$

5 $x - 5 = 20$

6 $36 = x - 3$

7 $\dfrac{x}{5} = 20$

8 $3 = \dfrac{x}{36}$

9 $3 = \dfrac{36}{x}$

10 $\dfrac{5}{x} = 20$

11 $20 - x = 5$

12 $36 = 3 - x$

13 $5x = 12$

14 $26 = 4x$

15 $x - 3.8 = 9.7$

16 $11.6 = x - 7.9$

17 $3.8 = \dfrac{x}{7}$

18 $\dfrac{x}{8} = 2.7$

19 $x + 9.7 = 11.1$

20 $13.1 = 17.9 + x$

21 $13.085 - x = 12.1$

22 $17 = 6.9 - x$

23 $\dfrac{34}{x} = 5$

24 $7 = \dfrac{44.1}{x}$

Solve these for $x$, giving each answer correct to 3 significant figures.

25 $23.5 + x = 123.4$

26 $34.5 = x + 167.8$

27 $7.6x = 39$

28 $50.2 = 4.7x$

29 $39.6 = x - 1.064$

30 $x - 0.987 = 3.6$

31 $45.7 = \dfrac{x}{12.7}$

32 $\dfrac{x}{0.93} = 34.8$

33 $7.89 = \dfrac{67}{x}$

34 $\dfrac{0.234}{x} = 5$

35 $40.9 - x = 2.06$

36 $90 = 0.0567 - x$

### Example 8

$3x - 5 = 7$   (Add 5 to both sides)
$3x = 12$   (Divide both sides by 3)
$x = 4$   (Check: $3 \times 4 - 5 = 7$)

### Example 9

$4(x + 3) = 20$   (Divide both sides by 4)
$x + 3 = 5$   (Subtract 3 from both sides)
$x = 2$   (Check: $4(2 + 3) = 20$)

### Example 10

$2(x + 3) = 9$ (Multiply out the bracket)
$2x + 6 = 9$ (Subtract 6 from both sides)
$2x = 3$ (Divide both sides by 2)
$x = \frac{3}{2}$   (Check: $2\left(\frac{3}{2} + 3\right) = 9$)

## Exercise 11

Solve these for $x$.

**1** $2x + 4 = 10$

**2** $3x + 2 = 14$

**3** $4x + 5 = 1$

**4** $6x + 9 = -9$

**5** $12x - 8 = -32$

**6** $15x - 11 = -41$

**7** $2(x + 3) = 10$

**8** $4(x + 2) = 24$

**9** $5(x - 2) = 30$

**10** $7(x - 4) = 35$

**11** $5 - x = 4$

**12** $13 - x = 7$

**13** $9 = 3 - x$

**14** $12 = 2 - x$

**15** $2(6 - 3x) = 6$

**16** $3(6 - 2x) = 12$

**17** $4(2 - x) = 16$

**18** $6(3 - x) = 24$

**19** $3(x - 5) = -13$

**20** $5(x - 7) = -34$

**21** $9(x + 4) = 41$

**22** $8(x + 3) = 29$

**23** $5(10 - 3x) = 30$

**24** $8(5 - 2x) = 24$

**25** $7(2 - 5x) = 49$

**26** $6(4 - 7x) = 36$

**27** The sum of two consecutive numbers is 477.
What are the numbers? (Let the first number be $x$.)

**28** The sum of three consecutive numbers is 219.
What are the numbers? (Let the first number be $x$.)

**29** Find $x$ and the size of each angle in this triangle.
(Remember that the angles of a triangle sum to 180°.)

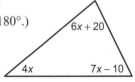

**30** AB is a straight line.
Find $x$ and the size of each angle.

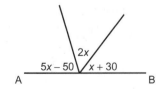

31 The area of a triangle is given by the formula $A = 0.5bh$.
   If $A = 12.2$ and $h = 6.1$, find $b$.

32 The area of a trapezium is given by the formula $A = 0.5h(a + b)$.
   If $A = 10.8$, $a = 2.1$ and $h = 4.8$, find $b$.

33 The formula for converting degrees Fahrenheit ($F$) to degrees Celsius ($C$) is $F = 32 + 1.8C$.
   Find $C$ when $F$ is 5.

34 The formula for summing an arithmetic progression is $S = 0.5n(2a + (n - 1)d)$.
   If $a = 8$, $n = 20$ and $S = 255$, find $d$.

# Exercise 11★

Solve these for $x$.

1  $5x - 3 = 17$      2  $7x - 12 = 9$      3  $27 = 3(x - 2)$

4  $30 = 5(x - 4)$      5  $7(x - 3) = -35$      6  $8(x - 6) = -32$

7  $12(x + 5) = 0$      8  $9(x + 4) = 0$      9  $-7 = 9 + 4x$

10  $-5 = 13 + 3x$      11  $5 - 4x = -15$      12  $8 - 7x = -6$

13  $34 = 17(2 - x)$      14  $39 = 13(4 - x)$

15 The sum of three consecutive even numbers is 222. Find the numbers.

16 The sum of four consecutive odd numbers is 504. Find the numbers.

17 John and Amelia have a baby daughter, Sonia. John is 23 kg heavier than Amelia, who is four times heavier than Sonia. Their combined weight is 122 kg. How heavy is each person?

18 Emma buys some cans of cola at 28p each, and twice as many cans of orange at 22p each. She also buys ten fewer cans of lemonade than orange at 25p each. She spends £14.58. How many cans of cola did she buy?

19 Solve for $x$: $0.3(6 - x) + 0.4(x + 8) = 0$.

20 Solve for $x$: $1.4(x - 3) + 0.2(2x - 1) = 0.1$.

21 Carly is training by running to a post across a field and then back. She runs the outward leg at $7\,\text{ms}^{-1}$ and the return leg at $5\,\text{ms}^{-1}$. She takes 15.4 s. Find the distance to the post.

22 A piece of wire 30 cm long is cut into two pieces.
   One of these is bent into a circle, and the other is bent into
   a square enclosing the circle, as shown in the diagram.
   What is the diameter of the circle? (Remember that the
   circumference of a circle = $2\pi r$.)

23 Sophie sets off on a walk at $6\,\text{kmh}^{-1}$. Ten minutes later, her brother Dave sets off after her on his bicycle at $15\,\text{kmh}^{-1}$. How far must Dave go to catch up with Sophie?

## Equations with *x* on both sides

Sometimes *x* appears on both sides of an equation.

**Example 11**

Solve this for *x*.

$$7x - 3 = 3x + 5 \quad \text{(Subtract } 3x \text{ from both sides)}$$
$$7x - 3 - 3x = 5 \quad \text{(Add 3 to both sides)}$$
$$4x = 5 + 3$$
$$4x = 8 \quad \text{(Divide both sides by 4)}$$
$$x = 2 \quad \text{(Check: } 7 \times 2 - 3 = 3 \times 2 + 5\text{)}$$

# Exercise 12

Solve these for *x*.

1  $8x - 3 = 4x + 1$

2  $5x - 6 = 3x + 2$

3  $2x + 5 = 5x - 1$

4  $4x + 3 = 6x - 7$

5  $7x - 5 = 9x - 13$

6  $4x - 3 = 8x - 15$

7  $2x + 7 = 5x + 16$

8  $3x + 5 = 9x + 17$

9  $5x + 1 = 8 - 2x$

10  $4x + 3 = 13 - x$

11  $14 - 3x = 10 - 7x$

12  $19 - 5x = 11 - 9x$

13  $6 + 2x = 6 - 3x$

14  $10 + 3x = 10 - 7x$

15  $8x + 9 = 6x + 8$

16  $6x + 6 = 3x + 5$

17 Find the value of *x*
and the perimeter of this rectangle.

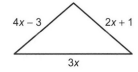

18 Find the value of *x*
and the perimeter of this isosceles triangle.

19 The result of adding 36 to a certain number is the same as multiplying that number by 5.
What is the number?

20 The result of doubling a certain number and adding 17 is the same as trebling that number and
adding 4. What is the number?

# Exercise 12★

Solve these for $x$.

**1** $3x + 8 = 7x - 8$  **2** $8x - 3 = 4x + 9$  **3** $7x + 5 = 5x + 1$

**4** $9x + 13 = 6x + 10$  **5** $5x + 7 = 9x + 1$  **6** $11x - 5 = 6x - 3$

**7** $4x + 3 = 7 - x$  **8** $3x + 5 = 8 - x$  **9** $15x - 4 = 10 - 3x$

**10** $22x - 8 = 10 - 5x$  **11** $5(x + 1) = 4(x + 2)$  **12** $7(x + 2) = 6(x + 3)$

**13** $8(x + 5) = 10(x + 3)$  **14** $9(x + 4) = 12(x + 2)$  **15** $3(x - 5) = 7(x + 4) - 7$

**16** $4(x - 3) = 8(x + 5) - 12$

**17** $3.1(4.8x - 1) - 3.9 = x + 1$

**18** $5.6(3.4x - 2) + 5.2 = x - 1$

**19** $8.9(x - 3.5) + 4.2(3x + 2.3) = 4.7x$

**20** $6.4(2x - 0.9) + 3.3(x + 4.1) = 6.8x - 3.6$

**21** If $\frac{x}{3} - 1$ is twice as large as $\frac{x}{4} - 3$, what is the value of $x$?

**22** A father is three times as old as his son. In 14 years' time, he will be twice as old as his son. How old is the father now?

## Negative signs outside brackets

> **Remember**
>
> $-(2x - 5)$ means $-1 \times (2x - 5) = (-1) \times (2x) + (-1) \times (-5) = -2x + 5$

> **Example 12**
>
> Solve this for $x$.
>
> | | |
> |---|---|
> | $2(3x + 1) - (2x - 5) = 15$ | (Remove brackets) |
> | $6x + 2 - 2x + 5 = 15$ | (Simplify) |
> | $6x - 2x = 15 - 2 - 5$ | (Subtract 2 and 5 from both sides) |
> | $4x = 8$ | (Divide both sides by 4) |
> | $x = 2$ | (Check: $2(3 \times 2 + 1) - (2 \times 2 - 5) = 15$) |

# Exercise 13

Solve these for $x$.

**1** $3(x - 2) - 2(x + 1) = 5$  **2** $4(x - 1) - 3(x + 2) = -6$

**3** $3(2x + 1) - 2(2x - 1) = 11$  **4** $9(x - 2) - 3(2x - 3) = 12$

**5** $2(5x - 7) - 6(2x - 3) = 0$  **6** $3(3x + 2) - 4(3x - 3) = 0$

**7** $4(3x - 1) - (x - 2) = 42$  **8** $2(2x - 1) - (x + 5) = 5$

**9** $4(3 - 5x) - 7(5 - 4x) + 3 = 0$  **10** $5(3x - 2) - 9(2 + 4x) - 7 = 0$

# Exercise 13★

Solve these for $x$.

**1** $5(x - 3) - 4(x + 1) = -11$

**2** $9(x - 2) - 7(x + 1) = -15$

**3** $4(3x + 5) - 5(2x + 6) = 0$

**4** $3(5x - 4) - 3(2x - 1) = 0$

**5** $3(3x + 1) - 8(2x - 3) + 1 = 0$

**6** $5(6x + 2) - 7(3x - 5) - 72 = 0$

**7** $-2(x + 3) - 6(2x - 4) + 108 = 0$

**8** $-3(x - 2) - 5(3x - 2) + 74 = 0$

**9** $7(5x - 3) - 10 = 2(3x - 5) - 3(5 - 7x)$

**10** $4(7 + 3x) - 5(6 - 7x) + 1 = 8(1 + 4x)$

**11** Lauren is shooting at a target at a fair. If she hits the target she receives 50p, but if she misses she has to pay 20p for the shot. After 15 shots, Lauren finds she has made a profit of £1.20.
How many hits has she had?

**12** Aidan is doing a multiple-choice test with 20 questions. He scores 3 marks for a correct answer and loses 1 mark if the answer is incorrect. Aidan answers all the questions and scores 40 marks.
How many questions has he got right?

**13** Freddie the frog is climbing up a well. Every day he climbs up 3 m but some nights he falls asleep and slips back 4 m. At the start of the sixteenth day, he has climbed a total of 29 m.
On how many nights was he asleep?

# Exercise 14 (Revision)

Simplify these as much as possible.

**1** $x + 2x + 3 - 5$

**2** $3ba - ab + 3ab - 4ba$

**3** $2a \times 3$

**4** $2a \times a$

**5** $a^2 \times a$

**6** $2a^2 \times a^2$

**7** $2a \times 2a \times a^2$

**8** $7a - 4a(b + 3)$

**9** $4(x + y) - 3(x - y)$

Solve these equations.

**10** $2(x - 1) = 12$

**11** $7x - 5 = 43 - 3x$

**12** $5 - (x + 1) = 3x - 4$

**13** Find three consecutive numbers whose sum is 438.

**14** The perimeter of a rectangle is 54 cm. One side is $x$ cm long, and the other is 6 cm longer.
**a** Form an equation involving $x$.
**b** Solve the equation, and write down the length of each of the sides.

# Exercise 14★ (Revision)

Simplify these as much as possible.

**1** $6xy^2 - 3x^2y - 2y^2x$

**2** $2xy^2 \times x^2y$

**3** $p - (p - (p - (p - 1)))$

**4** $xy(x^2 + xy + y^2) - x^2(y^2 - xy - x^2)$

Solve these equations.

**5** $4 = \dfrac{x}{5}$

**6** $4 = \dfrac{5}{x}$

**7** $43 - 2x = 7 - 8x$

**8** $1.3 - 0.3x = 0.2x + 0.3$

**9** $0.6(x + 1) + 0.2(6 - x) = x - 0.6$

**10** The length of a conference room is one and a half times its width. A carpet that is twice as long as it is wide is placed in the centre of the room, leaving a 3 m wide border round the carpet. Find the area of the carpet.

**11** Two years ago, my age was four times that of my son. Eight years ago, my age was ten times that of my son. Find the age of my son now.

**12** A river flows at $2\,\text{ms}^{-1}$. What is the speed through the water of a boat that can go twice as fast downstream as upstream?

**13** Matt goes to buy a television. If he pays cash, he gets a discount of 7%. If he pays by instalments, he has to pay an extra 10% in interest. The difference between the two methods is $16.66. Find the cost of the television.

# Graphs 1

## Gradient of a straight line

The slope of a line is its **gradient**. This is usually represented by the letter $m$.

For a straight line,

$$m = \frac{\text{change in the } y \text{ coordinates}}{\text{change in the } x \text{ coordinates}} = \frac{\text{'rise'}}{\text{'run'}}$$

---

### Example 1

Find the gradient of the straight line joining A(1, 2) to B(3, 6).

First draw a diagram.
Mark in the changes in coordinates.

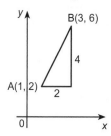

For AB, the change in the $y$ coordinate is $6 - 2 = 4$, and the change in the $x$ coordinate is $3 - 1 = 2$. The gradient is $\frac{4}{2} = 2$ (a positive gradient, running *uphill*).

### Example 2

Find the gradient of the straight line joining C(6, 4) to D(12, 1).

First draw a diagram.
Mark in the changes in coordinates.

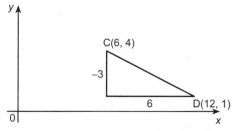

For CD, the change in the $y$ coordinate is $1 - 4 = -3$, and the change in the $x$ coordinate is $12 - 6 = 6$. The gradient is $\frac{-3}{6} = -\frac{1}{2}$ (a negative gradient, running *downhill*).

---

## Key Points

- Gradient $= \dfrac{\text{'rise'}}{\text{'run'}}$.

- Lines like this ⟋ have a *positive* gradient.

- Lines like this ⟍ have a *negative* gradient.

- Parallel lines have the same gradient.

- Always draw a diagram.

## Investigate

♦ Find the gradient of the line AB.

A(1, 1) B(11, 2) C(11, 1)

♦ Investigate the gradient of AB as point B moves closer and closer to point C. Tabulate your results. What is the gradient of the horizontal line AC?

What is the gradient of any horizontal line?

♦ Investigate the gradient of AB as point A moves closer and closer to point C. Tabulate your results. What is the gradient of the vertical line BC?

What is the gradient of any vertical line?

# Exercise 15

For questions 1–10, find the gradient of the straight line joining A to B when

**1** A is (2, 1) and B is (5, 4)      **2** A is (1, 2) and B is (5, 6)

**3** A is (2, 3) and B is (4, 4)      **4** A is (2, 1) and B is (6, 3)

**5** A is (1, 3) and B is (2, 6)      **6** A is (1, 1) and B is (3, 5)

**7** A is (−4, −1) and B is (4, 1)      **8** A is (−3, 1), and B is (6, 4)

**9** A is (−2, 2) and B is (2, 1)      **10** A is (−3, 2), and B is (3, −4)

**11** A hill has a gradient of 0.1. What is the value of $h$?

**12** A steep cliff has a gradient of 9. What is the value of $h$?

gradient = 0.1

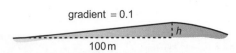

100 m

gradient = 9

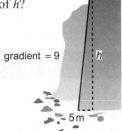

h

5 m

**13** A ladder reaches 6 m up a vertical wall and has a gradient of 4. How far is the foot of the ladder from the wall?

**14** After take-off, an aeroplane climbs in a straight line with a gradient of $\frac{1}{5}$.

When it has reached a height of 2000 m, how far has it travelled horizontally?

**15** The roof of a lean-to garden shed has a gradient of 0.35. Find the height of the shed.

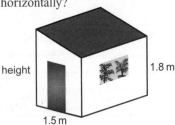

height               1.8 m

1.5 m

**16** The seats at a football stadium are on a slope with a gradient of $\frac{1}{2}$.

What is the height $h$ of the bottom seats?

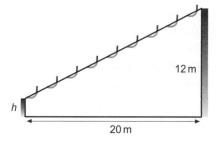

**17** A road has a gradient of $\frac{1}{15}$ for 90 m. Then there is a horizontal section 130 m long.

The final section has a gradient of $\frac{1}{25}$ for 200 m.

**a** Find the total height gained from start to finish.

**b** What is the average gradient from start to finish?

**18** The masts for London's Millennium Dome were held up during erection by wire ropes. The top of a mast, A, is 106 m above the ground, and C is vertically below A. The gradient of one wire rope, AB, is 1, and CD is 53 m.

**a** Find the gradient of AD.

**b** Find the length of BD.

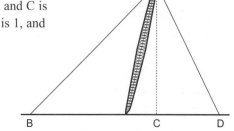

# Exercise 15★

For Questions 1–4, find the gradients of the straight lines joining A to B when

**1** A is (−4, −1), B is (4, 2)

**2** A is (−3, 1), B is (1, 6)

**3** A is (−2, 4), B is (2, 1)

**4** A is (−3, 2), B is (4, −4)

**5** The line joining A(1, 4) to B(5, $p$) has a gradient of $\frac{1}{2}$. Find the value of $p$.

**6** The line joining C(3, 1) to D(5, $q$) has a gradient of 2. Find the value of $q$.

For Questions 7–10, find, by calculating gradients, whether or not the opposite sides of quadrilateral ABCD are parallel.

**7** A is (2, 1), B is (14, 9), C is (24, 23), D is (10, 13)

**8** A is (0, 8), B is (12, 4), C is (22, 12), D is (11, 16)

**9** A is (2, 1), B is (14, 7), C is (20, 19), D is (8, 13)

**10** A is (1, 18), B is (16, 16), C is (17, 0), D is (2, 2)

**11** Brendan enjoys mountain biking. He has found that the maximum gradient up which he can cycle is 0.3, and the maximum gradient that he can safely descend is 0.5. His map has a scale of 2 cm to 1 km, with contours every 25 m.

**a** What is the minimum distance between the contours on his map that allows him to go uphill?

**b** What is the minimum distance between the contours on his map that allows him to go downhill?

**12** One of the world's tallest roller coasters is in Blackpool, England.
It has a maximum height of 72 m, and gives white-knuckle rides
at up to 140 km per hour.
The maximum drop is 65 m over a horizontal distance
of 65 m in two sections. The first section has a
gradient of 3, and the second section has a gradient of $\frac{1}{2}$.
How high above the ground is point A?

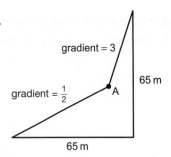

**13** The line joining $(3, p)$ to $(7, -4p)$ is parallel to the line joining $(-1, -3)$ to $(3, 7)$. Find $p$.

**14** The line joining $(-2, 1)$ to $(6, 4)$ is parallel to the line joining $(-q, 5)$ to $(4, q)$. Find $q$.

# Straight-line graphs

## Graphs of the form $y = mx + c$

### Activity 3

◆ For each of these equations, copy and complete this table of values.

| $x$ | $-2$ | $0$ | $2$ |
|-----|------|-----|-----|
| $y$ |      |     |     |

$y = x + 1$      $y = -x + 1$      $y = 2x - 1$

$y = -2x + 1$      $y = 3x - 1$      $y = \frac{1}{2}x + 2$

◆ Draw **one** set of axes, with the $x$-axis labelled from $-2$ to $2$ and the $y$-axis from $-7$ to $5$.
Plot the graphs of all six equations on this set of axes.

◆ Copy and complete this table.

The **$y$ intercept** is the value of $y$
where the line crosses the
$y$-axis. Can you see a connection
between the number in front of $x$
and the gradient?

Can you see a connection
between the number at the end of
the equation and the $y$ intercept?

| Equation | Gradient | $y$ intercept |
|----------|----------|---------------|
| $y = x + 1$ | | |
| $y = -x + 1$ | | |
| $y = 2x - 1$ | | |
| $y = -2x + 1$ | | |
| $y = 3x - 1$ | | |
| $y = \frac{1}{2}x + 2$ | | |
| $y = mx + c$ | | |

## Key Point

The graph of $y = mx + c$ is a straight line with gradient $m$ and $y$ intercept $c$.

Sketching a straight line means showing the approximate position and slope of the line *without* plotting the line. If you know the gradient and intercept you can sketch the straight line easily.

**Example 3**

Sketch these two lines.

$y = 2x - 1$          $y = -\frac{1}{2}x + 3$

$y = 2x - 1$ is a straight line with gradient 2 and intercept −1.

$y = -\frac{1}{2}x + 3$ is a straight line with gradient $-\frac{1}{2}$ and intercept 3.

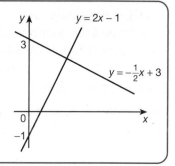

# Exercise 16

For Questions 1–14, write down the gradient and *y* intercept and then sketch the graph of the equation.

**1** $y = 3x + 5$          **2** $y = 4x + 1$          **3** $y = x - 7$          **4** $y = 2x - 3$

**5** $y = \frac{1}{3}x + 2$          **6** $y = \frac{1}{2}x + 4$          **7** $y = -\frac{1}{2}x + 5$          **8** $y = -\frac{1}{4}x + 3$

**9** $y = -\frac{1}{3}x - 2$          **10** $y = -\frac{1}{5}x - 1$          **11** $y = 4 - 2x$          **12** $y = 5 - x$

**13** $y = -2$          **14** $y = 3$

For Questions 15–18, write down the equations of the lines with gradient

**15** 3, passing through (0, −2)          **16** −1, passing through (0, 4)

**17** $\frac{1}{3}$, passing through (0, 10)          **18** −0.2, passing through (0, −5)

For Questions 19–22, write down the equations of the lines that are parallel to

**19** $y = 2x - 7$, passing through (0, 4)          **20** $y = 2 - x$, passing through (0, −1)

**21** $y = 4 - 5x$, passing through (0, −1)          **22** $y = 3x - 4$, passing through (0, 0)

**23** Write down possible equations for these sketch graphs.

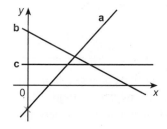

**24** Sanjay's biology project involves heating a tank of water and taking temperature readings every 5 minutes after turning on the heater.

| Time $m$ (minutes) | 5 | 10 | 15 | 20 |
|---|---|---|---|---|
| Temperature $t$ (°C) | 12 | 15 | 18 | 21 |

  **a** Plot this data on a graph of temperature against time for $0 \le t \le 20$.
     Draw a straight line to join the points.

b What was the temperature when Sanjay switched on the heater?

c Find the gradient and intercept of the line, and write down the equation of the line.

d Use your equation to find when the temperature was first 30 °C. Check this using your graph.

e Sanjay goes away and returns after 3 hours.
According to your equation, what will the temperature be after 3 hours? Is this value sensible?
What restrictions should you place on your equation?

# Exercise 16★

For Questions 1–12, write down the gradient and $y$ intercept and then sketch the graph of each equation.

**1** $y = 5x + \frac{1}{2}$      **2** $y = -4x - \frac{3}{4}$      **3** $y = -\frac{3}{4}$      **4** $y = 2\frac{1}{2}$

**5** $y = -3x + \frac{5}{2}$      **6** $y = \frac{4}{3}x + \frac{2}{3}$      **7** $y = 6x - \frac{3}{2}$      **8** $y = \frac{5}{4}x + \frac{2}{3}$

**9** $x = -1.5$      **10** $x = 3.5$      **11** $y = -\frac{2}{3}x - \frac{5}{3}$      **12** $y = -\frac{3}{4}x + \frac{5}{4}$

For Questions 13–16, write down the equations of the lines with gradient

**13** 2.5, passing through (0, −2.3)      **14** $-\frac{3}{5}$, passing through (0, 6.3)

**15** $\frac{1}{4}$, passing through (4, 2)      **16** −0.7, passing through (1, −4)

For Questions 17–20, write down the equations of the lines that are parallel to

**17** $2y = 5x + 7$, passing through (0, −3.5)      **18** $y = 8x − 14$, passing through (0, 8.4)

**19** $7x + 6y = 13$, passing through (6, 7)      **20** $9x − 5y = −3$, passing through (9, −3)

**21** Write down possible equations for these sketch graphs.

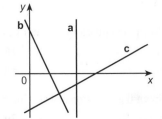

**22** Jack plants a magic bean at noon on 1 April. He measures the height of the beanstalk every hour until it becomes too high for him to measure. These are his results.

| Time ($t$) | 1pm | 2pm | 3pm | 4pm | 5pm | 6pm | 7pm |
|---|---|---|---|---|---|---|---|
| Height $h$ (m) | 0 | 0 | 0.4 | 0.7 | 1.0 | 1.3 | 1.6 |

a Plot this data on a graph of height against time for $0 \leqslant t \leqslant 7$.
Draw a straight line to join the last five points.

b Find the gradient and intercept of this line. Write down its equation.

c According to your equation, what is the height at 1pm? Is this a sensible value?
What restrictions should you place on your equation?

d The giant's castle is 100 m up in the air. When will the beanstalk reach the castle?
What assumptions have you made?

e Jack is very impatient. He decides that the beanstalk is growing too slowly. At noon on 2 April, he gives the beanstalk some fertiliser, which makes the beanstalk grow twice as fast as before.
Find an equation for $h$ in terms of $t$ for this. Use it to find the time when the beanstalk will now reach the castle.

# Activity 4

Equipment needed: a cylinder with a diameter of 5–10 cm (a baked-bean tin or cardboard tube is ideal), a length of string about 30 times as long as the diameter of the cylinder, a ruler and graph paper.

◆ Wrap the string tightly around the cylinder, keeping the turns close together. Ask a friend to draw a straight line across the string while you hold the ends of the string.

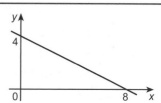

◆ Unwind the string. Measure the distance of each mark from the first mark. Enter your results in a table.

◆ Plot these points on a graph of $D$ against $M$. (Plot $D$ on the vertical axis and $M$ on the horizontal axis.) Draw the best straight line through these points.

◆ Calculate the gradient of the line, and then write down the equation of the line.

◆ The gradient should equal $\pi d$, where $d$ is the diameter of the cylinder. Use your gradient to work out an estimate for $\pi$.

◆ Repeat the activity with various cylinders, and obtain further estimates for $\pi$.

| Mark $M$ | Distance $D$ (cm) |
|---|---|
| 1 | 0 |
| 2 | |
| 3 | |
| 4 | |
| 5 | |
| 6 | |

## Graphs of the form $ax + by = c$

The graph of $3x + 4y = 12$ is a straight line. The equation can be rearranged as $y = -\frac{3}{4}x + 3$, showing that the graph is a straight line with gradient $-\frac{3}{4}$ and $y$ intercept $(0, 3)$.

An easy way to draw or sketch this graph is to find where the graph crosses the axes.

---

**Example 4**

Sketch the graph of $x + 2y = 8$.

Substituting $y = 0$ gives $x = 8$, which shows that $(8, 0)$ lies on the line.

Substituting $x = 0$ gives $y = 4$, which shows that $(0, 4)$ lies on the line.

---

# Exercise 17

For Questions 1–10, find where the graph crosses the axes and sketch the graph.

1  $2x + y = 6$      2  $x + 3y = 9$      3  $3x + 2y = 12$      4  $4x + 2y = 8$
5  $4x + 5y = 20$      6  $3x + 8y = 24$      7  $x - 2y = 4$      8  $x - 3y = 9$
9  $4y - 3x = 24$      10  $2y - 5x = 10$

11  A firm selling CDs finds that the number sold ($N$ thousand) is related to the price ($£P$) by the formula $6P + N = 90$.

    a  Draw the graph of $N$ against $P$ for $0 \le P \le 90$ (the vertical axis should be the $N$ axis, and the horizontal axis should be the $P$ axis).

    b  Use your graph to find the price when 30 000 CDs are sold.

    c  Use your graph to find the number sold if the price of a CD is set at £8.

    d  Use your graph to find the price if 90 000 CDs are sold. Is this a sensible value?

# Exercise 17★

For Questions 1–10, find where the graph crosses the axes and sketch the graph.

**1**  $6x + 3y = 36$     **2**  $4x + 7y = 56$     **3**  $6x + 4y = 21$     **4**  $8x + 5y = 12$

**5**  $4x - 5y = 30$     **6**  $5x - 8y = 32$     **7**  $7y - 2x = 21$     **8**  $5y - 4x = 12$

**9**  $6x - 7y = -21$    **10**  $4x - 9y = -27$

**11** Courtney has started playing golf. To try to reduce her handicap she has lessons with a professional. She keeps a record of her progress.

| Week (W)     | 5  | 10 | 20 | 30 |
|--------------|----|----|----|----|
| Handicap (H) | 22 | 21 | 20 | 19 |

   **a** Plot these points on a graph of $H$ against $W$. Draw in the best straight line.
   **b** What was Courtney's handicap before she started having lessons?
   **c** Find the gradient and intercept of the line.
      Write down the equation of the line in the form $ax + by = c$.
   **d** To have a trial for the youth team, Courtney needs to have a handicap of less than 12.
      Use your equation to find how many weeks it will take Courtney to reduce her handicap to 12.
      Do you think this is a reasonable time?

## Activity 5

Your aim is to find the equation of the straight line joining two points.

◆ Plot the points A(1, 3) and B(5, 5) on a graph. Find the gradient of AB.

◆ Calculate where the straight line passing through AB will intercept the $y$-axis.

◆ Write down the equation of the straight line passing through A and B.

◆ Use this method to find the equation of the straight line joining these pairs of points:

   (−2, 1) and (−1, 4)       (−3, 4) and (6, 1)       (−2, −1) and (4, 3)

# Exercise 18 (Revision)

**1** Find the gradient of the straight line joining A to B when
   **a** A is (3, 4), B is (5, 8)           **b** A is (−1, 2), B is (1, 0)

**2** The foot of a ladder is 1.5 m from the base of a vertical wall. The gradient of the ladder is 3. How far does the ladder reach up the wall?

**3** Write down the gradient and $y$ intercept of the graph of
   **a** $y = 3x - 2$          **b** $y = -2x + 5$

**4** Write down the equations of the lines with
   **a** gradient 2, passing through (0, −1)         **b** gradient −3, passing through (0, 2)

**5** Sketch the following graphs.
   **a** $y = 2x - 3$          **b** $y = 4 - x$          **c** $2x + 5y = 10$

6  Which of these lines are parallel?

$y = 2x + 4$          $x - 3y = 1$          $4x = 2y + 7$           $9y = 3x + 4$

$4x - 3y = 12$          $3x - 4y = 12$          $3y = 4x - 1$          $4y = 3x + 7$

7  A temperature $F$ in degrees Fahrenheit is related to the temperature $C$ in degrees Celsius by the
formula $F = \frac{9}{5}C + 32$.

   **a**  Draw a graph of $F$ against $C$ for $-50 \leqslant C \leqslant 40$.

   **b**  Use your graph to estimate $80\,°F$ and $-22\,°F$ in degrees Celsius and $25\,°C$ in degrees Fahrenheit.

   **c**  Use your graph to find which temperature has the same value in both degrees Fahrenheit and
degrees Celsius.

Write down the equations that will produce these patterns.

**8**                                        **9**

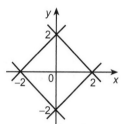          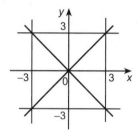

# Exercise 18★ (Revision)

1  Find the gradient of the straight line joining A to B when

   **a**  A is $(-4, -1)$, B is $(-1, -2)$

   **b**  A is $(-2, -8)$, B is $(1, -2)$

2  The Leaning Tower of Pisa is 55 m high, and the gradient
of its lean is 11. By how much does the top overhang the bottom?

3  Sketch the following graphs

   **a**  $y = 3x - 2$          **b**  $y = 3 - 2x$          **c**  $2y = 5 - x$          **d**  $5x + 3y = 10$

4  Show that A$(-5, 2)$, B$(-1, 5)$, C$(5.6, -0.5)$, D$(1, -3)$ is a trapezium.

5  Find $b$ such that the line from the origin to $(3, 4b)$ is parallel to the line from the origin to $(b, 3)$.

6  Find the equation of the line passing through $(6, 4)$ that is parallel to $3y = x + 21$.

7  A recipe book gives the time $F$ for fast roasting in a hot oven as 20 minutes plus 20 minutes per
pound. The time $S$ for slow roasting in a moderate oven is given as 35 minutes plus 35 minutes per
pound. The total weight is $P$ pounds.

   **a**  Write down the equations relating $F$ to $P$ and $S$ to $P$.

   **b**  Draw both graphs on the same axes for $0 \leqslant P \leqslant 12$.

   **c**  Use your graphs to estimate the cooking time for 5.5 pounds (slow roasting) and 7 pounds (fast
roasting).

   **d**  What is the weight of a cut of meat that takes 3 hours 45 minutes to cook in a hot oven?

   **e**  A 4-pound roast is put in the oven at 11am. The temperature is set midway between hot and
moderate by mistake. By drawing a third line on your graph, estimate when the roast will be
cooked.

8  A spider is descending over Little Miss Muffet, aiming for her bowl of whey. The spider's heights

above the ground are in the table.

| Time $t$ (s) | 5 | 10 | 15 | 20 |
|---|---|---|---|---|
| Height $h$ (m) | 2.6 | 2.1 | 1.7 | 1.3 |

**a** Plot this data on a graph of $h$ against $t$. Draw in the best straight line.

**b** Find the gradient and intercept of the line. Write down the equation of the line.

**c** Use your equation to find when the spider was first 2 m above the ground. Check your answer using your graph.

**d** According to your equation, how far above the ground is the spider after it has been descending for 1 minute? Is this a sensible value? What restrictions should you place on your equation?

**e** The bowl of whey is 60 cm above the ground. When will the spider reach the whey? What assumptions have you made?

# Shape and space 1

## Basic principles

### Triangles

(A broken line indicates an axis of symmetry.)

**Isosceles triangle**

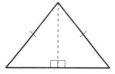

Acute, obtuse and right angles are possible.

**Equilateral triangle**

The rotational symmetry is of order 3.

### Angle properties

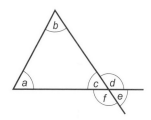

$$a + b + c = 180° \quad \text{(Angle sum of triangle)}$$
$$c + d = 180° \quad \text{(Angles on straight line)}$$
$$c = e \quad \text{(Vertically opposite angles)}$$
$$c + d + e + f = 360° \quad \text{(Angles at a point)}$$

Since $\quad d = 180° - c \quad$ (Angles on straight line)

and $\quad a + b = 180° - c \quad$ (Angle sum of triangle)

$$d = a + b \quad \text{(Exterior angle of triangle)}$$

### Parallel lines

Alternate angles are equal.

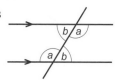

Corresponding angles are equal.

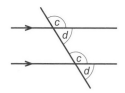

### Quadrilaterals

**Square**

Rotational symmetry of order 4

**Rhombus**

Rotational symmetry of order 2

**Rectangle**

Rotational symmetry of order 2

**Parallelogram**

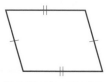

Rotational symmetry of order 2

**Arrowhead**

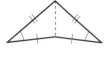

**Kite**

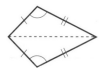

Acute, obtuse and right angles are possible.

**Trapezium**

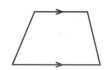

There is no symmetry. Right angles are possible.

**Isosceles trapezium**

## Activity 6

Copy and complete this table to show which properties are true for each type of quadrilateral.

| Property | Square | Rectangle | Rhombus | Parallelogram | Arrowhead | Trapezium | Kite |
|---|---|---|---|---|---|---|---|
| The diagonals are equal in length. | | | | No | | | |
| The diagonals bisect each other. | | | | Yes | | | |
| The diagonals are perpendicular. | | | | No | | | |
| The diagonals bisect the angles at the corners. | | | | No | | | |
| Both pairs of opposite angles are equal. | | | | Yes | | | |

## Angles of a regular polygon

**Interior angles of an *n*-sided polygon**

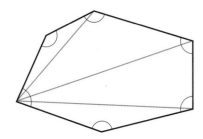

**Exterior angles of an *n*-sided polygon**

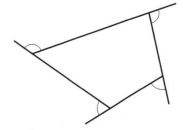

The polygon can be divided into $(n - 2)$ triangles.
Therefore the angle sum $= (n - 2) \times 180°$.

The angles add up to one complete turn.
Therefore they sum to $360°$.

In a *regular* polygon, all the interior angles are equal and all the exterior angles are equal.

Each interior angle $= \dfrac{(n - 2) \times 180°}{n}$

Each exterior angle $= \dfrac{360°}{n}$

Interior and exterior angles add up together to $180n°$.

---

### Example 1

Find the angle sum of a polygon with seven sides.

$$n = 7$$
$$\text{Angle sum} = (7 - 2) \times 180°$$
$$= 5 \times 180° = 900°$$

---

**Example 2**

A regular polygon has ten sides. Find the size of each interior and each exterior angle.

$$n = 10$$

Interior angle $= \dfrac{(10 - 2) \times 180°}{10}$

Or, find the exterior angle first.

$$= 8 \times 18° = 144°$$

Exterior angle $= \dfrac{360°}{10} = 36°$

Exterior angle $= 180° - 144° = 36°$

Interior angle $= 180° - 36° = 144°$

# Exercise 19

Calculate the size of each lettered angle.

**1**

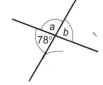

**2**

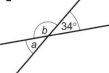

**3**

**4**

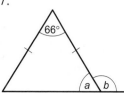

**5**

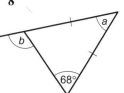

**6**

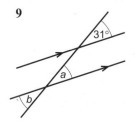

**7.**

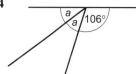

**8**

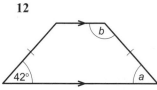

**9**

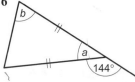

**10**

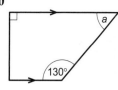

**11**

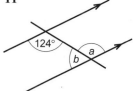

**12**

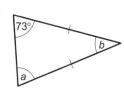

**13**

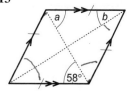

**14**

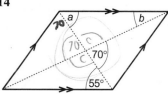

**15** A regular polygon has eight sides.

    **a** Calculate the size of the exterior angles.

    **b** Calculate the size of the interior angles.

    **c** Calculate the sum of the interior angles.

**16** A heptagon has seven sides.

    **a** Calculate the sum of the interior angles.

    **b** Calculate the size of each interior angle.

**17** The angle sum of an irregular polygon is 1260°. How many sides has it?

**18** The interior angle of a regular polygon is 150°. How many sides has it?

**19** Calculate the size of the two unknown angles.

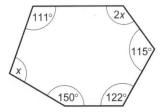

**20** The exterior angle of a regular polygon is 24°.

    **a** Calculate the number of sides.

    **b** Calculate the sum of the interior angles.

# Exercise 19★

Calculate the size of each lettered angle.

**1**

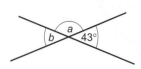

**2**

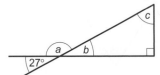

**3**

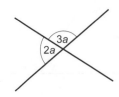

**4**

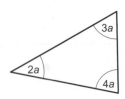

**5**

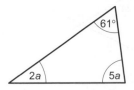

**6**

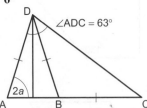

**7** Express ∠ABC in terms of *x*.

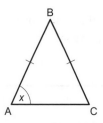

**8** Express ∠EDF in terms of *y*.

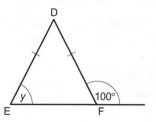

**9** Find the size of angle *x*.
Write out a 'solution' giving a reason for each step.

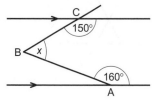

**10** DEFG is an isosceles trapezium.
Calculate the size of angle *x*, giving a reason for each step.

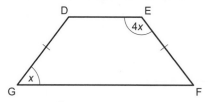

**11** ABCD is a rectangle. Find angles *a* and *b*, giving reasons with each step.

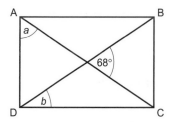

**12** Find, giving reasons, the size of angle *a*.

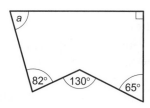

**13** Find angles *a* and *b*, giving reasons with each step.

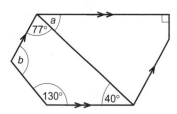

**14** Find angles *a* and *b*, giving reasons with each step.

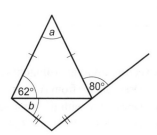

**15** The interior angle of a regular polygon is nine times larger than the exterior angle. How many sides has the polygon?

**16** There are two polygons. The larger one has three times as many sides as the smaller one, and its angle sum is four times bigger. How many sides has the smaller polygon?

## Investigate

◆ How might the 'exterior angle' of a polygon be defined when the interior angle is reflex? Try to define it in such a way that the exterior angles can still be made to add up to 360°.

◆ A star of David is an intersection of two equilateral triangles. What are the exterior angles for the star of David?

# Constructions

The properties of triangles and quadrilaterals are used in the standard ruler and compass constructions.

**Remember**

### Constructing a 60° angle (equilateral triangle)

◆ Draw an arc from A to intersect AB at P.
◆ With the same radius, draw an arc from P to intersect the first arc at Q.
◆ Draw the line AQ. ∠**BAQ = 60°**.

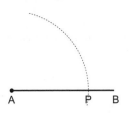 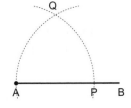 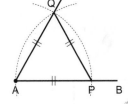 

### Bisecting an angle (diagonals of a rhombus or kite)

◆ Draw an arc from A to intersect the lines at P and Q.
◆ Draw arcs, with the same radius, from P and Q to intersect each other at R.
◆ Draw AR. ∠**PAR = ∠QAR**.

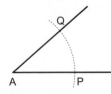 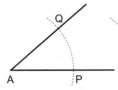 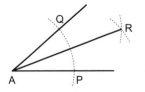 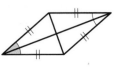

**Constructing a perpendicular bisector of a line (diagonals of a rhombus)**
- Draw arcs from A, with the same radius, above and below the line.
- With the same radius, draw arcs from B to intersect those from A, above and below the line. Label these two intersections P and Q.
- Draw the line PQ. **PQ is the perpendicular bisector of AB**.
  (Note: R is the mid-point of AB.)

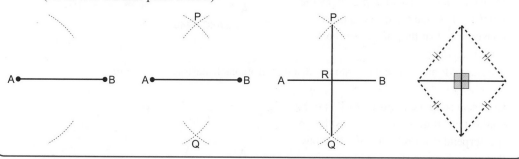

**Remember**

**Constructing a perpendicular from a point X on the line**
- With the same radius, draw arcs from the point X to cut the line at either side. Label these points A and B.
- The perpendicular bisector of this part AB of the line will pass through X.

# Loci

**A locus is the position of a set of points that obey a particular rule.** It can be a line, curve or region, depending on the rule.

**Remember**

**Common loci**

Points on the **angle bisector** of ∠BAC are equidistant from the lines AB and AC.

Points on the **perpendicular bisector** of PQ are equidistant from P and Q.

Points on the **circle** with centre at X are equidistant from X.

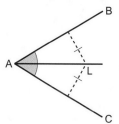

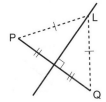

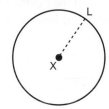

**Example 3**

A 10 km road race attracts so many runners that it is decided to split up the beginning of the race and have three different starts in a park.

The top diagram shows the three starting positions, A, B and C. On an accurate scale drawing, show the point P where the three routes must converge so that they are all of the same distance.

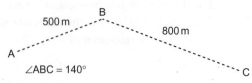

Measure AP, and hence calculate the distance from each starting position to P.

Choose a scale 1 cm : 50 m. As P is to be equidistant from A and B, it must be on the perpendicular bisector of AB. Draw this.

As P is also to be equidistant from B and C, it must be on the perpendicular bisector of BC. Draw this.

The point of intersection is equidistant from A and B and C.

AP = 19 cm. Therefore the distance from each starting position to P is 950 m.

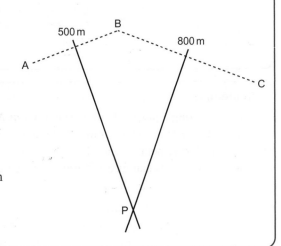

# Exercise 20

Questions 1–12 should be done on plain paper. Protractors can be used, and all construction arcs are to be shown. Make a rough sketch of the figure before you begin a construction.

1   Construct triangle ABC, where AB = 8 cm, ∠A = 60°, and ∠B = 45°. Measure the length of AC.

2   Construct triangle DEF, where DE = 7 cm, ∠D = 75° and ∠E = 30°. Measure the length of DF.

3   Construct triangle FGH, where FG = 5 cm, GH = 6 cm and FH = 7 cm.
    Construct the perpendicular from G to the line FH.
    Measure the length of this perpendicular, and hence calculate the area of the triangle.

4   Construct the figure JKLM and
    measure the lengths of JK and JM.

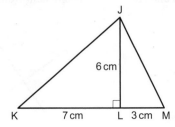

**5 a** Construct triangle PQR, where PQ = 7 cm, QR = 8 cm and RP = 9 cm.

  **b** Construct the perpendicular bisectors of QR and RP. These two lines intersect at the point S.

  **c** Measure the lengths of PS, QS and RS. Comment on your answers.

**6** Construct triangle XYZ, where ∠Z = 90°, XY = 8 cm and XZ = 6 cm. Measure the length of YZ.

**7** A goat is tied in a field to a horizontal rail 30 m long by a chain which is 20 m long. The chain is free to slide along the rail. Draw a scale diagram that shows the area in which the goat can graze.

**8** A fierce dog is tethered by a rope 10 m long to a post 6 m from a straight path. If the path is 2 m wide, draw a scale diagram to illustrate the area of path along which a walker would be in danger.

**9** P, Q and R represent the positions of three radio beacons. Signals from P have a range of 300 km, Q has a range of 350 km and R has a range of 200 km.

  **a** Reproduce the diagram and shade the region in which all three signals can be received.

  **b** Measure the shortest distance from Q to this region.

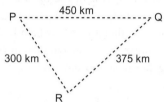

**10** Gas rig Beta is 7 km from gas rig Gamma on a bearing of 210°. Bearings are measured from North in a clockwise direction. The region less than 4 km from gas rig Beta is an exclusion zone for ships.

  **a** Using a scale of 1 cm to 1 km, draw a scale diagram showing the positions of the gas rigs, and shade the region that represents the exclusion zone.

  **b** A boat sails so that it is always the same distance from Gamma and Beta. Draw the route taken by the boat.

  **c** For what distance is the boat within 4 km of oil rig Beta?

**11** PQ is a breakwater, 750 m long, with a lighthouse at Q. Using a scale drawing, find the distance from P of a ship that is 190 m from the breakwater and 280 m from the lighthouse.

**12** Some treasure is hidden in a field in which there are three trees: an ash A, a beech B and a chestnut C. BC = 300 m, CA = 210 m and AB = 165 m.

  The treasure is the same distance from the chestnut as from the beech, and it is 60 m from the ash. Use a scale drawing to find out how far the treasure is from the beech tree.

**13** The diagram shows a sheep pen that is in the middle of a field. A sheepdog is tethered at the corner C by a rope 6 m long.

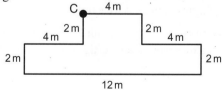

  **a** Draw a scale diagram of the pen, and shade the region that the dog can cover if he is outside the pen.

  **b** Shade the region that he can cover if he is inside the pen.

# Exercise 20★

**1** Construct a rhombus with diagonals of length 9 cm and 6 cm. Measure the length of the side.

**2** Construct a parallelogram with diagonals that intersect at 60° and have lengths of 7 cm and 8 cm. Measure the length of the longer side.

**3** Construct the isosceles trapezium TUVW, where TU = 8.5 cm, VW = 5 cm and ∠UTW = 75°. Measure the length of TW.

**4** Construct the arrowhead ABCD, where BD = 6 cm and is the line of symmetry, ∠ABC = 30° and ∠ADC = 90°. Measure the length of BC.

**5** The block ABCD is tipped over to the flat position by rotation about C. It is then put into an upright position again by rotation about B (which is then on the floor). Draw a horizontal line, and then draw the locus of A during these two movements.

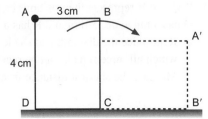

**6** The front wheel of a bicycle has a radius of 30 cm. It has a reflector that is positioned 20 cm from the centre.
Sketch the locus of the reflector as the bicycle moves forward and the wheel completes one revolution. Start with the reflector in its highest position.

**7** A ladder is 15 m long. It is resting almost vertically against a wall.
The bottom of the ladder is pulled out from the wall and allowed to slide into the horizontal position.
Draw x- and y-axes from 0 to 15, and make a scale drawing of the locus of the middle rung of the ladder. (A 15 cm ruler may be useful.)

**8** A and B are two points that are exactly 5 cm apart. Plot the locus of the point P such that AP + PB = 9 cm. (A short piece of string may be useful.)

**9** In a sailing race, boats follow a triangular course around three buoys A, B and C.
B is 1 km due West of A, and C is 800 m from A on a bearing of 240°.
The wind is from the East, and boats cannot sail on a bearing between 050° and 130°.
  **a** Make a scale drawing, showing the position of the three buoys.
  **b** Draw on the diagram the shortest clockwise route.
  **c** Draw on the diagram the shortest anticlockwise route.

**10** An athletics track can be described as a locus: 'point P is x metres from a fixed line of length y metres'. Show this locus on a diagram, and investigate the values of x and y that would produce a 400 m track.

# Exercise 21 (Revision)

**1** Find angles *a*, *b* and *c*.

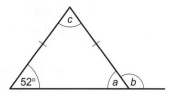

**2** A regular polygon has nine sides.
  **a** Calculate the size of the exterior angles.
  **b** Calculate the size of the interior angles.
  **c** Calculate the sum of the interior angles.

**3** The angle sum of a regular polygon is 1800°.
  **a** Find the number of its sides.
  **b** Find the exterior angle.

**4** Find angles *a* and *b*.

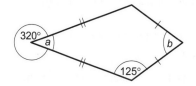

**5** Find angles *a* and *b*.

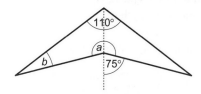

**6** Construct triangle PQR such that PQ = 8 cm , ∠PQR = 60°, and ∠RPQ = 75°.
Construct the perpendicular from R to intersect PQ at S.
Measure RS, and hence calculate the area of the triangle PQR.

**7** The diagram represents a rectangular lawn. There is a water sprinkler at the point E, halfway between C and D. The sprinkler wets the area within 15 m from E.
  **a** Using a scale of 1 cm to 5 m, draw a diagram of the garden, and shade the area wetted by the sprinkler.
  **b** A child is playing on the lawn. She starts at A, and then runs across the lawn, keeping the same distance from the sides AD and AB until she is 10 m from the side DC. She then runs straight to the corner B.
  Draw the path that the child takes onto your diagram.
  **c** What length of her path is wet?

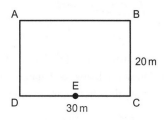

# Exercise 21★ (Revision)

**1** Find angles *a*, *b*, *c* and *d*.

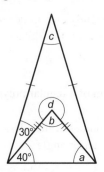

**2** Find angle *x*, giving a reason with each step of your working.

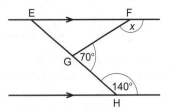

**3** **a** Construct a rhombus with diagonals of length 5 cm and 12 cm.
   **b** Measure the length of the side of the rhombus.
   **c** Calculate the area of the rhombus.

**4** What do you call a regular polygon if
   **a** its interior angle is twice the exterior angle?
   **b** its exterior angle is twice the interior angle?

**5** HIJK is a parallelogram. ∠KJL = 49°, ∠JLI = 64°, ∠HKL = 3 × ∠JKL. Show that ∠JKH = 60°.

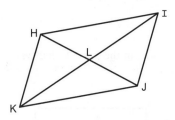

**6** LNMP is a quadrilateral. LO = OM and PO = ON. Find the size of angle *x*.

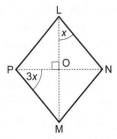

**7** **a** Construct triangle ABC such that AB = 10 cm, BC = 9 cm and AC = 8 cm.
   **b** Draw the locus of P such that P is equidistant from AB and AC.
   **c** Draw the circumcircle and measure its radius.

**8** Three radio tracking stations are at A, B and C. B is 120 km North of A and C is 100 km East of A. All three stations are able to track the direction, to within 5°, of a radio transmission. A ship issues a mayday call. It is tracked by A at 045° ± 5°, by B at 120° ± 5°, and by C at 340° ± 5°. Draw a scale diagram to show the positions of A, B and C, and shade the area that gives the possible position of the ship. From your diagram, estimate the greatest and least distance of the ship from C.

# SETS 1

The concept of a set is a simple but powerful idea. The theory of sets is mainly due to the work of the German mathematician, Cantor. It led to arguments and controversy, but by the 1920s his ideas were generally accepted and led to great advances in mathematics.

## Basic ideas

> **Remember**
>
> A **set** is a collection of objects, which are called the elements or members of the set.

The objects can be numbers, animals, ideas, colours, in fact anything you can imagine.
A set can be described by **listing** all the members of the set, or by giving a **rule** to describe the members. The list or rule is enclosed by **braces** { }.

> **Example 1**
>
> *A set described by a list:*
> {Abigail, Ben, Carmel} is the set consisting of the three people called Abigail, Ben and Carmel.

> **Example 2**
>
> *A set described by a rule:*
> {even numbers between 1 and 11} is the set consisting of the five numbers 2, 4, 6, 8, 10.

Sets are often called by a single capital letter. A = {odd numbers between 2 and 10} means A is the set consisting of the four numbers 3, 5, 7, 9.

> **Remember**
>
> The **number of elements** in the set A is written as $n(A)$.

Sets can be infinite in size, for example the set of prime numbers.

> **Remember**
>
> **Membership** of a set is indicated by the symbol $\in$ and non-membership by the symbol $\notin$.

> **Example 3**
>
> If E = {2, 8, 4, 6, 10} and F = {even numbers between 1 and 11}, then:
>
> $n(E) = 5$, $n(F) = 5$; in other words both E and F have the same number of elements
>
> $3 \notin E$ means 3 is not a member of the set E
>
> $6 \in F$ means 6 is a member of the set F
>
> E = F because both E and F have the same members. The order in which the members are listed does not matter.

> **Remember**
>
> The **empty set** is the set with no members. It is denoted by the symbol $\varnothing$ or { }.

The concept of the empty set might seem strange, but it is very useful.

> **Example 4**
>
> Give two examples of the empty set.
>
> **a** The set of people you know over 4 m tall.    **b** The set of odd numbers divisible by two.

## Exercise 22

**1** Write down two more members of each of these sets.
   **a** {carrot, potato, pea, …}          **b** {red, green, blue, …}
   **c** {a, b, c, d, …}                    **d** {1, 3, 5, 7, …}

**2** Write down two more members of each of these sets.
   **a** {apple, pear, cherry, …}          **b** {sock, tie, blouse, …}
   **c** {+, ×, √, …}                       **d** {2, 4, 6, 8, …}

**3** List these sets.
   **a** {days of the week}                 **b** {square numbers less than 101}
   **c** {subjects you study at school}     **d** {prime numbers less than 22}

**4** List these sets.
   **a** {continents of the world}          **b** {odd numbers between 2 and 10}
   **c** {TV programmes you enjoy}          **d** {all factors of 12}

**5** Describe these sets by a rule.
   **a** {a, b, c, d}                       **b** {Tuesday, Thursday}
   **c** {1, 4, 9, 16}                      **d** {2, 4, 6, 8, …}

**6** Describe these sets by a rule.
   **a** {u, v, w, x, y, z}                 **b** {January, June, July}
   **c** {1, 2, 3, 4, 5}                    **d** {2, 3, 5, 7, 11, …}

**7** Which of these statements are true?
   **a** cat $\in$ {animals with two legs}   **b** Square $\notin$ {parallelograms}
   **c** 1 $\in$ {prime numbers}             **d** 2 $\notin$ {odd numbers}

**8** Which of these statements are true?
   **a** Jupiter $\notin$ {Solar System}     **b** Triangle $\in$ {polygons}
   **c** 3 $\notin$ {odd numbers}            **d** 17 $\in$ {prime factors of 357}

**9** Which of these are examples of the empty set?
   **a** The set of men with no teeth
   **b** The set of months of the year with 32 days
   **c** The set of straight lines drawn on the surface of a sphere
   **d** The set of prime numbers between 35 and 43.

**10** Which of these are examples of the empty set?

  **a** The set of spiders with six legs

  **b** The set of days of the week that start with the letter F

  **c** The set of triangles with four sides

  **d** The set of even numbers that give an odd number when divided by two.

# Exercise 22★

**1** Write down two more members of each of these sets.

  **a** {Venus, Earth, Mars, ...}      **b** {triangle, square, hexagon, ...}

  **c** {hydrogen, iron, aluminium, ...}      **d** {1, 4, 9, 16, ...}

**2** Write down two more members of the following sets.

  **a** {tennis, basketball, cricket, ...}      **b** {cube, sphere, icosahedron, ...}

  **c** {sin, DEL, ×, STO, ...}      **d** {1, 3, 6, 10, ...}

**3** List these sets.

  **a** {all possible means of any two elements of 1, 3, 5}

  **b** {different digits of $11^4$}

  **c** {all factors of 35}

  **d** {powers of 10 less than one million}

**4** List these sets.

  **a** {all arrangements of the letters SET}      **b** {factors of $x^3 - x$}

  **c** {Fibonacci numbers less than 20}      **d** {multiples of 3 less than 13}

**5** Describe these sets by a rule.

  **a** {spring, summer, autumn, winter}      **b** {circle, ellipse, parabola, hyperbola}

  **c** {1, 2, 4, 8, 16}      **d** {(3, 4, 5), (5, 12, 13), (7, 24, 25), ...}

**6** Describe these sets by a rule.

  **a** {red, amber, green}      **b** {α, β, γ, δ, ε}

  **c** {1, 5, 10, 10, 5, 1}      **d** {123, 132, 213, 231, 312, 321}

**7** Which of these statements are true?

  **a** Everest ∈ {mountains over 2000 m high}      **b** 2000 ∉ {leap years}

  **c** $2x + 3y = 5$ ∈ {straight-line graphs}      **d** $-2$ ∈ {solutions of $x^3 - 2x^2 = 0$}

**8** Which of these statements are true?

  **a** shark ∈ {fish}      **b** C ∉ {Roman numerals}

  **c** π ∈ {integers}      **d** $x - y$ ∉ {factors of $x^2 - y^2$}

**9** Which of these are examples of the empty set?

  **a** The set of three-legged kangaroos

  **b** The set that has the numeral zero as its only member

  **c** The set of common factors of 11 and 13

  **d** The set of solutions of $x^2 = -1$.

**10** Which of these are examples of the empty set?

  **a** The set of mermaids

  **b** The set of prime numbers greater than one million

  **c** The set of square numbers between 150 and 170

  **d** The set of English words with five consecutive vowels.

# Venn diagrams

Sets can be shown in a diagram called a **Venn diagram** after the English mathematician John Venn (1834–1923). The members of the set are shown within a closed curve.

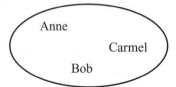   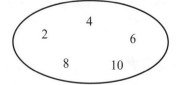

When the number of elements in a set is so large that they cannot all be shown, then a simple closed curve is drawn to indicate the set. If T = {all tabby cats} then this is shown on a Venn diagram as

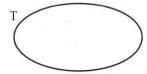

If C = {all cats in the world}, then T and C can be shown on a Venn diagram as

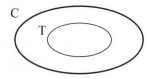

The set T is shown inside the set C because every member of T is also a member of C.

---

**Remember**

If T is inside C, then T is called a **subset** of C. This is written as T ⊂ C.

---

**Example 5**

A = {1, 2, 3, 4, 5, 6, 7, 8, 9}

**a**  List the subset O = {odd numbers}.

**b**  List the subset P = {prime numbers}.

**c**  Is Q = {8, 4, 6} a subset of A?

**d**  Is R = {0, 1, 2, 3} a subset of A?

*Answers*

**a**  O = {1, 3, 5, 7, 9}

**b**  P = {2, 3, 5, 7}

**c**  Q is a subset of A (Q ⊂ A) because every member of Q is also a member of A.

**d**  R is not a subset of A (R ⊄ A) because the element 0 is a member of R but is not a member of A.

If the problem was only about cats in this world and wasn't concerned about cats outside this world, then it is more usual to call the set C the **universal set**, denoted by $\mathscr{E}$. The universal set contains all the elements being discussed in a particular problem, and is shown as a rectangle.

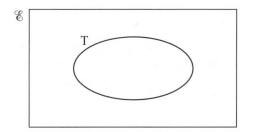

If the problem was only about cats in Rome then $\mathscr{E}$ = {all cats in Rome}; the Venn diagram does not change. If we knew there were 10 000 cats in Rome, and 1000 were tabby cats, then these numbers can be entered on the Venn diagram. The diagram shows that there are 9000 cats outside T.

The cats outside T are all non-tabby cats. This set is denoted by T′ and is known as the **complement** of T.

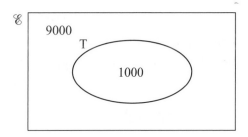

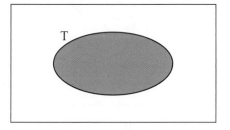

T shown shaded

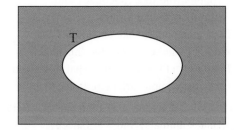

T′ shown shaded

## Intersection and union

Sets can overlap. Let M = {all male cats}. T and M overlap because some cats are both tabby and male. T and M are shown on this Venn diagram:

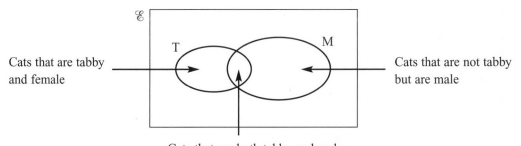

Cats that are tabby and female

Cats that are not tabby but are male

Cats that are both tabby and male

The set of cats that are both tabby and male is where the sets T and M overlap.

---

**Remember**

Where T and M overlap is called the **intersection** of the two sets T and M, and is written $T \cap M$.

---

## Example 6

$\mathscr{E}$ = {all positive integers less than 10}, P = {prime numbers less than 10}
and O = {odd numbers less than 10}.

**a** Illustrate these sets on a Venn diagram.

**b** Find the set P∩O and $n$(P∩O).

**c** List P′.

*Answers*

**a** The set P∩O is shown shaded on the Venn diagram.

**b** From the Venn diagram, P∩O = {3, 5, 7}
and $n$(P∩O) = 3.

**c** P′ is every element not in P, so P′ = {1, 4, 6, 8, 9}.

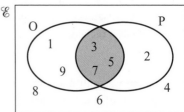

# Exercise 23

**1** On the Venn diagram, $\mathscr{E}$ = {pupils in a class},
C = {pupils who like chocolate} and
T = {pupils who like toffee}.

**a** How many pupils like chocolate?

**b** Find $n$(T) and express what this means in words.

**c** Find $n$(C∩T) and express what this means in words.

**d** How many pupils are there in the class?

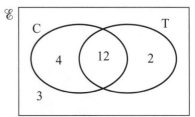

**2** On the Venn diagram, $\mathscr{E}$ = {animals in a field},
B = {black animals} and S = {sheep}.

**a** How many animals are there in the field?

**b** How many non-black sheep are there?

**c** Find $n$(B∩S) and express what this means in words.

**d** How many black animals are there?

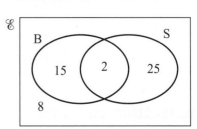

**3** $\mathscr{E}$ = {a, b, c, d, e, f, g, h, i, j}, A = {a, c, e, g, i},
B = {c, d, e, f}.

**a** Illustrate this information on a Venn diagram.

**b** List A∩B and find $n$(A∩B).

**c** Does A∩B = B∩A?

**d** Is B a subset of A? Give a reason for your answer.

**4** $\mathscr{E}$ = {a, b, c, d, e, f, g, h, i, j}, C = {c, d, e}, D = {b, c, d, e, f, g, h}.

**a** Illustrate this information on a Venn diagram.

**b** List C∩D and D∩C.

**c** Does C = C∩D?

**d** Is C a subset of D? Give a reason for your answer.

**5** $\mathscr{E}$ = {all cars in the world}, P = {pink cars}, R = {Rolls-Royce cars}.

**a** Describe the set P∩R in words.

**b** If P∩R = ∅, describe what this means.

**6** $\mathscr{E}$ = {all the clothes in a shop}, J = {set of jeans}, Y = {yellow clothes}.

**a** Describe the set J∩Y in words.

**b** If J∩Y ≠ ∅, describe what this means.

**7** On the Venn diagram, $\mathscr{E}$ = {people at a disco},
P = {people who like pop music}, C = {people who
like classical music} and J = {people who like jazz}.
   **a** How many people liked pop music only?
   **b** How many liked pop music and classical music?
   **c** How many liked jazz and classical music, but not
   pop music?
   **d** How many liked all three types of music?
   **e** How many people were at the disco?

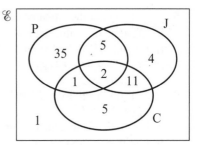

**8** On the Venn diagram, $\mathscr{E}$ = {ice-creams in a shop},
C = {ice-creams containing chocolate},
N = {ice-creams containing nuts} and
R = {ice-creams containing raisins}.
   **a** How many ice-creams contain both chocolate and nuts?
   **b** How many ice-creams contain all three ingredients?
   **c** How many ice-creams contain just raisins?
   **d** How many ice-creams contain chocolate and raisins
   but not nuts?
   **e** How many different types of ice-creams are there
   in the shop?

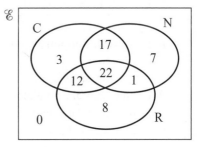

# Exercise 23★

**1** $\mathscr{E}$ = {all positive integers less than 12}, A = {2, 4, 6, 8, 10}, B = {4, 5, 6, 7, 8}.
   **a** Illustrate this information on a Venn diagram.
   **b** List A∩B and find $n(A∩B)$.
   **c** Does A∩B = B∩A?
   **d** List (A∩B)′.
   **e** Is A∩B a subset of A?

**2** $\mathscr{E}$ = {all positive integers less than 12}, C = {2, 3, 4}, D = {1, 2, 3, 4, 5, 6, 7}.
   **a** Illustrate this information on a Venn diagram.
   **b** List C∩D and D∩C.
   **c** Does C = C∩D?
   **d** Is C⊂D?
   **e** If C = C∩D, does this imply that C⊂D?

**3** $\mathscr{E}$ = {all positive integers less than 12}, E = {1, 2, 3, 4}, F = {5, 6, 7, 8}.
   **a** Illustrate this information on a Venn diagram.
   **b** List E∩F.
   **c** If E∩F = ∅, what does this imply about the sets E and F?

**4** R is the set of roses in a flower shop and W is the set of white flowers in the same shop.
   **a** Illustrate this information on a Venn diagram.
   **b** Describe the set R∩W in words.
   **c** If R∩W = ∅, what can you say?

**5** $\mathscr{E}$ = {letters of the alphabet}, V = {vowels}, A = {a, b, c, d, e}, B = {d, e, u}.

    **a** Illustrate this information on a Venn diagram.

    **b** List the sets V∩A, V∩B′, A′∩B.

    **c** List the set V∩A∩B.

**6** $\mathscr{E}$ = {positive integers less than 10}, P = {prime numbers}, E = {even numbers}, F = {factors of 6}.

    **a** Illustrate this information on a Venn diagram.

    **b** List P′∩E, E∩F, P∩F′.

    **c** Describe the set P∩E∩F.

**7** $\mathscr{E}$ = {all positive integers}, F = {4, 8, 12, 16, 20, 24}, S = {6, 12, 18, 24}.

    **a** Illustrate this information on a Venn diagram.

    **b** List F∩S.

    **c** What is the smallest member of F∩S?

    **d** F is the set of the multiples of 4, S is the set of the multiples of 6. What is the LCM of 4 and 6? How is this related to the set F∩S?

    **e** Use this method to find the LCM of **i)** 6 and 8       **ii)** 8 and 10.

**8** $\mathscr{E}$ = {all positive integers}, T = {factors of 24}, E = {factors of 18}.

    **a** Illustrate this information on a Venn diagram.

    **b** List T∩F.

    **c** What is the name usually given to the largest member of T∩F?

    **d** Use this method to find the highest common factor of **i)** 48 and 32  **ii)** 80 and 45.

**9** Show that a set of three elements has eight subsets, including ∅. Find a rule giving the number of subsets (including ∅) for a set of *n* elements.

**10** A = {multiples of 2}, B = {multiples of 3}, C = {multiples of 5}. If $n$(A∩B∩C) = 1, what can you say about $\mathscr{E}$?

---

**Remember**

The **union** of two sets A and B is the set of elements that belong to both sets, and is written A∪B.

---

The union of two sets is the set of elements that belong to A or to B or to both A and B.

---

**Example 7**

$\mathscr{E}$ = {all positive integers less than 10},
P = {prime numbers less than 10} and
O = {odd numbers less than 10}.

    **a** Illustrate these sets on a Venn diagram.

    **b** Find the set P∪O and $n$(P∪O).

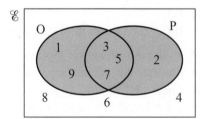

*Answers*

    **a** The set P∪O is shown shaded in the Venn diagram.

    **b** From the Venn diagram, P∪O = {1, 2, 3, 5, 7, 9} and $n$(P∪O) = 3.

# Exercise 24

**1** $\mathscr{E}$ = {all positive integers less than 10}, A = {1, 3, 5, 7, 9}, B = {3, 4, 5, 6}.

  **a** Illustrate this information on a Venn diagram.

  **b** List A∪B and find $n$(A∪B).

  **c** Does A∪B = B∪A?

  **d** List (A∪B)′.

  **e** Is A∪B a subset of A?

**2** $\mathscr{E}$ = {all positive integers less than 12}, A = {1, 3, 5, 7, 9}, B = {3, 5, 7}.

  **a** Illustrate this information on a Venn diagram.

  **b** List A∪B and B∪A.

  **c** Does A = B∪A?

  **d** Is B a subset of A?

  **e** If A = B∪A, does this imply that B⊂A?

**3** $\mathscr{E}$ = {pack of 52 playing cards}, B = {black cards}, C = {clubs}, K= {kings}.

  **a** Draw a Venn diagram to show the sets B, C and K.

  **b** Describe the set B∪K.

  **c** Describe the set B∪K∪C.

  **d** Describe the set B′∪K.

**4** $\mathscr{E}$ = {letters of the alphabet}, V = {vowels}, A = {a, b, c, d, e}, B = {d, e, u}.

  **a** Draw a Venn diagram to show the sets V, A and B.

  **b** List the set V∪A.

  **c** Describe the set V′.

  **d** Is B⊂V∪A?

**5** $\mathscr{E}$ = {all triangles}, E = {equilateral triangles}, I = {isosceles triangles} and R = {right-angled triangles}.

  **a** Draw a Venn diagram to show the sets E, I and R.

  **b** Sketch a member of I∩R.

  **c** Describe the sets I∪E and I∪R.

  **d** Describe the sets I∩E and E∩R.

**6** $\mathscr{E}$ = {polygons}, F = {polygons with 4 sides}, R = {regular polygons} and S = {squares}

  **a** Draw a venn diagram to show the sets F, R and S.

  **b** Sketch a member of F∩R.

  **c** Describe the sets S∪F and S∪R.

  **d** Describe the sets S∩F and F∩R.

# Exercise 24★

1 $\mathscr{E}$ = {all positive integers less than 10}, E = {2, 4, 6, 8}, O = {1, 3, 5, 7, 9}.
   a Illustrate this information on a Venn diagram.
   b List E∪O.
   c If $n$(E) + $n$(O) = $n$(E∪O), what does this imply about the sets E and O?
   d If (E∪O)′ = ∅, what does this tell you about E and O?

2 In Joe's Pizza Parlour, H is the set of pizzas containing ham and C is the set of pizzas containing cheese.
   a Describe the set H∪C in words.
   b Describe the set H∩C in words.
   c If (H∪C)′ = ∅, what can you say?

3 Draw a Venn diagram to show the intersections of three sets A, B and C. Given that $n$(A) = 18, $n$(B) = 15, $n$(C) = 16, $n$(A∪B) = 26, $n$(B∪C) = 23, $n$(A∩C) = 7 and $n$(A∩B∩C) = 1, find $n$(A∪B∪C).

4 Draw a Venn diagram to show the intersections of three sets A, B and C. Given that $n$(A) = 21, $n$(B) = 20, $n$(C) = 19, $n$(A∩B) = 5, $n$(B∪C) = 31, $n$(A∩C) = 7 and $n$(A∩B∩C) = 3, find $n$(A∪B∪C).

5 If $n$(A) = $n$(A∪B), what can you say about the sets A and B?

6 Does $n$(A) + $n$(B) = $n$(A∪B) + $n$(A∩B) for all possible configurations of sets A and B?

# Exercise 25 (Revision)

1 Write down two more members of these sets.
   a {salt, pepper, thyme, …}
   b {cat, dog, rabbit, …}
   c {apple, banana, orange, …}
   d {red, black, blue, …}

2 List these sets.
   a {square numbers between 2 and 30}
   b {all factors of 24}
   c {vowels in the word 'mathematics'}
   d {months of the year containing 30 days}

3 Describe these sets by a rule.
   a {2, 3, 5, 7}
   b {32, 34, 36, 38}
   c {Saturday, Sunday}
   d {a, e, i, o, u}

4 $\mathscr{E}$ = {all positive integers}, P = {prime numbers}, E = {even numbers}, O = {odd numbers}.
   Say which of these are true or false.
   a 51 ∈ P    b P is a subset of O    c E∩O = ∅    d E∪O = $\mathscr{E}$

**5** $\mathscr{E}$ = {positive integers less than 15}, A = {5, 7, 11, 13}, B = {6, 7, 9}, C = {multiples of 3}.

   **a** List C.

   **b** Draw a Venn diagram to illustrate the sets $\mathscr{E}$, A, B and C.

   **c** List A ∪ B.

   **d** List B ∩ C.

   **e** What is A ∩ C?

**6** $\mathscr{E}$ = {members of an expedition to the South Pole}, A = {people born in Africa},
F = {females}, C = {people born in China}.

   **a** Describe A ∩ F.

   **b** What is A ∩ C?

   **c** Jordan ∈ A ∪ C. What can you say about Jordan?

   **d** Illustrate the sets $\mathscr{E}$, A, F and C on a Venn diagram.

# Exercise 25★ (Revision)

**1** List these sets.

   **a** {multiples of 4 less than 20}

   **b** {colours of the rainbow}

   **c** {arrangements of the letters CAT}

   **d** {all pairs of products of 1, 2, 3}

**2** Describe these sets by a rule.

   **a** {1, 2, 3, 4, 6, 12}

   **b** {1, 1, 2, 3, 5}

   **c** {hearts, clubs, diamonds, spades}

   **d** {tetrahedron, cube, octahedron, dodecahedron, icosahedron}

**3** **a** A and B are two sets. A contains 12 members, B contains 17 members and A ∪ B contains 26 members. How many members of A are not in A ∩ B?

   **b** Draw a Venn diagram with circles representing three sets A, B and C, such that these are true:
A ∩ C = ∅, B ∩ C ≠ ∅, A ∩ B = ∅.

**4** $\mathscr{E}$ = {pack of 52 playing cards}, A = {aces}, B = {black cards}, D = {diamonds}.

   **a** Describe A ∩ D.

   **b** Describe B ∩ D.

   **c** Describe A ∪ D.

   **d** Find $n(A ∩ B)$.

   **e** Illustrate the sets $\mathscr{E}$, A, B and D on a Venn diagram.

**5** $\mathscr{E}$ = {triangles}, R = {right-angled triangles}, I = {isosceles triangles}, E = {equilateral triangles}.

   **a** Describe I ∩ R.

   **b** Describe I ∩ E.

   **c** Describe R ∩ E.

   **d** Draw a Venn diagram to illustrate the sets $\mathscr{E}$, R, I and E.

**6** $\mathscr{E}$ = {positive integers less than 30}, P = {multiples of 4}, Q = {multiples of 5},
R = {multiples of 6}.

   **a** List P ∩ Q.

   **b** $x ∈ P ∩ R$. List the possible values of $x$.

   **c** Is it true that Q ∩ R = ∅? Explain your answer.

# Summary 1

Number

## Fractions

To change $\frac{1}{8}$ into a decimal: divide 1 by 8          0.125

To change $\frac{7}{20}$ into a percentage: multiply by 100          $\frac{7}{20} \times 100 = 35\%$

To simplify $\frac{15}{1.2}$: multiply top and bottom by 10          $\frac{150}{12} = \frac{25}{2} = 12.5$

## Directed numbers

$3 + (-4) = 3 - 4 = -1$          $6 \times (-2) = -12$

$3 - (-4) = 3 + 4 = 7$          $(-6) \times (-2) = 12$

$(-3) + (-4) = -3 - 4 = -7$          $6 \div (-2) = -3$

$(-3) - (-4) = -3 + 4 = 1$          $(-6) \div (-2) = 3$

## Order of operations

**BIDMAS**:   Brackets,  Indices,  Division/Multiplication,  Addition/Subtraction.

## Percentages

**Multiplying factor method**

If the original price is $60:

For a **loss** of 30%, selling price $x = 60 - \left(60 \times \frac{30}{100}\right) = 60\left(1 - \frac{30}{100}\right) = 60 \times 0.7 = \$42$

For a **profit** of 30%, selling price $y = 60 + \left(60 \times \frac{30}{100}\right) = 60\left(1 + \frac{30}{100}\right) = 60 \times 1.30 = \$78$

## Standard form

$67\,000 = 6.7 \times 10^4$          To display $6.7 \times 10^4$ on a calculator: 6.7  4

## Significant figures

34.779 to 3 s.f. = 34.8

0.0659 to 2 s.f. = 0.066

## Decimal places

2.0765 to 2 d.p. = 2.08

0.052 96 to 3 d.p. = 0.053

# Algebra

## Simplifying

$4ac$ means $4 \times a \times c$.

$2s \times 3t = 6st$

$y^2 \times y = y^3$

$3ab \times 5b = 15ab^2$

Only add or subtract **like terms**.

$a - 3b + 2a - 2b = a + 2a - 3b - 2b = 3a - 5b$

$2c^2b + cb^2 - c^2b = c^2b + cb^2$

Multiply each term inside **brackets** by the term outside.

$4(2a - 3b) = 8a - 12b$

$2d - (3 - 4d) = 2d - 3 + 4d = 6d - 3$

## Solving equations

Do the same to both sides.

| | |
|---|---|
| $3x + 4 = 2(x + 3)$ | (expand brackets) |
| $3x + 4 = 2x + 6$ | (collect like terms) |
| $3x - 2x = 6 - 4$ | (simplify) |
| $x = 2$ | |
| Check: $3 \times 2 + 4 = 2(2 + 3)$ | |

| | |
|---|---|
| $3(x + 4) - (x - 7) = 25$ | (expand brackets) |
| $3x + 12 - x + 7 = 25$ | (simplify) |
| $2x + 19 = 25$ | (subtract 19) |
| $2x = 6$ | (divide by 2) |
| $x = 3$ | |
| Check: $3(3 + 4) - (3 - 7) = 25$ | |

# Graphs

## Graphs of the form $y = mx + c$

The equation of any straight-line graph can always be written in the form $y = mx + c$, where $m$ is the gradient and $c$ is the $y$ intercept.

Line **a** has a gradient of $\frac{4}{2} = 2$ (so $m = 2$), and a $y$ intercept of $+4$.

Therefore its equation is $y = 2x + 4$.

Line **b** has a gradient of $-\frac{2}{4} = -\frac{1}{2}$ $\left(\text{so } m = -\frac{1}{2}\right)$, and a $y$ intercept of $+4$.

Therefore its equation is $y = -\frac{1}{2}x + 4$.

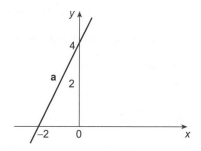

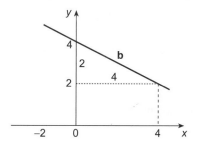

## Graphs of the form $ax + by = c$

The graph of $3x + 4y = 12$ can be rearranged to give $y = -\frac{3}{4}x + 3$, showing that it is a straight line with gradient $-\frac{3}{4}$ and $y$ intercept at 3. To **sketch** the graph of $3x + 4y = 12$, substitute $x = 0$ to get the $y$ intercept of 3, and substitute $y = 0$ to get the $x$ intercept of 4.

## Shape and space

### Compass constructions

**60° angle (equilateral triangle)**

Draw arc from A to intersect AB at P. Keeping same radius, draw arc from P to intersect arc at Q.

∠BAQ = 60°

**Bisecting an angle**

Draw arc from A to intersect lines at P and Q. Keeping the same radius, draw arcs from P and Q to intersect at R. Draw AR.

∠PAR = ∠QAR.

**Perpendicular bisector of a line**

Draw arc from A, with same radius, above and below line. Keeping the same radius, draw arcs from B to intersect those from A above and below the line. These points are P and Q. Draw line PQ, the perpendicular bisector of AB.

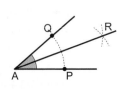

## Sets

| | | |
|---|---|---|
| A **set** is a collection of objects, described by a list or a rule. | A = {1, 3, 5} | |
| Each object is an **element** or a **member** of the set. | 1∈ A, 2∉ A | |
| Sets are **equal** if they have exactly the same elements. | B = {5, 3, 1}, B = A | |
| The **number of elements** of set A is given by $n(A)$. | $n(A) = 3$ | |
| The **empty set** is the set with no members. | { } or ∅ | |
| The **universal set** contains all the elements being discussed in a particular problem. | ℰ | ℰ |
| B is a **subset** of A if every member of B is a member of A. | B⊂A | ℰ 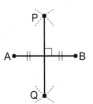 |
| The **complement** of set A is the set of all elements not in A. | A′ | ℰ  |
| The **intersection** of A and B is the set of elements that are in both A and B. | A∩B | ℰ  |
| The **union** of A and B is the set of elements that are in A or B or both. | A∪B | ℰ  |

**1** Simplify these.

  **a** $3a^2 + 4a^3 + 4a^2$      **b** $2a \times a^2$

  **c** $5(x + 4) - 7(5 - x)$

**2** Solve these for $x$.

  **a** $3x = 27$          **b** $\dfrac{x}{8} = 5$

  **c** $4(x - 6) = 2(x + 3)$

**3** There are about 100 million million cells in the human body. Write this number in

  **a** index form          **b** standard form

**4** $r = 6 \times 10^4$ and $s = 5 \times 10^5$.

  **a** Find the value of $r \times s$ and give your answer in standard form.

  **b** Find the value of $s - r$ and give your answer as a decimal number.

**5** Calculate these, and give your answers as a decimal number.

  **a** $97.6 - 4.82$     **b** $45.7 \times 4$     **c** $\dfrac{28.3}{0.4}$

**6** There were 420 girls at St Bright's School in 2001. Three years later, the numbers have risen to 504. What percentage increase is this?

**7** Find the value of these when $x = -2$, $y = 6$.

  **a** $y^2 - x$          **b** $2y + 3x - 3y - 2x$

  **c** $(y - x)^2$

**8** On squared paper, plot the points A(0, 4) and B(2, 0), and draw the line AB.

  Write down

  **a** the gradient of AB

  **b** the $y$ intercept

  **c** the equation of the line AB

**9** On squared paper, draw the graphs of $y = 3x + 2$, $y = -3x - 3$, $y = 3x - 3$, and $y = \dfrac{1}{3}x + 2$.

  Write down the equations of the lines that are parallel.

**10** Solve these.

  **a** $\dfrac{3.6}{x} = 0.6$      **b** $3 - 2(x + 10) = 4$

  **c** $2(x - 4) + 27 = 3(3 - x)$

**11** Calculate these.

  **a** $\dfrac{14}{70}$     **b** $17 \div 0.25$     **c** $\dfrac{6}{4.2} \times 0.14$

**12** Calculate these, and write them in standard form, correct to 2 significant figures.

  **a** $(8.4 \times 10^4) \times (9.5 \times 10^4)$

  **b** $(8.4 \times 10^4) - (9.5 \times 10^3)$

  **c** $\dfrac{3.99 \times 10^3}{4.2 \times 10^2}$

**13** Simplify these.

  **a** $4a^2b - ba^2 + a$     **b** $(2a)^2 \times a$

  **c** $5(c + 1) - (c + 1)$

**14** A computer is bought for $380, and then sold at a profit of $45.60. Find the profit as a percentage of the original purchase price.

**15** Expressed as a percentage, how much steeper is the steeper hill?

**16** Find the value of these when $a = -2$ and $b = -3$.

  **a** $ab^2$         **b** $(a - b)^2$       **c** $(ab)^2 - b$

**17** $\mathscr{E} = \{$positive integers less than 13$\}$, $E = \{$even integers$\}$, $F = \{$multiples of 4$\}$ and $T = \{$multiples of 3$\}$.

  **a** List the sets $E'$, $E \cap T$ and $F \cap T$.

  **b** Give descriptions of the sets $E'$, $E \cap T$ and $F \cap T$.

  **c** Draw a Venn diagram showing these sets.

**18** In the village of Cotterstock, not all the houses are connected to electricity, mains water or gas.

$\mathscr{E}$ = {houses in Cotterstock},
E = {houses with electricity},
W = {houses with mains water} and
G = {houses with gas}.

The relationship of these sets is shown in the Venn diagram.

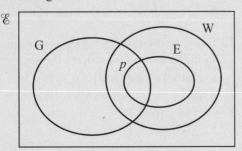

**a** Express both in words and in set notation the relationship between E and W.

**b** Give all the information you can about house $p$ shown on the Venn diagram.

**c** Copy the diagram and shade the set $G \cap E$.

**d** Mark on the diagram house $q$ that is connected to mains water but not to gas or electricity.

**19** A straight line passes through A(0, 2) and B(3, 0).

**a** Find the equation of the line.

**b** Find the equation of the line drawn parallel to AB that passes through (−3, 1).

**20** When 2 is added to a number and the answer trebled, it gives the same result as doubling the number and adding 10.

**a** Form an equation with $x$ representing the number.

**b** Solve your equation to find the number.

**21** A kind teacher gives you 20 cents for every question you get right, but you have to pay the teacher 10 cents for every question you get wrong. After 30 questions you have made a profit of $1.80.

**a** Form an equation with $x$ representing the number of equations you got right.

**b** Solve your equation to find how many questions you got right.

**22** A cup of tea costs 10 cents less than a cup of coffee, while a cup of hot chocolate costs 20 cents more than a cup of coffee. Three cups of coffee, five cups of tea and two cups of hot chocolate cost $8.90.

**a** Form an equation with $x$ representing the price of a cup of coffee.

**b** Solve your equation to find the price of a cup of coffee.

# Number 2

## Standard form with negative indices

We can write numbers, however small, in **standard form.**

### Activity 7

Copy and complete the table.

| Decimal form | Multiples of 10 or fraction form | Standard form |
|---|---|---|
| 100 | $10 \times 10$ | $1 \times 10^2$ |
| 10 | 10 | |
| 1 | | |
| 0.1 | | $1 \times 10^{-1}$ |
| | $\dfrac{1}{100} = \dfrac{1}{10^2}$ | |
| 0.001 | $=$ | |
| 0.0001 | $=$ | |
| | $=$ | $1 \times 10^{-5}$ |

## Key Point

$$10^{-2} = \frac{1}{10^2} = \frac{1}{100} = 0.01$$

## Activity 8

House mouse

$10^{-2}$ kg

Pigmy shrew

$10^{-3}$ kg

Grain of sand

$10^{-7}$ kg

Staphylococcus bacterium

$10^{-15}$ kg

Write down the mass of each of the first three objects in grams

◆ in ordinary numbers

◆ in standard form.

Copy and complete these statements:

◆ A house mouse is … times heavier than a pigmy shrew.

◆ A pigmy shrew is … times heavier than a grain of sand.

◆ A grain of sand is 100 000 times lighter than a ….

◆ A pigmy shrew is 10 000 times heavier than a ….

◆ A … is 100 million times heavier than a ….

◆ A house mouse is … 10 000 billion* times heavier than a ….

*1 billion can equal $10^9$ (US, France) or $10^{12}$ (UK, Germany). In this equation please take 1 billion to be $10^9$.

---

**Example 1**

Write 0.987 in standard form.

$$0.987 = 9.87 \times \frac{1}{10} = 9.87 \times 10^{-1}$$

To display this on your calculator, press

---

# Exercise 26

For Questions 1–8, write each number in standard form.

**1** 0.1      **2** 0.01      **3** 0.001      **4** 0.0001

**5** $\frac{1}{1000}$      **6** $\frac{1}{100}$      **7** 10      **8** 1

For Questions 9–16, write each one as an ordinary number.

**9** $10^{-3}$      **10** $10^{-5}$      **11** $1.2 \times 10^{-3}$      **12** $8.7 \times 10^{-1}$

**13** $10^{-6}$      **14** $10^{-4}$      **15** $4.67 \times 10^{-2}$      **16** $3.4 \times 10^{-4}$

For Questions 17–24, write each number in standard form.

**17** 0.543      **18** 0.0708      **19** 0.007      **20** 0.0009

**21** 0.67      **22** 0.00707      **23** 100      **24** 1000

For Questions 25–34, write each one as an ordinary number. Check your answers with a calculator.

**25** $10^{-2} \times 10^{4}$      **26** $10^{3} \times 10^{-1}$

**27** $10^{2} \div 10^{-2}$      **28** $10^{3} \div 10^{-3}$

**29** $(3.2 \times 10^{-2}) \times (4 \times 10^{3})$      **30** $(2.4 \times 10^{-2}) \div (8 \times 10^{-1})$

**31** $(6 \times 10^{-1}) \div (2 \times 10^{-2})$      **32** $(4 \times 10^{-3}) \times (9 \times 10^{2})$

**33** $(2 \times 10^{-2}) \times (9 \times 10^{-1})$      **34** $(3.6 \times 10^{-2}) \div (1.2 \times 10^{-3})$

# Exercise 26★

For Questions 1–8, write each answer as an ordinary number.

**1** $10^{3} \times 10^{-2}$      **2** $10^{-1} \times 10^{-2}$      **3** $10^{-2} + 10^{-3}$      **4** $10^{-1} - 10^{-3}$

**5** $10^{-4} \times 10^{2}$      **6** $10^{-3} \times 10^{-1}$      **7** $10^{-3} + 10^{-4}$      **8** $10^{-3} - 10^{-1}$

For Questions 9–16, write each answer in standard form.

**9** $10 \div 10^{-2}$      **10** $10^{2} \div 10^{-2}$      **11** $10^{-1} \div 10^{-2}$      **12** $10^{-4} \div 10^{-3}$

**13** $10^{3} \div 10^{-1}$      **14** $10^{-1} \div 10^{3}$      **15** $10^{-2} \div 10^{-4}$      **16** $10^{-5} \div 10^{-2}$

For Questions 17–24, write each answer in standard form. Check your answers with a calculator.

**17** $(4 \times 10^{2})^{-2}$      **18** $(4 \times 10^{-2})^{2}$      **19** $(6.9 \times 10^{3}) \div 10^{-4}$      **20** $10^{-3} \div (2 \times 10^{-2})$

**21** $(5 \times 10^{2})^{-2}$      **22** $(5 \times 10^{-2})^{2}$      **23** $(4.8 \times 10^{2}) \div 10^{-3}$      **24** $10^{-2} \div (5 \times 10^{-3})$

You will need this information to answer Questions 25 and 26.

Cough virus            Human hair            Pin

$9.144 \times 10^{-6}$ mm diameter      $5 \times 10^{-2}$ mm diameter      $6 \times 10^{-1}$ mm diameter

**25** How many viruses, to the nearest thousand, can be placed in a straight line across the width of a human hair?

**26** How many viruses, to the nearest thousand, can be placed in a straight line across the width of a pin?

**27** The radius of the nucleus of a hydrogen atom is $1 \times 10^{-12}$ mm. How many would fit in a straight line across a human hair of diameter 0.06 mm?

**28** The average mass of a grain of sand is $10^{-4}$ g. How many grains of sand are there in 2 kg?

**29** Devise a sensible method to work out $(3.4 \times 10^{23}) + (3.4 \times 10^{22})$.

**30** A molecule of water is a very small thing, so small that its volume is $10^{-27}$ m$^3$.

  **a** How many molecules are there in 1 m$^3$ of water?
    If you wrote your answer in full, how many zero digits would there be?

  **b** If you assume that a water molecule is in the form of a cube, show that its side length is $10^{-9}$ m.

  **c** If a number of water molecules were placed touching each other in a straight line, how many would there be in a line 1 cm long?

  **d** The volume of a cup of tea is 200 cm$^3$.
    How many molecules of water would the cup hold?
    If all these were placed end to end in a straight line, how long would the line be?

Take the circumference of the Earth to be 40 000 km.
How many times would the line of molecules go around the Earth?

# Four rules of fractions

This section will give you practice in using fractions. The questions will help you to understand much of the algebra in this unit.

## Addition and subtraction

A common denominator is required.

**Example 2** Addition

$$\frac{3}{4} + \frac{1}{6}$$
$$= \frac{9}{12} + \frac{2}{12} = \frac{9+2}{12} = \frac{11}{12}$$

**Example 3** Subtraction

$$\frac{3}{4} - \frac{2}{5}$$
$$= \frac{15}{20} - \frac{8}{20} = \frac{15-8}{20} = \frac{7}{20}$$

**Example 4** With mixed fractions

$$3\frac{1}{3} - 1\frac{3}{4}$$
$$= (3 + \frac{1}{3}) - (1 + \frac{3}{4})$$
$$= 3 - 1 + \frac{1}{3} - \frac{3}{4} = 2 + \frac{1}{3} - \frac{3}{4}$$
$$= 2 + \frac{4}{12} - \frac{9}{12}$$
$$= 1 + \frac{12}{12} + \frac{4}{12} - \frac{9}{12} = 1 + \frac{12+4-9}{12}$$
$$= 1\frac{7}{12}$$

## Multiplication and division

Convert mixed fractions into improper fractions.

**Example 5** Multiplication

$$1\frac{3}{4} \times \frac{3}{5}$$
$$= \frac{7}{4} \times \frac{3}{5} = \frac{7 \times 3}{4 \times 5} = \frac{21}{20} = 1\frac{1}{20}$$

**Example 6** Cancelling

$$1\frac{5}{9} \times 2\frac{1}{7}$$
$$= \frac{14}{9} \times \frac{15}{7} = \frac{14 \times 15}{9 \times 7}$$

*Divide top and bottom by 7, and by 3.*

$$= \frac{2 \times 5}{3 \times 1} = \frac{10}{3} = 3\frac{1}{3}$$

Much easier than $\frac{14 \times 15}{9 \times 7} = \frac{210}{63}$ !

**Example 7** Division

$$\frac{3}{4} \div \frac{5}{11}$$

*Turn the divisor upside down and multiply.*

$$= \frac{3}{4} \times \frac{11}{5} = \frac{3 \times 11}{4 \times 5} = \frac{33}{20} = 1\frac{13}{20}$$

Because $\frac{3}{4} \div \frac{5}{11} \equiv \dfrac{\frac{3}{4}}{\frac{5}{11}} \equiv \dfrac{\frac{3}{4} \times \frac{11}{5}}{\frac{5}{11} \times \frac{11}{5}} \equiv \dfrac{\frac{3}{4} \times \frac{11}{5}}{1}$

**Example 8** Multiplying with a whole number.

$$\frac{3}{4} \times 7$$

*Change the whole number into a fraction.*

$$= \frac{3}{4} \times \frac{7}{1} = \frac{3 \times 7}{4 \times 1} = \frac{21}{4} = 5\frac{1}{4}$$

**Example 9** Dividing into a whole number.

$$8 \div 1\frac{1}{2}$$
$$= \frac{8}{1} \div \frac{3}{2} = \frac{8}{1} \times \frac{2}{3} = \frac{8 \times 2}{1 \times 3} = \frac{16}{3} = 5\frac{1}{3}$$

## Exercise 27

Work these out.

1 $\frac{2}{7} + \frac{4}{7}$

2 $\frac{1}{9} + \frac{4}{9}$

3 $\frac{3}{10} + \frac{1}{10}$

4 $\frac{3}{8} + \frac{1}{8}$

5 $\frac{7}{9} + \frac{4}{9}$

6 $\frac{3}{8} + \frac{7}{8}$

7 $\frac{5}{6} - \frac{1}{3}$

8 $\frac{11}{20} - \frac{3}{10}$

9 $\frac{3}{8} + \frac{7}{12}$

10 $\frac{5}{6} - \frac{3}{4}$

11 $3\frac{1}{4} + 1\frac{1}{6}$

12 $4\frac{3}{5} - 2\frac{1}{2}$

13 $\frac{5}{6} \times \frac{1}{3}$

14 $\frac{3}{8} \times \frac{4}{7}$

15 $\frac{3}{4} \div \frac{7}{8}$

16 $\frac{3}{10} \div \frac{4}{5}$

17 $4 \times \frac{3}{20}$

18 $\frac{2}{3} \times 5$

19 $\frac{12}{25} \div 4$

20 $6 \div \frac{3}{4}$

21 $2\frac{1}{7} \times 1\frac{2}{5}$

22 $1\frac{1}{4} \times 1\frac{1}{5}$

23 $1\frac{1}{8} \div \frac{3}{4}$

24 $3\frac{3}{8} \div 1\frac{1}{4}$

25 $2\frac{5}{6} + 1\frac{3}{4}$

26 $3\frac{7}{8} + 4\frac{1}{4}$

27 $5\frac{3}{10} - 2\frac{11}{20}$

28 $36\frac{3}{8} - 32\frac{7}{12}$

29 $1\frac{3}{5} \times 3$

30 $1\frac{7}{8} \times 4$

31 $1\frac{4}{5} \div 6$

32 $8 \div 1\frac{1}{3}$

## Exercise 27★

Work these out.

1 $\frac{1}{3} + \frac{5}{12}$

2 $\frac{1}{4} + \frac{9}{20}$

3 $\frac{5}{6} - \frac{7}{30}$

4 $\frac{11}{15} - \frac{3}{20}$

5 $\frac{1}{5} + \frac{3}{10} + \frac{9}{20}$

6 $\frac{1}{4} + \frac{3}{20} - \frac{1}{40}$

7 $\frac{1}{8} + \frac{1}{8} + \frac{1}{8} + \frac{1}{8} + \frac{1}{8} + \frac{1}{8}$

8 $\frac{5}{6} \times \frac{9}{25}$

9 $\frac{5}{18} \times \frac{8}{25}$

10 $6 \times \frac{1}{8}$

11 $\frac{2}{3} \div \frac{20}{21}$

12 $\frac{8}{15} \div \frac{5}{6}$

13 $\frac{2}{3} \div 3$

14 $\frac{2}{3} \div 2$

15 $\frac{1}{2} \times \frac{15}{16} \times \frac{4}{5}$

16 $\frac{2}{3} \times \frac{3}{5} \times \frac{9}{10}$

17 $4\frac{1}{2} + 3\frac{1}{6}$

18 $6\frac{2}{5} + 7\frac{1}{3}$

19 $7\frac{2}{3} - 1\frac{1}{6}$

20 $4\frac{7}{9} - 3\frac{1}{3}$

21 $7\frac{2}{3} - \frac{8}{9}$

22 $6\frac{1}{12} - 4\frac{7}{10}$

23 $3\frac{1}{7} \times \frac{7}{15}$

24 $2\frac{1}{2} \times 4\frac{2}{3}$

25 $4 \times \frac{2}{3}$

26 $4\frac{1}{2} \div \frac{3}{4}$

27 $2\frac{1}{3} \div 2\frac{4}{5}$

28 $3\frac{1}{9} \div 14$

29 $14 \div 3\frac{1}{9}$

30 $\left(2\frac{1}{8} + 2\frac{1}{4}\right) \div 2\frac{1}{3}$

# Ratio

The ratio $3:5$ ('three to five') can be written and used in several different forms:

$3:5 \equiv 1:\frac{5}{3} \equiv 1:1.67$  or  $\frac{3}{5}:1 \equiv 0.6:1$  or  $\frac{3}{8}:\frac{5}{8} \equiv 0.375:0.625$.

The last form is the most useful when dividing up quantities in a given ratio, since the two parts of the ratio add up to 1 ($\frac{3}{8} + \frac{5}{8} = 1$).

**Example 10**

A marinade in a recipe contains rice wine and soy sauce in the ratio $2:3$.

How much of each ingredient is needed for 100 ml of the mixture?

(Add the ratios together: $2 + 3 = 5$.)

Then the parts are $\frac{2}{5}$ and $\frac{3}{5}$.

Amount of rice wine $= \frac{2}{5}$ of $100 = 40$ ml.  Amount of soy sauce $= \frac{3}{5}$ of $100 = 60$ ml.

## Exercise 28

1   Divide \$392 in the ratio of $3:4$.

2   Divide 637 m in the ratio of $2:5$.

3   Divide 752 kg in the ratio of $1:7$.

4   Divide 243 miles in the ratio of $4:5$.

5   Divide 984 in the ratio of $7:5$.

6   Divide 405 in the ratio of $7:2$.

7   Divide 13.5 in the ratio of $3:2$.

8   Divide 5.6 in the ratio of $4:3$.

**Example 11**

Divide £1170 in the ratio of $2:3:4$.        (Add the ratios together: $2 + 3 + 4 = 9$.)

Then the first part  $= \frac{2}{9}$ of £1170 $=$ £260.

The second part  $= \frac{3}{9}$ of £1170 $=$ £390.

The third part  $= \frac{4}{9}$ of £1170 $=$ £520.    (Note that £260 + £390 + £520 = £1170.)

## Exercise 28*

1   Divide \$120 in the ratio $3:5$.

2   Divide €350 in the ratio $1:6$.

3   The fuel for a lawn mower is a mixture of 8 parts petrol to one part oil.  How much oil is required to make 1 litre of fuel?

4   A chicken stock contains chicken bones and chicken pieces in the ratio $20:7$.  What weight of chicken pieces is needed to make 2 kg of stock?

5   Divide 702 in the ratio $1:2:3$.

6   Divide 576 tonnes in the ratio $4:3:2$.

7   Mr Chan has three daughters, An, Lien and Tao, aged 7, 8 and 10 years respectively.  He shares \$100 between them in the ratio of their ages.  How much does Lien receive?

8   A breakfast cereal contains the vitamins thiamin, riboflavin and niacin in the ratio $2:3:25$.  A bowl of cereal contains 10 mg of these vitamins.  Calculate the amount of riboflavin in a bowl of cereal.

# Direct proportion

In mathematics, there are many ways to relate two quantities together. Here are a few examples.

**Changes of units:** 1 mile = 1.609 km    **Velocities:** 30 miles travelled in 1 hour (30 miles/hour)

**Gradients:** 1 in 5    **Scales:** 1:50

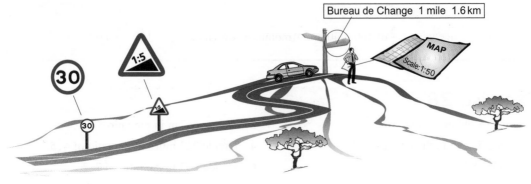

Bureau de Change  1 mile  1.6 km

**Exchange rates:** £1 = $1.60    **Equations:** $3x + 5 = 16$

**Ratios:** $\frac{4}{5}$ means $4 \div 5$

**Densities:** 13 g is the mass of 1 cm³ (13 g/cm³)    **Graphs:** (two axes)

**Problem solving:** 3.4 m of timber costs $6.80 ($2/m)

Two different ways of doing the same calculation are shown in Example 12.

---

## Example 12

Since July 2000, the Olympic organising committee ATHENS 2004 has recycled 108 tonnes of paper, saving 1836 trees and cutting energy consumption by 442 800 kW. Calculate the amount of energy saved by recycling one tonne of paper.

**Method 1** 'unitary method'

| | | |
|---|---|---|
| 108 tonnes | save | 442 800 kW |
| ∴  1 tonne | saves | $\dfrac{442\,800}{108}$ kW |
| | = | 4100 kW |

**Method 2** 'per method'

Energy saved per tonne $= \dfrac{\text{Total energy saved}}{\text{Total tonnage}}$

$= \dfrac{442\,800}{108} = 4100$ kW per tonne

---

## Remember

'Per' means divide. For example, 'miles *per* hour' means miles divided by hours, or miles/hour.

---

# Exercise 29

For Questions 1–6, find the exchange rate in the form £1 = ... .

**1** £6 = ¥1080

**2** US$15 = £9

**3** Aus$62 = £26

**4** £7 = US$13

**5** NZ$22 = £8

**6** Can$54 = £24

**7** 4m of timber costs £33.60.

   **a** What is the cost of 1m?   **b** What is the cost of 9m?

**8** Three CDs cost $33.75.

   **a** What is the cost of one CD?   **b** What is the cost of seven CDs?

**9** In one of the strongest hurricanes to sweep across South America, 63 cm of rain fell in 6 hours.

   **a** Find the amount of rain that fell in millimetres per hour.

   **b** Find the amount of rain that fell per minute.

**10** At blast off, the US Saturn rocket uses 104.5 tonnes of fuel in 9.5 s.

   **a** Find the amount of fuel used per second.

   **b** Find the amount of fuel used per minute.

**11** A military jet uses 3000 litres of fuel on a 45-minute flight.

   **a** For how long would it travel using 1 litre?

   **b** How many litres does it use in 1 minute?

**12** A spacecraft travelled 22 billion miles between 1977 and 1999.

   **a** Find, correct to 3 significant figures, its speed in miles per day.

   **b** Find, correct to 3 significant figures, its speed in miles per hour.

# Exercise 29★

**1** 6 Hong Kong dollars can be exchanged for 80 Japanese yen.

   **a** How many dollars can be exchanged for 200 yen?

   **b** How many yen can be exchanged for 200 dollars?

**2** When 18 g of peanuts are burned, 390 kJ (kilojoules) of energy are produced.

   **a** How much energy is produced if 24 g of peanuts are burned?

   **b** How many grams of peanuts must be burned to produce 130 kJ?

**3** Red blood cells are replaced by the bone marrow at the rate of $8.28 \times 10^9$ per hour. How many are replaced per second?

**4** There are $25 \times 10^{11}$ red blood cells in the human body. Suppose that the total mass of red blood cells is 2.5 kg and the total volume is 2.75 litres.

   **a** Find the number of cells in 1 g.

   **b** Find the density of red blood cells in grams per cubic centimetre.

   (Give your answer as a fraction.)

**5** In its 12 years of life, it is estimated that a bird called the chimney swift flies 1.25 million miles and can sleep while flying.

Approximately how far would you expect it to fly in 1 hour?

**6** It is estimated that by the age of 18, the average American child has seen 350 000 commercials on television. How many is this per day, approximately?

**7** At full speed, the cruise ship QE 2 uses 40 000 litres of fuel in 55 minutes.

   **a** Find the time, in seconds, taken to use 1 m³ of fuel.

   **b** Find the fuel consumption in litres per second.

**8** A supersonic aircraft uses 12 000 litres of fuel every 25 minutes.

   **a** Find the time, in seconds, taken to use $1\,\text{m}^3$.

   **b** Find the fuel consumption in litres per second.

**9** Air bags in a car 'explode' at 340 km/h.

   **a** Convert this to metres per second.

   **b** How long would the air bag take to move 10 cm?
     Give your answer to the nearest thousandth of a second.

**10** In Portuguese (standard of Lisbon) wine measures, 1 quartilo = 397 ml and
    1 almude = 48 quartilos. Convert 1000 litres into almudes.

## Exercise 30 (Revision)

**1** Write 0.001 in standard form.

**2** Write 0.036 89 in standard form correct to 3 significant figures.

For Questions 3–5, write each answer as an ordinary number.

**3** $10^{-2} \times 10^{4}$                             **4** $10^{-2} \div 10$                   **5** $10^{-2} + 10^{-1}$

For Questions 6–8, write each answer as a fraction in its simplest form.

**6** $2\frac{2}{5} + \frac{1}{4}$                             **7** $\frac{4}{9} \div \frac{2}{3}$                   **8** $2\frac{1}{6} \times \frac{3}{26}$

**9** Divide \$275 in the ratio of $2:3$.

**10** If 240 Swiss Francs can be exchanged for 312 Brunei Dollars, find the number of dollars for
    1 Swiss Franc, and for 15 Swiss Francs.

**11** At lift off, the US Apollo rocket uses 150 gallons of fuel in 0.15 s.
    How many gallons per second is this?

**12** The average adult reads at the rate of 5 words per second. How long would it take someone to
    read 3500 words? Give your answer to the nearest whole number of minutes.

## Exercise 30★ (Revision)

**1** Write one millionth in standard form.

**2** Write $\dfrac{2.7}{540}$ in standard form.

For Questions 3–5, write each answer as a fraction in its simplest form.

**3** $2\frac{3}{4} + 1\frac{1}{6}$                        **4** $5\frac{1}{3} \div 4\frac{12}{13}$                  **5** $16\frac{1}{3} - 1\frac{3}{4}$

**6** What is the reciprocal of 0.125?

**7** Divide \$936 in the ratio $2:3:4$.

**8** Calculate this, and write your answer as an ordinary number.

$$\frac{(3 \times 10^{-4}) + (3 \times 10^{-2})}{3 \times 10^{-2}}$$

**9** Comment on the statement 'during the summer in the Scottish Highlands, there are, on average, 20 million midges per hectare'. (1 ha = 10 000 m$^2$)

**10** An athlete runs 1 mile along a normal 400 m athletics track and finishes exactly on the finish line. Where did he start from? (1 mile = 1.609 km.)

**11** The density of lead is 11.4 g/cm$^3$. What mass of lead has a volume of 1 m$^3$?

**12** Take 830 cm$^3$ of air to weigh 1 g.
Find, to 2 significant figures, the weight of air in a classroom of dimensions 4 m by 10 m by 8 m.

**13** A camel can drink 45 litres of liquid in 5 minutes. How long, in seconds, would it take to drink a quartilho? Give your answer correct to 3 significant figures. (1 quartilho = 0.397 litres.)

# Algebra 2

## Simplifying fractions

Algebraic fractions are simplified in the same way as arithmetic fractions.

## Multiplication and division

**Example 1**

Simplify $\dfrac{4x}{6x}$.    $\dfrac{\overset{2}{\cancel{4}}\overset{1}{\cancel{x}}}{\underset{3}{\cancel{6}}\underset{1}{\cancel{x}}} = \dfrac{2}{3}$

**Example 2**

Simplify $\dfrac{3x^2}{6x}$.    $\dfrac{3x^2}{6x} = \dfrac{\cancel{3} \times x \times \overset{1}{\cancel{x}}}{\underset{2}{\cancel{6}} \times \underset{1}{\cancel{x}}} = \dfrac{x}{2}$

**Example 3**

Simplify $(27xy^2) \div (60x)$.    $(27xy^2) \div (60x) = \dfrac{27xy^2}{60x} = \dfrac{\overset{9}{\cancel{27}} \times \overset{1}{\cancel{x}} \times y \times y}{\underset{20}{\cancel{60}} \times \underset{1}{\cancel{x}}} = \dfrac{9y^2}{20}$

## Exercise 31

Simplify these.

**1** $\dfrac{4x}{x}$    **2** $\dfrac{3x}{3}$    **3** $\dfrac{6y}{2}$    **4** $\dfrac{8y}{4}$

**5** $(6x) \div (3x)$    **6** $(10y) \div (5y)$    **7** $\dfrac{12a}{4b}$    **8** $\dfrac{15x}{5y}$

**9** $\dfrac{3ab}{6a}$    **10** $\dfrac{2xy}{4y}$    **11** $(9a) \div (3b)$    **12** $(20x) \div (5y)$

**13** $\dfrac{12c^2}{3c}$    **14** $\dfrac{15y^2}{3y}$    **15** $\dfrac{4a^2}{8a}$    **16** $\dfrac{5b^2}{15b}$

**17** $\dfrac{12x}{3x^2}$    **18** $(21a) \div (7a^2)$    **19** $\dfrac{8ab^2}{4ab}$    **20** $\dfrac{12xy^2}{3x}$

**21** $\dfrac{3a}{15ab^2}$    **22** $\dfrac{2x}{12xy^2}$    **23** $(3a^2b^2) \div (12ab^2)$    **24** $(5xy^2) \div (15x^2y)$

# Exercise 31★

Simplify these.

**1** $\dfrac{5y}{10y}$      **2** $\dfrac{4x}{16x}$      **3** $\dfrac{12a}{6ab}$      **4** $\dfrac{15b}{5ab}$

**5** $(3xy) \div (12y)$      **6** $(5ab) \div (20a)$      **7** $\dfrac{3a^2}{6a}$      **8** $\dfrac{2x^2}{10x}$

**9** $\dfrac{10b}{5b^2}$      **10** $\dfrac{12x}{6x^2y}$      **11** $(18a) \div (3ab^2)$      **12** $(25a^2) \div (5ab^2)$

**13** $\dfrac{3a^2b^2}{6ab^3}$      **14** $\dfrac{14x^3y^2}{21x^2y^3}$      **15** $\dfrac{15abc}{5a^2b^2c^2}$      **16** $\dfrac{9xyz^2}{12x^2yz^2}$

**17** $(3a^2) \div (12ab^2)$      **18** $(4x)^3 \div (12x^2y)$      **19** $\dfrac{abc^3}{(abc)^3}$      **20** $\dfrac{a(bc)^4}{(ab)^4c}$

**21** $\dfrac{150a^3b^2}{400a^2b^3}$      **22** $\dfrac{52ab^2c^3}{65a^3b^2c}$      **23** $\dfrac{45x^3y^4z^5}{150x^5y^4z^3}$      **24** $\dfrac{221(2a^5)^3b^{10}}{34(13b^3)^3a^{14}}$

---

**Example 4**

Simplify $\dfrac{3x^2}{y} \times \dfrac{y^3}{x}$.    $\dfrac{3x^2}{y} \times \dfrac{y^3}{x} = \dfrac{3 \times x \times \cancel{x}}{\cancel{y}} \times \dfrac{\cancel{y} \times y \times y}{\cancel{x}} = 3xy^2$

**Example 5**

Simplify $\dfrac{2x^2}{y} \div \dfrac{2x}{5y^3}$.    $\dfrac{2x^2}{y} \div \dfrac{2x}{5y^3} = \dfrac{\cancel{2} \times x \times \cancel{x}}{\cancel{y}} \times \dfrac{5 \times \cancel{y} \times y \times y}{\cancel{2x}} = 5xy^2$

---

**Remember**

To divide by a fraction, turn the fraction upside down and multiply.

---

# Exercise 32

Simplify these.

**1** $\dfrac{3x}{4} \times \dfrac{5x}{3}$      **2** $\dfrac{2x}{5} \times \dfrac{5x}{3}$      **3** $\dfrac{x^2y}{z} \times \dfrac{xz^2}{y^2}$      **4** $\dfrac{xy^3}{z^2} \times \dfrac{yz}{x}$

**5** $\dfrac{x^2}{y} \times \dfrac{z}{x^2} \times \dfrac{y}{z}$      **6** $\dfrac{a}{b^3} \times \dfrac{ab}{c} \times \dfrac{b^2c}{a^2}$      **7** $\dfrac{4c \times 7c^2}{7 \times 5c}$      **8** $\dfrac{5d^2 \times 6d}{6d \times 6d}$

**9** $\dfrac{3x}{4} \div \dfrac{x}{8}$      **10** $\dfrac{2a}{5} \div \dfrac{a}{10}$      **11** $4 \div \dfrac{8}{ab}$      **12** $9 \div \dfrac{12}{xy}$

**13** $\dfrac{2b}{3} \div 4$      **14** $\dfrac{6x}{7} \div 3$      **15** $\dfrac{2x}{3} \div \dfrac{2x}{3}$      **16** $\dfrac{5a}{4} \div \dfrac{5a}{4}$

**17** $\dfrac{2x}{y^2} \div \dfrac{x}{y}$      **18** $\dfrac{3a}{b} \div \dfrac{a^2}{b}$      **19** $\dfrac{5ab}{c^2} \div \dfrac{10a}{c}$      **20** $\dfrac{3x^2}{y^2} \div \dfrac{x^2}{z}$

# Exercise 32★

Simplify these.

**1** $\dfrac{4a}{3} \times \dfrac{5a}{2} \times \dfrac{3a}{5}$

**2** $\dfrac{x}{6} \times \dfrac{5x}{2} \times \dfrac{6x}{5}$

**3** $\dfrac{3x^2y}{z^3} \times \dfrac{z^2}{xy}$

**4** $\dfrac{4ab^3}{c} \times \dfrac{c^2}{a^2b}$

**5** $\dfrac{45}{30} \times \dfrac{p^2}{q} \times \dfrac{q^3}{p}$

**6** $\dfrac{24}{abc^2} \times \dfrac{a^2bc^3}{12}$

**7** $\dfrac{3x}{y} \div \dfrac{6x^2}{y}$

**8** $\dfrac{2y^2}{x^3} \div \dfrac{4y}{x}$

**9** $\dfrac{15x^2y}{z} \div \dfrac{3xz}{y^2}$

**10** $\dfrac{12xy^3}{z^2} \div \dfrac{4zy}{x}$

**11** $\dfrac{2x}{y} \times \dfrac{3y}{4x} \times \dfrac{2y}{3}$

**12** $\dfrac{a}{3} \times \dfrac{3b}{2a} \times \dfrac{2a}{b}$

**13** $\left(\dfrac{x}{2y}\right)^3 \times \dfrac{2x}{3} \div \dfrac{2}{9y^2}$

**14** $\dfrac{3x}{4y} \times \left(\dfrac{2x^2}{y^3}\right)^2 \div \dfrac{6x^3}{5y^5}$

**15** $\dfrac{\sqrt{a^3b^2}}{6a^3} \times \dfrac{3a^5b}{(a^3b^2)^2} \div \dfrac{ab}{\sqrt{a^3b^2}}$

**16** $\dfrac{13\sqrt{x^2+y^3}}{7x^5y^6} \div \dfrac{36\sqrt{x^2+y^3}}{35(x^2+y)^2} \times \dfrac{216(x^2y)^2}{52(y+x^2)^3}$

## Addition and subtraction

**Example 6**

Simplify $\dfrac{a}{4} + \dfrac{b}{5}$.

$$\dfrac{a}{4} + \dfrac{b}{5} = \dfrac{5a+4b}{20}$$

**Example 7**

Simplify $\dfrac{3x}{5} - \dfrac{x}{3}$.

$$\dfrac{3x}{5} - \dfrac{x}{3} = \dfrac{9x-5x}{15} = \dfrac{4x}{15}$$

**Example 8**

Simplify $\dfrac{2}{3b} + \dfrac{1}{2b}$.

$$\dfrac{2}{3b} + \dfrac{1}{2b} = \dfrac{4+3}{6b} = \dfrac{7}{6b}$$

**Example 9**

Simplify $\dfrac{3+x}{7} - \dfrac{x-2}{3}$.     Remember to use brackets here.  Note sign change.

$$\dfrac{3+x}{7} - \dfrac{x-2}{3} = \dfrac{3(3+x)-7(x-2)}{21} = \dfrac{9+3x-7x+14}{21} = \dfrac{23-4x}{21}$$

# Exercise 33

Simplify these.

**1** $\dfrac{x}{3} + \dfrac{x}{4}$

**2** $\dfrac{a}{5} + \dfrac{a}{4}$

**3** $\dfrac{a}{3} - \dfrac{a}{4}$

**4** $\dfrac{x}{4} - \dfrac{x}{6}$

**5** $\dfrac{a}{3} + \dfrac{b}{4}$

**6** $\dfrac{x}{5} + \dfrac{y}{6}$

**7** $\dfrac{2x}{3} - \dfrac{x}{4}$

**8** $\dfrac{4y}{5} - \dfrac{y}{4}$

**9** $\dfrac{2a}{7} + \dfrac{3a}{14}$

**10** $\dfrac{3b}{9} + \dfrac{5b}{18}$

**11** $\dfrac{a}{4} + \dfrac{b}{3}$

**12** $\dfrac{x}{2} - \dfrac{y}{5}$

**13** $\dfrac{2a}{3} - \dfrac{a}{2}$

**14** $\dfrac{4b}{5} - \dfrac{3b}{10}$

**15** $\dfrac{a}{4} + \dfrac{2b}{3}$

**16** $\dfrac{x}{9} + \dfrac{2y}{3}$

# Exercise 33★

Simplify these.

**1** $\dfrac{x}{6} + \dfrac{2x}{9}$

**2** $\dfrac{a}{5} - \dfrac{2a}{15}$

**3** $\dfrac{2a}{3} - \dfrac{3a}{7}$

**4** $\dfrac{4y}{5} - \dfrac{2y}{9}$

**5** $\dfrac{2x}{5} + \dfrac{4y}{7}$

**6** $\dfrac{6a}{11} - \dfrac{2b}{5}$

**7** $\dfrac{3a}{4} + \dfrac{a}{3} - \dfrac{5a}{6}$

**8** $\dfrac{x}{3} + \dfrac{3x}{5} + \dfrac{x}{15}$

**9** $\dfrac{3}{2b} + \dfrac{4}{3b}$

**10** $\dfrac{5}{4c} + \dfrac{3}{10c}$

**11** $\dfrac{2}{d} + \dfrac{3}{d^2}$

**12** $\dfrac{1}{x^2} - \dfrac{2}{x}$

**13** $\dfrac{2-x}{5} + \dfrac{3-x}{10}$

**14** $\dfrac{x+4}{3} + \dfrac{x-3}{4}$

**15** $\dfrac{y+3}{5} - \dfrac{y+4}{6}$

**16** $\dfrac{3-a}{2} - \dfrac{3-a}{7}$

**17** $\dfrac{x-3}{3} + \dfrac{x+5}{4} - \dfrac{2x-1}{6}$

**18** $\dfrac{2y+1}{6} - \dfrac{y-2}{5} - \dfrac{y+4}{10}$

**19** $\dfrac{a}{a-1} - \dfrac{a-1}{a}$

**20** $\dfrac{4}{x-4} + \dfrac{3}{x+3}$

# Solving equations

> **Remember**
>
> To solve equations, do the same operations to both sides.

---

**Example 10**

Solve $3x^2 + 4 = 52$.

$3x^2 + 4 = 52$   (Subtract 4 from both sides)

$3x^2 = 48$   (Divide both sides by 3)

$x^2 = 16$   (Square root both sides)

$x = \pm 4$

Check: $3 \times 16 + 4 = 52$

(Note that −4 is also an answer because $(-4) \times (-4) = 16$.)

**Example 11**

Solve $5\sqrt{x} = 50$.

$5\sqrt{x} = 50$   (Divide both sides by 5)

$\sqrt{x} = 10$   (Square both sides)

$x = 100$

Check: $5 \times \sqrt{100} = 50$

**Example 12**

Solve $\dfrac{\sqrt{x+5}}{3} = 1$.

$\dfrac{\sqrt{x+5}}{3} = 1$   (Multiply both sides by 3)

$\sqrt{x+5} = 3$   (Square both sides)

$x + 5 = 9$   (Subtract 5 from both sides)

$x = 4$

Check $\dfrac{\sqrt{4+5}}{3} = 1$

# Exercise 34

Solve these equations.

**1** $4x^2 = 36$

**2** $3x^2 = 12$

**3** $\dfrac{x^2}{3} = 12$

**4** $\dfrac{x^2}{2} = 32$

**5** $x^2 + 5 = 21$

**6** $x^2 - 5 = 59$

**7** $\dfrac{x^2}{2} + 5 = 37$

**8** $\dfrac{x^2}{7} - 3 = 4$

**9** $2x^2 + 5 = 23$

**10** $2x^2 + 13 = 63$

**11** $5x^2 - 7 = -2$

**12** $3x^2 - 20 = 55$

**13** $\dfrac{x + 12}{5} = 5$

**14** $\dfrac{x - 15}{8} = 4$

**15** $\dfrac{x^2 + 4}{5} = 4$

**16** $\dfrac{x^2 - 11}{7} = 10$

**17** $\sqrt{x} + 27 = 31$

**18** $\sqrt{x} - 4 = 8$

**19** $4\sqrt{x} + 4 = 40$

**20** $\dfrac{\sqrt{x + 4}}{4} = 1$

# Exercise 34★

Solve these equations.

**1** $4x^2 + 26 = 126$

**2** $38 = 5x^2 - 7$

**3** $\dfrac{x^2}{7} - 3 = 4$

**4** $37 = \dfrac{x^2}{2} + 5$

**5** $\dfrac{x^2 - 11}{7} = 10$

**6** $4 = \dfrac{x^2 + 4}{5}$

**7** $4\sqrt{x} + 4 = 40$

**8** $1 = \dfrac{\sqrt{x + 4}}{4}$

**9** $\sqrt{\dfrac{x - 3}{4}} + 5 = 6$

**10** $8 = 4 + \sqrt{7 + \dfrac{x}{4}}$

**11** $\dfrac{40 - 2x^2}{2} = 4$

**12** $\dfrac{37 - 3\sqrt{x}}{5} = 2$

**13** $22 = 32 - \dfrac{2x^2}{5}$

**14** $-6 = 10 - \dfrac{4\sqrt{x}}{3}$

**15** $(3 + x)^2 = 169$

**16** $4(4 + x)^2 = 9$

**17** $\sqrt{\dfrac{3x^2 + 5}{2}} + 4 = 8$

**18** $2 = \sqrt{22 - 8x^2}$

**19** $\sqrt{3 + \dfrac{(4 + \sqrt{x + 3})^2}{6}} = 3$

**20** $\sqrt{1 + \dfrac{(1 + \sqrt{2x - 1})^2}{4}} = 2$

# Using formulae

A formula is a way of describing a relationship, using algebra. For example, the formula to calculate the volume of a cylindrical can is $V = \pi r^2 h$ where $V$ is the volume, $r$ is the radius and $h$ is the height.

---

**Example 13**

Find the volume of a cola can that has a radius of 3 cm and a height of 11 cm.

| | |
|---|---|
| **Facts** | $r = 3$ cm, $h = 11$ cm, $V = ?$ cm$^3$ |
| **Equation** | $V = \pi r^2 h$ |
| **Substitution** | $V = \pi \times 3^2 \times 11$ |
| **Working** | $\pi \times 3^2 \times 11 = \pi \times 9 \times 11 = 311$ cm$^3$ (3 s.f.) |
| | Volume $= 311$ cm$^3$ (3 s.f.) |

---

**Remember**

When using any formula: Write down the **facts** with the correct units. Then write down the **equation**, **substitute** the facts, and do the **working**.

**Some commonly-used formulae**

You will need the following formulae to complete Exercise 35. The formulae are covered more fully later in this book.

In a right-angled triangle, Pythagoras's theorem:
$$a^2 = b^2 + c^2$$

Area of parallelogram
$$= bh$$

Area of trapezium
$$= \frac{h}{2}(a+b)$$

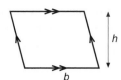

Area of a triangle $= \frac{1}{2}$ base $\times$ height

Area of a circle $= \pi r^2$

Circumference of a circle $= 2\pi r$

Speed $= \dfrac{\text{distance}}{\text{time}}$

**Challenge**

Prove the formulae for the area of a parallelogram and the area of a trapezium.

# Exercise 35

1  The area of a parallelogram is $31.5\,\text{cm}^2$ and its base is $7\,\text{cm}$ long. Find its height.

2  The area of a parallelogram is $57.6\,\text{cm}^2$ and its base is $8\,\text{cm}$ long. Find its height.

3  The radius of a circle is $7\,\text{cm}$. Find the circumference of the circle, and its area.

4  The radius of a circle is $14\,\text{cm}$. Find the circumference of the circle, and its area.

5  The area of this triangle is $72\,\text{cm}^2$.

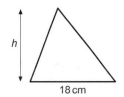

Find its height $h$.

6  The area of this triangle is $96\,\text{cm}^2$.

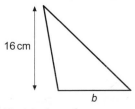

Find its base $b$.

**7** Find YZ.

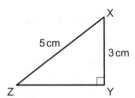

**8** Find BC.

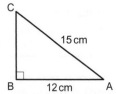

**9** The Earth, which is 150 million km from the Sun, takes 365 days to complete one circular orbit.
  **a** Find the length of one orbit, giving your answer in standard form correct to 3 significant figures.
  **b** Find the speed of the Earth around the Sun in km per hour correct to 2 significant figures.

**10** The planet Uranus, which is 3 billion km from the Sun, takes 84 years to complete one circular orbit.
  **a** Find the length of one orbit, giving your answer in standard form correct to 3 significant figures.
  **b** Find the speed of Uranus around the Sun in km per hour correct to 2 significant figures.

**11** The area of this trapezium is $21.62 \text{ cm}^2$.

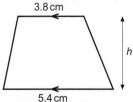

Find its height $h$.

**12** The area of this trapezium is $10.4 \text{ cm}^2$.

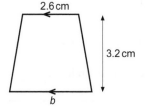

Find the length $b$.

**13** A ship travels 6 km North-West and then 10 km North-East.
How far is the ship from its starting point?

**14** It takes light $8\frac{1}{3}$ minutes to reach the Earth from the Sun.
Calculate the distance between the Sun and the Earth if light travels at $300\,000$ km/s.

**15** A bullet from a high-speed military machine gun travels at $108\,000$ km/h. Assuming that the bullet does not slow down, find the time, in seconds, that it takes to travel 30 km.

# Exercise 35★

1   The circumference of a circle is 88 cm. Find its radius.

2   The area of a circle is 154 cm². Find its radius.

3   The area of triangle ACD is 25.2 cm².

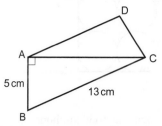

Find AC, and the perpendicular height of D above AC.

4   This arrowhead has an area of 35 cm².

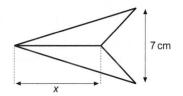

Find the length x.

5   The area of this trapezium is 30.8 cm².

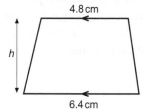

Find its height h.

6   The area of this trapezium is 51 cm².

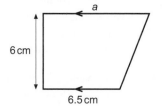

Find the length a.

7   Can a brick which measures 9 cm by 12 cm by 25 cm pass down a cylindrical pipe of diameter 14 cm?

8   ABCD is a rectangle. AB = 11 cm and AD = 6 cm. P is a point on DC such that DP = 8 cm, and Q is a point on BC such that BQ = 2 cm. Find the lengths of AP and PQ, and the area of APQ.

9   The diagram shows the cross-section of a metal pipe. Find the area of the shaded part correct to 3 significant figures.

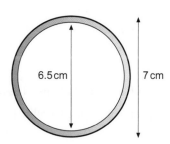

10  Calculate the distance between the points (3, 2) and (−2, −1). Give your answer correct to 3 significant figures.

11 Find the area of the shaded region in each of the diagrams correct to 3 significant figures.
Use your answers to work out the shaded region of a similar figure with 100 identical circles.

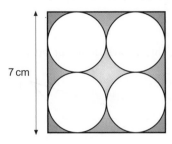

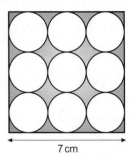

12 Owing to the rotation of the Earth, a point on the equator is moving at 1660 km/hour.
Find the circumference of the Earth at the equator, correct to 3 significant figures.

13 How long would the minute hand of a clock have to be if its end were to move at 100 km/hour?

14 A family goes on a cruise around the world. They have a pet hamster which remains in their cabin, which is at sea level, throughout the voyage. The captain has a parrot on the bridge. If the bridge is 35 m above sea level, calculate how much further the parrot travels than the hamster. (Assume that the voyage is a circular orbit.) Give your answer correct to 3 significant figures.

15 A triangle has sides of 5 cm, 3.3 cm and 6 cm.
What type of triangle is it: obtuse, acute or right-angled?

16 Calculate the height of an equilateral triangle which has the same area as a circle with a circumference of 10 cm. Give your answer correct to 3 significant figures.

17 The area of the shaded region is 20 cm². Find the value of $x$, correct to 3 significant figures.

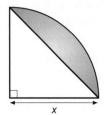

18 A sector of a circle of radius $r$ has a perimeter of $4r$. Show that its area is $r^2$.

19 This equation gives the monthly repayments £$S$ for a house mortgage over $n$ years on £$P$ borrowed at $R$%.

$$S = \frac{PR\left(1 + \dfrac{R}{100}\right)^n}{1200\left[\left(1 + \dfrac{R}{100}\right)^n - 1\right]}$$

Calculate, correct to 3 significant figures, the monthly repayments on £200 000 over 25 years at
a 8%
b 12%
c 17%

# Positive integer indices

UNIT 2 ◆ Algebra

> **Remember**
>
> $1000 = 10 \times 10 \times 10$, so it can be written as $10^3$.
>
> $1\,000\,000\,000$ is a billion, and can be written as $10^9$.
>
> This is because a billion is equal to $10 \times 10 \times 10 \times 10 \times 10 \times 10 \times 10 \times 10 \times 10$.
>
> Similarly, $3 \times 3 \times 3 \times 3 \times 3$ can be written as $3^5$, and, in algebra, $a \times a \times a \times a \times a \times a$ can be written as $a^6$.

To help you to understand how the rules of indices work, look carefully at these examples.

| Operation | Example | Rule |
|---|---|---|
| **Multiplying** | $a^4 \times a^2 = (a \times a \times a \times a) \times (a \times a)$ <br> $= a^6 = a^{4+2}$ | **Add** the indices <br> $(a^m \times a^n = a^{m+n})$ |
| **Dividing** | $a^4 \div a^2 = \dfrac{a \times a \times a \times a}{a \times a} = a^2 = a^{4-2}$ | **Subtract** the indices <br> $(a^m \div a^n = a^{m-n})$ |
| **Raising to a power** | $(a^4)^2 = (a \times a \times a \times a) \times (a \times a \times a \times a)$ <br> $= a^8 = a^{2 \times 4}$ | **Multiply** the indices <br> $(a^m)^n = a^{mn}$ |

> **Example 14**
>
> Use the rules of indices first to simplify $6^3 \times 6^4$. Then use your calculator to calculate the answer.
>
> $6^3 \times 6^4 = 6^7 = 279\,936$     (Add the indices)
>
> **Example 15**
>
> $9^5 \div 9^2 = 9^3 = 729$     (Subtract the indices)
>
> **Example 16**
>
> $(4^2)^5 = 4^{10} = 1\,048\,576$     (Multiply the indices)
>
> Notice that some answers become very large after only a few multiplications.

# Exercise 36

Use the rules of indices to simplify these. Then use your calculator to calculate the answer.

**1** $2^4 \times 2^6$      **2** $3^7 \times 3^2$      **3** $4^3 \times 4^4$      **4** $5^3 \times 5^4$

**5** $2^{10} \div 2^4$      **6** $3^{20} \div 3^{10}$      **7** $\dfrac{7^{13}}{7^{10}}$      **8** $\dfrac{8^{12}}{8^4}$

**9** $(2^3)^4$      **10** $(3^3)^3$      **11** $(6^2)^4$      **12** $(1^{10})^{10}$

Use the rules of indices to simplify these.

**13** $a^3 \times a^2$  **14** $b^4 \times b^3$  **15** $c^6 \div c^2$  **16** $d^5 \div d^3$

**17** $(e^2)^3$  **18** $(f^3)^2$  **19** $a^2 \times a^3 \times a^4$  **20** $b^3 \times b^4 \times b^5$

**21** $\dfrac{c^8}{c^3}$  **22** $\dfrac{d^9}{d^7}$  **23** $2 \times 6 \times a^4 \times a^2$  **24** $3 \times 4 \times b^6 \times b^2$

**25** $2a^3 \times 3a^2$  **26** $4b^3 \times 3b^4$  **27** $2(e^4)^2$  **28** $3(f^3)^2$

## Exercise 36★

Use the rules of indices to simplify these. Then use your calculator to calculate the answer.
Give your answers correct to 3 significant figures and in standard form.

**1** $6^6 \times 6^6$  **2** $7^7 \times 7^3$  **3** $7^{12} \div 7^6$  **4** $8^{10} \div 8^3$

**5** $(8^3)^4$  **6** $(9^3)^5$  **7** $4(4^4)^4$  **8** $(5^5)^5 \div 5$

Use the rules of indices to simplify these.

**9** $a^5 \times a^3 \times a^4$  **10** $b^5 \times b^4 \times b^5$  **11** $(12c^9) \div (4c^3)$  **12** $(16d^{10}) \div (2d^5)$

**13** $2(e^4)^2$  **14** $3(f^3)^2$  **15** $(2g^4)^3$  **16** $(3h^3)^3$

**17** $3(2j^3)^4$  **18** $4(3k^4)^2$  **19** $3m(2m^2)^3$  **20** $4n(3n^3)^2$

**21** $3a^2(3a^2)^2$  **22** $\dfrac{(3a^2)^2}{3a^2}$  **23** $\dfrac{2a^8 + 2a^8}{2a^8}$  **24** $\dfrac{2a^8 \times (2a)^8}{2a(2a)^8}$

**25** $\dfrac{12b^8}{6b^4} + 6b^4$  **26** $2a^8(2a^8 + 2a^8)$  **27** $\dfrac{b^4 + b^4 + b^4 + b^4 + b^4 + b^4}{b^4}$

**28** $16(d^{10} \div 2)d^5$

## Inequalities

### Number lines

**Remember**

These are examples of how to show inequalities on a number line.

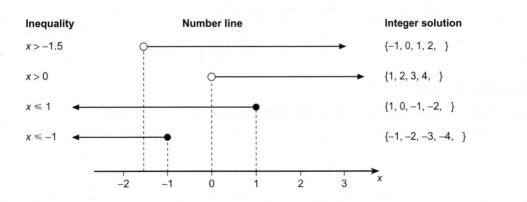

| Inequality | Number line | Integer solution |
|---|---|---|
| $x > -1.5$ | | $\{-1, 0, 1, 2,\ \}$ |
| $x > 0$ | | $\{1, 2, 3, 4,\ \}$ |
| $x \leqslant 1$ | | $\{1, 0, -1, -2,\ \}$ |
| $x \leqslant -1$ | | $\{-1, -2, -3, -4,\ \}$ |

# Solving linear inequalities

Inequalities are solved in the same way as algebraic equations, **except** that when multiplying or dividing by a negative number the inequality sign is reversed.

**Example 17**

Solve the inequality $4 < x \leqslant 10$. Show the result on a number line.

$4 < x \leqslant 10$          (Split the inequality into two parts)

$4 < x$    and    $x \leqslant 10$

$x > 4$    and    $x \leqslant 10$

Note that $x$ cannot be equal to 4.

**Example 18**

Solve the inequality $4 \geqslant 13 - 3x$. Show the result on a number line.

$4 \geqslant 13 - 3x$        (Add $3x$ to both sides)

$3x + 4 \geqslant 13$        (Subtract 4 from both sides)

$3x \geqslant 9$        (Divide both sides by 3)

$x \geqslant 3$

**Example 19**

Solve the inequality $5 - 3x < 1$. List the four smallest integers in the solution set.

$5 - 3x < 1$        (Subtract 5 from both sides)

$-3x < -4$        (Divide both sides by $-3$, so **reverse** the inequality sign)

$x > \dfrac{-4}{-3}$

$x > 1\dfrac{1}{3}$

Thus the four smallest integers are 2, 3, 4 and 5.

**Example 20**

Solve the inequality $x \leqslant 5x + 1 < 4x + 5$. Show the inequality on a number line.

$x \leqslant 5x + 1 < 4x + 5$        (Split the inequality into two parts)

**a**    $x \leqslant 5x + 1$        (Subtract $5x$ from both sides)

$-4x \leqslant 1$        (Divide both sides by $-4$, so **reverse** the inequality sign)

$x \geqslant -\dfrac{1}{4}$

**b**    $5x + 1 < 4x + 5$        (Subtract $4x$ from both sides)

$x + 1 < 5$        (Subtract 1 from both sides)

$x < 4$

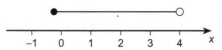

# Key Points

When finding the solution set of an inequality:

♦ Collect up the lettered terms on one side.

♦ When dividing or multiplying both sides by a negative number, **reverse** the inequality sign.

**Remember**

$x > 4$ means that $x$ cannot be equal to 4, while $x \geqslant 4$ means that $x$ can be equal to 4 or greater than 4.

## Exercise 37

For Questions 1–8, insert the correct symbol, <, > or =.

**1** $-3 \; \square \; 3$      **2** $\frac{1}{6} \; \square \; \frac{1}{7}$      **3** $30\% \; \square \; \frac{1}{3}$      **4** $0.01 \; \square \; 1\%$

**5** $-3 \; \square \; -4$      **6** $\frac{1}{9} \; \square \; \frac{1}{7}$      **7** $0.3 \; \square \; \frac{1}{3}$      **8** $\frac{1}{20} \; \square \; 0.5\%$

**9** Write down the inequalities represented by this number line.

**10** Write down the inequalities represented by this number line.

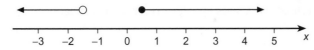

For Questions 11 and 12, write down the single inequality represented by each number line.

**11**

**12**

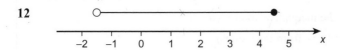

For Questions 13–28, solve the inequality, and show the result on a number line.

**13** $x - 3 > 2$      **14** $x + 5 > 8$      **15** $x - 3 \leqslant 1$

**16** $x - 6 \geqslant -8$      **17** $4 < 7 - x$      **18** $5 < 3 - x$

**19** $10 \geqslant 13 - x$      **20** $10 \geqslant 7 - x$      **21** $4x \geqslant 3x + 9$

**22** $3x \leqslant x - 6$      **23** $6x + 3 < 2x + 19$      **24** $5x - 7 > 2x + 5$

**25** $2(x + 3) < x + 6$      **26** $4(x + 2) \leqslant 2x + 6$      **27** $5(x - 1) > 2(x + 2)$

**28** $6(x - 2) \geqslant 4(x - 3)$

Solve these inequalities.

**29** $3 > x + 5$          **30** $2 < x - 4$          **31** $-2x \leqslant 10$

**32** $-3x \geqslant 12$          **33** $3 > 2x + 5$          **34** $4 \geqslant 3x + 5$

**35** $x - 4 \geqslant 3x$          **36** $x + 4 \leqslant 5x$          **37** $2(x - 1) \leqslant 5x$

**38** $3(x - 2) \leqslant 7x$          **39** $2(x - 3) \leqslant 5(x + 3)$          **40** $2(x - 2) \leqslant 4(x + 1)$

Solve these inequalities. List the integers in each solution set.

**41** $4 < x \leqslant 6$          **42** $4 \leqslant x < 6$          **43** $2 < x \leqslant 4.5$

**44** $-3 \leqslant x < 0$          **45** $-1 < x \leqslant 1.5$          **46** $-4.5 \leqslant x < -1.5$

**47** $2 \leqslant 2x < x + 5$          **48** $3 \leqslant 3x < x + 6$          **49** $4 < 2x + 1 \leqslant 7$

**50** $4 < 3x + 1 \leqslant 12$

# Exercise 37★

**1** Write down the inequalities represented by this number line.

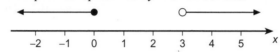

Explain why your two answers *cannot* be combined into a single inequality.

**2** Write down the single inequality represented by this number line.

What is the largest integer that satisfies your inequality?

For Questions 3–14, solve the inequality and show the result on a number line.

**3** $3x \leqslant x + 5$          **4** $4x > x - 11$          **5** $5x + 3 < 2x + 19$

**6** $5x - 7 \geqslant 2x + 4$          **7** $3(x + 3) < x + 12$          **8** $5(x + 1) \leqslant 2x - 20$

**9** $2(x - 1) > 7(x + 2)$          **10** $3(x - 2) \geqslant 5(x - 3)$          **11** $\dfrac{x}{2} - 3 \geqslant 3x - 8$

**12** $3 + \dfrac{x}{3} < 2(x + 4)$          **13** $x < 2x + 1 \leqslant 7$          **14** $2x < 3x + 1 \leqslant 13$

For Questions 15 and 16, solve the inequality, and then list the four largest integers in each solution set.

**15** $\dfrac{x + 1}{4} \geqslant \dfrac{x - 1}{3}$          **16** $\dfrac{x - 2}{7} \geqslant \dfrac{x + 1}{3}$

**17** Find the largest prime number $y$ that satisfies $4y \leqslant 103$.

**18** Find the largest prime number $z$ that satisfies $z^2 < 225$.

For Questions 19 and 20, list the integers that satisfy both the inequalities.

**19** $-3 \leqslant x < 4$   and   $x > 0$          **20** $-2.5 < x < 2$   and   $2x + 1 \leqslant x + 2$

# Exercise 38 (Revision)

Simplify these.

**1** $\dfrac{3y}{y}$

**2** $\dfrac{4x}{4}$

**3** $\dfrac{9x^2}{3x}$

**4** $\dfrac{2a}{3} \times \dfrac{6}{a}$

**5** $\dfrac{6b}{4} \div \dfrac{3b}{2a}$

**6** $\dfrac{10x^2}{3} \times \dfrac{9}{5x}$

**7** $\dfrac{y}{4} + \dfrac{y}{5}$

**8** $\dfrac{x}{3} - \dfrac{x}{5}$

**9** $\dfrac{2a}{5} + \dfrac{b}{10}$

Solve these.

**10** $\dfrac{x^2}{2} + 2 = 10$

**11** $\dfrac{x^2 + 2}{2} = 19$

**12** $\sqrt{\dfrac{4 + x}{6}} = 2$

Use the rules of indices to simplify these.

**13** $a^4 \times a^6$

**14** $b^7 \div b^5$

**15** $(c^4)^3$

**16** The area of the triangle is $7\,\text{cm}^2$, and the base is $2.8\,\text{cm}$ long.

2.8 cm

Find the height $h$.

**17** $AC = 20\,\text{cm}$ and $BC = 16\,\text{cm}$.

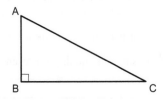

Find the length AB.

For Questions 18–21, rewrite each expression, replacing □ by the correct symbol, $<$, $>$ or $=$.

**18** $-2 \;\square\; -3$

**19** $\dfrac{1}{8} \;\square\; \dfrac{1}{7}$

**20** $0.009 \;\square\; 0.01$

**21** $0.1 \;\square\; 10\%$

**22** Write down the single inequality represented by this number line.

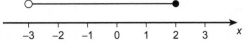

What is the smallest integer that $x$ can be?

For Questions 23–25, solve the inequality and show each result on a number line.

**23** $x - 4 > 1$

**24** $5x \leqslant 3x + 9$

**25** $5(x - 2) \geqslant 4(x - 2)$

**26** Solve the inequality $x + 5 \leqslant 6x$.

**27** List the integers in the solution set $3 \leqslant x < 5$.

**28** The area of a circle is $33\,\text{cm}^2$. Taking the area of the circle to be $\pi r^2$, find its radius correct to 3 significant figures.

**29** The Niagara Falls, one of the world's most spectacular waterfalls, is eroding the rockface over which the water cascades at the rate of $0.9\,\text{m/year}$.
Find the length of the erosion since the Falls were formed at the end of the Ice Age, $12\,600$ years ago.

# Exercise 38★ (Revision)

Simplify these.

**1** $\dfrac{20a}{5b}$

**2** $\dfrac{35x^2}{7xy}$

**3** $\dfrac{12ab^2}{48a^2b}$

**4** $\dfrac{2a}{b} \times \dfrac{b^2}{4a}$

**5** $\dfrac{30}{xy^2} \div \dfrac{6x^2}{x^2y}$

**6** $\dfrac{(3a)^2}{7b} \div \dfrac{a^3}{14b^2}$

**7** $\dfrac{3a}{2} + \dfrac{a}{10}$

**8** $\dfrac{2}{3b} + \dfrac{3}{4b} - \dfrac{5}{6b}$

**9** $\dfrac{x+1}{7} - \dfrac{x-3}{21}$

Solve these.

**10** $3x^2 + 5 = 32$

**11** $2 = \dfrac{\sqrt{2x} + 2}{2}$

**12** $\sqrt{100 - 4x^2} = 6$

For Questions 13–15, use the rules of indices to simplify each expression.

**13** $a^5 \times a^6 \div a^7$

**14** $(2b^3)^2$

**15** $3c(3c^2)^3$

**16** A 5 m ladder just reaches a window 4.5 m from the ground.
How far is the bottom of the ladder from the building?

**17** Write down the single inequality represented by the number line.

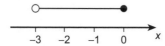

What is the smallest integer that satisfies the inequality?

For Questions 18–20, solve the inequality and show each result on a number line.

**18** $7x + 3 < 2x - 19$

**19** $2(x - 1) < 5(x + 2)$

**20** $\dfrac{x-2}{5} \geqslant \dfrac{x-3}{3}$

**21** Find the largest prime number $y$ which satisfies $3y - 11 \leqslant 103$.

**22** List the integers which satisfy both these inequalities simultaneously.
$$-3.5 < x < 3 \quad \text{and} \quad 4x + 1 \leqslant x + 2$$

**23** Find the circumference of a circle of area $200\,\text{cm}^2$.

**24** The fastest speed of a ball that has been served in tennis is 222 km/hour.
How long would it have taken the ball to travel 24 m, the length of a tennis court?
Give your answer in seconds correct to 2 significant figures.

**25** A pulsar (an imploding star) rotates at an incredible rate of 30 times a second.
If its diameter is 12 km, find the speed of a point on its equator.
Give your answer correct to 3 significant figures.

# Graphs 2

## Simultaneous equations

**Activity 9**

Viv is trying to decide between two Internet service providers, Pineapple and Banana. Pineapple charges $9.99/month plus 1.1 cents/minute online, while Banana charges $4.95/month plus 1.8 cents/minute online.

If $C$ is the cost in cents and $t$ is the time (in minutes) online per month then the cost of using Pineapple is $C = 999 + 1.1t$, and the cost of using Banana is $C = 495 + 1.8t$.

◆ Copy and complete this table to give the charges for Pineapple.

| Time online $t$ (minutes) | 0 | 500 | 1000 |
|---|---|---|---|
| Cost $C$ (cents) | | | |

◆ Draw a graph of this data with $t$ along the horizontal axis and $C$ along the vertical axis.
◆ Make a similar table for the Banana charges. Add the graph of this data to your previous graph.
◆ How many minutes online per month will result in both companies charging the same amount?

In Activity 9, you solved the simultaneous equations $C = 999 + 1.1t$ and $C = 495 + 1.8t$ graphically. From the graph you can also tell which is the cheaper option for any number of minutes online.

---

**Remember**

To solve simultaneous equations **graphically**:
◆ Draw the graphs for both equations on one set of axes.
◆ Only plot three points for a straight-line graph.
◆ The solution is where the graphs intersect.
◆ If the graphs do not intersect, there is no solution.
◆ If the graphs are the same, there is an infinite number of solutions.

---

**Example 1**

Solve the simultaneous equations $y = \frac{1}{2}x + 2$ and $y = 4 - x$ graphically.
First, make a table of values for each equation.

| $x$ | 0 | 2 | 4 |
|---|---|---|---|
| $y = \frac{1}{2}x + 2$ | 2 | 3 | 4 |

| $x$ | 0 | 2 | 4 |
|---|---|---|---|
| $y = 4 - x$ | 4 | 2 | 0 |

Next, draw accurate graphs for both equations on one set of axes.

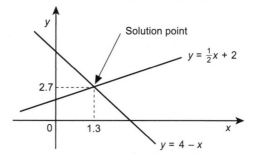

The solution point is approximately $x = 1.3$, $y = 2.7$.

# Exercise 39

**1** Copy and complete these tables, and then draw both graphs on one set of axes.

| $x$ | 0 | 2 | 4 |
|---|---|---|---|
| $y = x + 1$ | | | |

| $x$ | 0 | 2 | 4 |
|---|---|---|---|
| $y = 2x - 2$ | | | |

Solve the simultaneous equations $y = x + 1$, $y = 2x - 2$ using your graph.

**2** Copy and complete these tables, and then draw both graphs on one set of axes.

| $x$ | 0 | 2 | 4 |
|---|---|---|---|
| $y = x - 1$ | | | |

| $x$ | 0 | 2 | 4 |
|---|---|---|---|
| $y = 6 - x$ | | | |

Solve the simultaneous equations $y = x - 1$, $y = 6 - x$ using your graph.

**3** On one set of axes, draw the graphs of $y = 3x - 1$ and $y = 2x + 1$ for $0 \le x \le 6$.
Then, solve the simultaneous equations $y = 3x - 1$ and $y = 2x + 1$ using your graph.

**4** On one set of axes, draw the graphs of $y = 2x - 2$ and $y = x - 1$ for $0 \le x \le 6$.
Then, solve the simultaneous equations $y = 2x - 2$ and $y = x - 1$ using your graph.

For Questions 5–10, solve the simultaneous equations graphically, using $0 \le x \le 6$ in each question.

**5** $y = 2x + 2$      $y = 3x - 1$        **6** $y = 4x - 3$      $y = 3x + 1$

**7** $y = 4x + 3$      $y = 2x + 6$        **8** $y = x + 3$      $y = 3x - 2$

**9** $y = \frac{1}{2}x + 1$      $y = 4 - x$        **10** $y = x - 2$      $y = 4 - \frac{1}{2}x$

**11 a** A cross-Channel hovercraft ferry leaves Dover harbour at 9am travelling at 35 km/hour. Find an equation giving $d$ (the distance travelled in km) in terms of $t$ (the time in hours after 9am). Plot the graph of this equation for $0 \le t \le 1$.

   **b** As the hovercraft leaves the harbour, the captain sees a car ferry 5 km ahead travelling in the same direction at 15 km/hour. Find an equation for $d$ in terms of $t$ for the car ferry. Plot the graph of this equation, and find when the hovercraft catches up with the car ferry.

**12** The Purple Mobile Phone Company offers the following two pricing plans to customers. Plan A costs $15/month, with calls at 25c/minute, while Plan B costs $100/month with calls at 14c/minute.

   **a** Find an equation that gives $C$, the cost in dollars, in terms of $t$, the call time in minutes per month for each plan.

   **b** Plot the graphs of these equations on one set of axes.

   **c** What call time per month costs the same under both plans?

# Exercise 39★

1 On one set of axes, draw the graphs of $y = 2x + 1$ and $y = 3x - 5$ for $0 \leqslant x \leqslant 6$.
   Then, solve the simultaneous equations $y = 2x + 1$ and $y = 3x - 5$ using your graphs.

2 On one set of axes, draw the graphs of $y = x + 3$ and $y = 3x - 7$ for $0 \leqslant x \leqslant 8$.
   Then, solve the simultaneous equations $y = x + 3$ and $y = 3x - 7$ using your graphs.

For Questions 3–8, solve the simultaneous equations graphically, using $0 \leqslant x \leqslant 6$ for each pair.

3 $x + y = 6$ $\qquad$ $3x - y = 1$ $\qquad\qquad$ 4 $x + 2y = 6$ $\qquad$ $x - y = 1$

5 $2x + 3y = 6$ $\qquad$ $2y = x - 2$ $\qquad\qquad$ 6 $5x + 4y = 20$ $\qquad$ $3y = x + 6$

7 $6x - 5 = 2y$ $\qquad$ $3x - 7 = 6y$ $\qquad\qquad$ 8 $7y = x + 20$ $\qquad$ $3y = 6 + 2x$

9 McMountain Construction is digging a tunnel in the Alps. The mountain can be represented by $3y = 4x + 6000$ for $0 \leqslant x \leqslant 3000$ and $7y + 6x = 60\,000$ for $3000 \leqslant x \leqslant 10\,000$. The tunnel can be represented by the line $10y + x = 30\,000$. All the units are metres. Find the coordinates of the ends of the tunnel.

10 Copy and complete this table to show the angle that the minute hand of a clock makes with the number 12 for various times after 12 noon.

| Time after 12 noon (hours) | 0 | $\frac{1}{4}$ | $\frac{1}{2}$ | $\frac{3}{4}$ | 1 | $1\frac{1}{4}$ | $1\frac{1}{2}$ | $1\frac{3}{4}$ | 2 | $2\frac{1}{4}$ | $2\frac{1}{2}$ |
|---|---|---|---|---|---|---|---|---|---|---|---|
| Angle (degrees) | 0 | | 180 | | 90 | | | | | | |

a Use the table to draw a graph of angle against time. Show the time from 0 hours to 6 hours along the $x$-axis, and the angle from 0° to 360° along the $y$-axis.

b Draw another line on your graph to show the angle that the hour hand makes with the number 12 for various times after 12 noon.

c Use your graph to find the times between 12 noon and 6pm when the hour hand and the minute hand of the clock are in line.

# Inequalities

Inequalities in two variables can be represented on a graph.

**Example 2**

Find the region representing $x + y < 4$.

First draw the line $x + y = 4$.

This line divides the graph into two regions.

One of these regions satisfies $x + y < 4$, and the other satisfies $x + y > 4$.

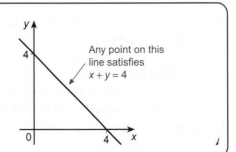

Any point on this line satisfies $x + y = 4$

To decide which region satisfies $x + y < 4$, take any point in one of the regions, for example (1, 1).

Substitute $x = 1$ and $y = 1$ into $x + y$ to see if the result is less than 4.

$$1 + 1 < 4$$

So (1, 1) is in the required region, because it satisfies $x + y < 4$.

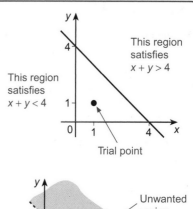

Therefore the required region is **below** the line $x + y = 4$.

Therefore this is the solution.

The line $x + y = 4$ is drawn as a broken line, to show that points on the line are *not* required. (Draw a solid line if points on the line *are* required.)

Notice that the **unwanted** region is always shaded.

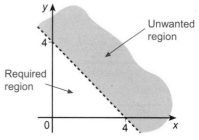

# Key Points

♦ Find the line representing the equality.

♦ If points on the line are required, draw a solid line. Otherwise, draw a broken line.

♦ Find the required region by using a trial point that is *not* on the line.

♦ Shade the **unwanted** region.

Inequalities in one variable can also be represented on a graph.

## Example 3

Find the region that represents $y \leqslant 3$.

Draw the line $y = 3$ as a solid line because points on the line satisfy $y \leqslant 3$.

Points below the line $y = 3$ have $y$ values less than 3, so the required region is below the line, and the unwanted region, above the line, is shaded.

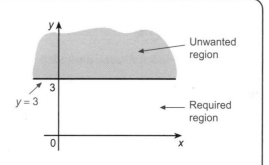

# Exercise 40

For Questions 1–8, describe the **unshaded** region in each graph.

**1**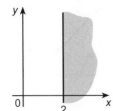

**2**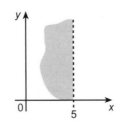

**3**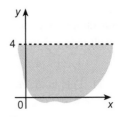

**4**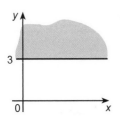

**5**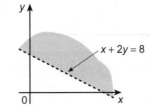

**6**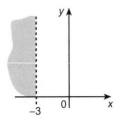

**7**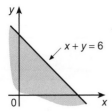

**8**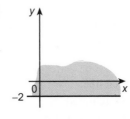

For Questions 9–16, illustrate each inequality on a graph.

**9** $x < 1$

**10** $x \geqslant -1$

**11** $y \geqslant -2$

**12** $y < 3$

**13** $y \geqslant 8 - x$

**14** $y < 10 - x$

**15** $y < 6 - 2x$

**16** $y \geqslant 3 - \dfrac{x}{3}$

# Exercise 40★

For Questions 1–8, describe the **unshaded** region in each graph.

**1**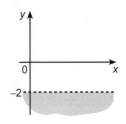

**2**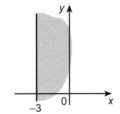

**3**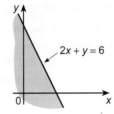

**4**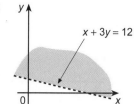

**5**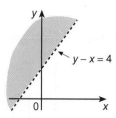

**6**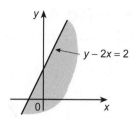

**7**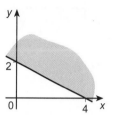

**8**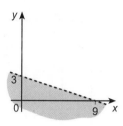

For Questions 9–16, illustrate each inequality on a graph.

**9** $x \leqslant -5$

**10** $y > -4$

**11** $3x + 4y > 12$

**12** $2x + 5y \leqslant 10$

**13** $y - 3x > 4$

**14** $y - 2x \leqslant -2$

**15** $2y - x \geqslant 4$

**16** $3y - 6x < 2$

Simultaneous inequalities can also be represented on a graph.

---

**Example 4**

Find the region representing $1 \leqslant x < 4$.

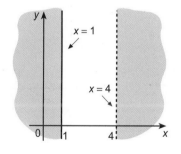

**Example 5**

Find the region representing $x + y \leqslant 4$ and $y - x < 2$.

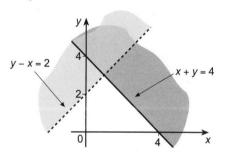

---

**Investigate**

What would several inequalities drawn on one graph look like if the **wanted** region was shaded?

# Exercise 41

For Questions 1–12, describe the **unshaded** region in the graph.

**1**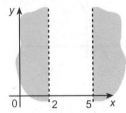

**2**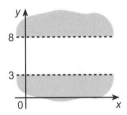

**3**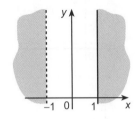

**4**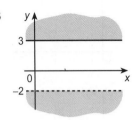

**5**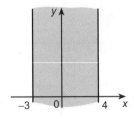

**6**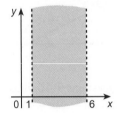

**7**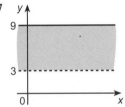

**8**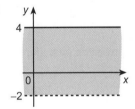

**9**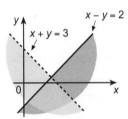

**10**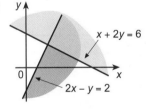

**11**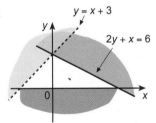

**12**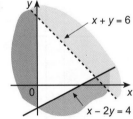

For Questions 13–20, illustrate each inequality on a graph.

**13** $2 \leqslant y < 5$

**14** $-1 < x \leqslant 3$

**15** $x < -1$ or $x \geqslant 4$

**16** $y < 1$ or $y \geqslant 2$

**17** $y > 5 - x$ and $y \geqslant 2x - 2$

**18** $y < 2 - \dfrac{x}{3}$ and $y < x$

**19** $x \geqslant 0$, $y > 2x - 3$ and $y \leqslant 2 - \dfrac{x}{2}$

**20** $y \geqslant 0$, $y < 4 + x$ and $y < 3 - \dfrac{3}{4}x$

# Exercise 41★

For Questions 1–8, describe the **unshaded** region in the graph.

**1**
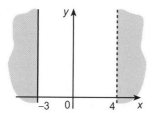

**2**
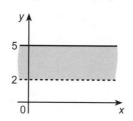

**3**
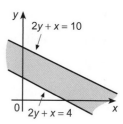

**4**
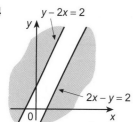

**5**
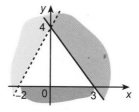

**6**
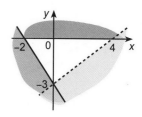

**7**
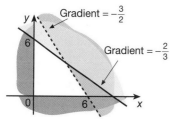

**8**
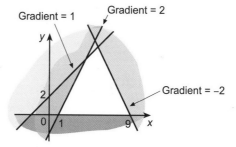

For Questions 9–12, illustrate each inequality on a graph.

**9** $y \geqslant 2x$, $x + 2y \leqslant 4$ and $y + 2x > 1$

**10** $2y > x$, $y + 2x \leqslant 4$ and $y > 2x + 2$

**11** $x \geqslant 0$, $y > 0$, $y < \dfrac{x}{2} + 4$ and $y \leqslant 6 - 2x$

**12** $x > 0$, $y \geqslant 0$, $3x + 4y \leqslant 12$ and $5x + 2y \leqslant 10$

**13 a** On a graph, draw the triangle with vertices $(-2, 0)$, $(0, 2)$ and $(2, -2)$.
   **b** Find the three inequalities that define the region inside the triangle.
   **c** What is the smallest integer value of $y$ that satisfies all three inequalities?

**14 a** On a graph, draw the triangle with vertices $(3, 4)$, $(1, 0)$ and $(0, 2.5)$.
   **b** Find the three inequalities that define the region inside the triangle.
   **c** What is the smallest integer value of $x$ that satisfies all three inequalities?

**15** Illustrate on a graph the region that satisfies $y > x^2 - 4$ and $y \leqslant 0$.

**16** Illustrate on a graph the region that satisfies both of the inequalities $y \geqslant x^2 - x - 2$ and $y < x + 6$.

**17** Two numbers have a sum that is less than 16 and a product that is more than 36.
   Illustrate on a graph all the number pairs that satisfy these conditions, and list all the positive integers that satisfy these conditions.

**18** A line is drawn on a graph to represent $y = mx + c$. Which side of the line always illustrates $y \geqslant mx + c$?

# Exercise 42 (Revision)

1  Solve $y = 3 - x$ and $y = 2x - 3$ graphically using $0 \leqslant x \leqslant 4$.

2  Solve $y = x - 1$ and $y = \frac{1}{2}x + 1$ graphically using $0 \leqslant x \leqslant 6$.

3  Solve $y = x + 4$ and $y = 1 - 2x$ graphically using $-2 \leqslant x \leqslant 2$.

4  Solve $y = 2x + 2$ and $y = -x - 4$ graphically using $-4 \leqslant x \leqslant 2$.

5  Write down the four pairs of inequalities that describe the regions A, B, C and D.

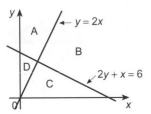

6  Write down the four pairs of inequalities that describe the regions A, B, C and D.

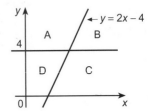

7  Illustrate the region that satisfies the inequalities $x \geqslant 0$, $y < 6 - x$ and $y \leqslant x + 1$.

8  Illustrate the region that satisfies the inequalities $y > 0$, $y \leqslant \frac{1}{2}x + 3$ and $y < 5 - x$.

# Exercise 42★ (Revision)

1  Solve $3x + 8y = 24$ and $2y = x + 2$ graphically using $0 \leqslant x \leqslant 4$.

2  Solve $6x + 5y = 30$ and $3y = 12 - x$ graphically using $0 \leqslant x \leqslant 4$.

3  Solve $y - 3x - 6 = 0$ and $y + 2x + 2 = 0$ graphically using $-4 \leqslant x \leqslant 2$.

4  Solve $2y + x + 3 = 0$ and $3y - x + 1 = 0$ graphically using $-4 \leqslant x \leqslant 2$.

5  Write down the inequalities that define the interior of a triangle with vertices at $(-1, -1)$, $(3, 1)$ and $(1, 5)$.

6  Write down the four pairs of inequalities that describe the regions A, B, C and D.

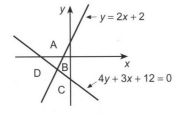

7  Write down the inequalities that define the unshaded region marked A on the graph.

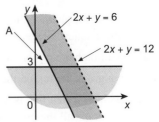

8  Illustrate the region that satisfies the inequalities $y \leqslant \frac{x}{2} + 3$, $y > 3x - 4$ and $y + 2x + 4 > 0$.

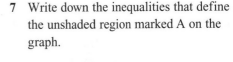

# Shape and space 2

## Tangent ratio

The history of mankind is full of examples of how we used features of **similar triangles** to build structures (for example Egyptian pyramids), survey the sky (as Greek astronomers did), and survey the land (for instance to produce Eratosthenes' map of the world).

 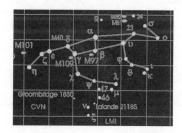 

Trigonometry (triangle measurement) is used today by architects, engineers, surveyors and scientists. It allows us to solve **right-angled triangles** without the use of scale drawings.

The sides of a right-angled triangle are given special names which must be easily recognised.

**Remember**

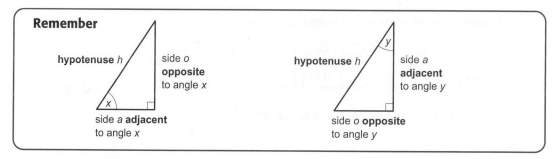

hypotenuse *h*  side *o* **opposite** to angle *x*  side *a* **adjacent** to angle *x*

hypotenuse *h*  side *a* **adjacent** to angle *y*  side *o* **opposite** to angle *y*

## Activity 10

Triangles X, Y and Z are similar.

 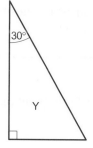 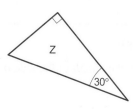

- ♦ For each triangle, measure the sides opposite (*o*) and adjacent (*a*) to the 30° angle in millimetres.
- ♦ Calculate the ratio of $\dfrac{o}{a}$ to 2 decimal places for X, Y and Z.
- ♦ What do you notice?

You should have found that the ratio *o* : *a* for the 30° angle is the same for all three triangles.
This is the case for *any* similar right-angled triangle with a 30° angle, and should not surprise you because you were calculating the **gradient** of the same slope each time.

The actual value of $\dfrac{\text{opposite}}{\text{adjacent}}$ for $30°$ is $0.577350$ (to 6 d.p.).

The ratio $\dfrac{\text{opposite}}{\text{adjacent}}$ for a given angle $x$ is a fixed number. It is called the **tangent of $x$,** or **tan $x$.**

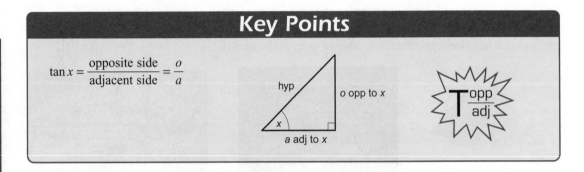

# Key Points

$$\tan x = \frac{\text{opposite side}}{\text{adjacent side}} = \frac{o}{a}$$

You can find the tangent ratio on your calculator.

♦ Make sure your calculator is in degree mode.

♦ Press the `tan` button followed by the value of the angle. Then press `=`.

Copy and complete the table, correct to 3 significant figures.

| $x$ (°) | 0° | 15° | 30° | 45° | 60° | 75° | 89° | 90° |
|---------|-----|------|------|------|------|------|------|------|
| $\tan x$ |     |      | 0.577 |    |     | 3.73 |      |      |

Why is $\tan 89°$ so large? Why can $\tan 90°$ not be found?

## Calculating sides

If you know an angle – e.g. an angle of elevation or angle of depression – and one side, you can find the length of another side by using the tangent ratio.

**Example 1**

Find the length of the side $p$ correct to 3 significant figures.

$$\tan 30° = \frac{p}{12\,\text{cm}}$$

$$12\,\text{cm} \times \tan 30° = p$$

$$p = 6.93\,\text{cm} \quad\text{(to 3 s.f.)}$$

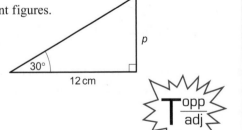

`1` `2` `×` `tan` `3` `0` `=` `6.92820` (to 6 s.f.)

## Example 2

PQ represents a 25 m tower, and R is a surveyor's mark $p$ m away from Q.

The **angle of elevation** of the top of the tower from the surveyor's mark R on level ground is 60°. Find the distance RQ correct to 3 significant figures.

$$\tan 60° = \frac{25\,\text{m}}{p\,\text{m}}$$

$$p \times \tan 60° = 25$$

$$p = \frac{25}{\tan 60°}$$

$$p = 14.4 \text{ (to 3 s.f.)}$$

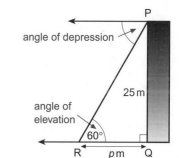

2 5 ÷ tan 6 0 = `14.4338` (to 6 s.f.)

## Remember

Bearings are measured

♦ clockwise

♦ from North.

A is 310° from B.
B is 130° from A.

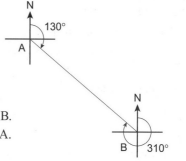

# Exercise 43

In this exercise, give answers correct to 3 significant figures.

In Questions 1–3, which sides are the hypotenuse, opposite and adjacent to the given angle $a$?

**1**

**2**

**3**

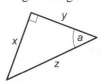

In Questions 4–6, what is the value of $\tan b$ for each of the triangles?

**4**

**5**

**6**

For Questions 7–12, find the length of side *x*.

**7**

**8**

**9**

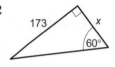

**10**

**11**

**12**

For Questions 13–18, find the length of side *x*.

**13**

**14**

**15**

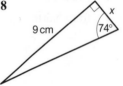

**16**

**17**

**18**

**19** The angle of elevation of the top of a cliff from a boat 125 m away from its foot is 35°. Find the height of the cliff.

**20** The angle of elevation of a radio mast from Remi is 60°. She is 50 m away from the base of the mast. Find its height.

**21** Find the cross-sectional area of the pitched roof WXY if WX = WY.

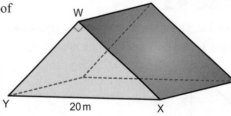

# Exercise 43★

Give answers correct to 3 significant figures.

For Questions 1–7, find the length of the side marked *x*.

**1**

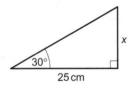

**2**

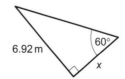

**3**

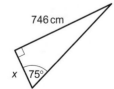

**4**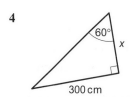
60°
x
300 cm

**5**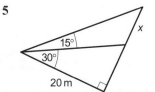
15°
30°
x
20 m

**6**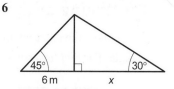
45°
30°
6 m
x

**7**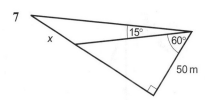
15°
60°
x
50 m

**8** The angle of depression from a window 10 m above the ground to a coin on the ground is 15°. Find the distance of the coin from the base of the building.

**9** The gnomon (central pin) of the giant equatorial sundial with observatory tower in Jaipur, India, is an enormous right-angled triangle. The angle the hypotenuse makes with the base is 27°. The base is 44 m. Find the height of the gnomon.

For Questions 10–13, find the lengths of $x$ and $y$.

**10**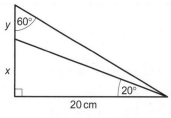
35°
60°
x
12 cm
y

**11**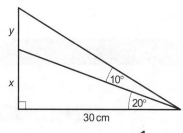
y
x
10°
20°
30 cm

**12**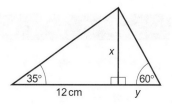
y
60°
x
20°
20 cm

**13**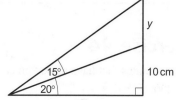
15°
20°
y
10 cm
x

**14** From the top of a 25 m high cliff, the angle of depression of a jet ski at point A is 20°. The jet ski moves in a straight line towards the base of the cliff so that its angle of depression 5 s later at point B is 30°.

  **a** Calculate the distance AB.
  **b** Find the average speed of the jet ski in kilometres/hour.

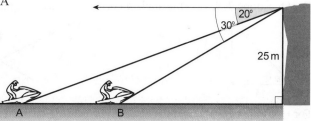
20°
30°
25 m
A
B

**15** In a triangle ABC, the angle at B is 62°, the angle at C is 75°, and the altitude AX = 5 m. Calculate BX and BC.

**16** A regular pentagon has sides of 10 cm. Find the radius of the largest circle which can be drawn inside the pentagon.

## Calculating angles

If you know the adjacent and opposite sides of a right-angled triangle, you can find the angles in the triangle. For this 'inverse' operation, you need to use the INV tan buttons on your calculator.

---

**Example 3**

The diagram shows a child on a slide.
Find angles $x$ and $y$ to the nearest degree.

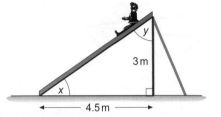

$$\tan x = \frac{3}{4.5} \text{ and } \tan y = \frac{4.5}{3}$$

INV tan ( 3 ÷ 4 . 5 ) = 33.6901 (to 6 s.f.)

INV tan ( 4 . 5 ÷ 3 ) = 56.3099 (to 6 s.f.)

$$T\frac{opp}{adj}$$

So $x = 34°$ and $y = 56°$ (to the nearest degree).

---

## Key Points

To calculate an angle from a tangent ratio, use the

INV tan or SHIFT tan buttons.

---

# Exercise 44

Find the angles that have these tangents, giving your answers correct to 2 significant figures.

**1** 1.000        **2** 0.577        **3** 0.268

Find the angles that have these tangents, giving your answers correct to 3 significant figures.

**4** 1.732        **5** 2.747        **6** 3.732

For Questions 7–9, find the angle $a$ correct to 1 decimal place.

**7**

**8**

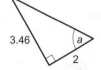

**9**

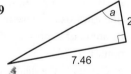

For Questions 10–15, find the angle $x$ correct to 1 decimal place.

**10**

**11**

**12**

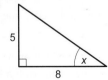

**13**

**14**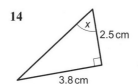

**15**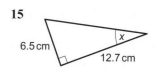

**16** Calculate the angles $a$ and $b$ of this dry ski slope correct to 1 decimal place.

**17** A bell tower is 65 m high. Find the angle of elevation, to 1 decimal place, of its top from a point 150 m away on level ground.

**18** Ollie is going to walk in the rain forest. He plans to go up a slope along a straight footpath from point P to point Q. The hill is 134 m high, and distance PQ *on the map* is 500 m.
Find the angle of the hill.

## Exercise 44★

**1** ABCD is a rectangle.
Find angles $a$ and $b$ to the nearest degree.

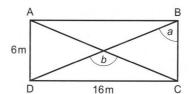

**2** The rhombus PQRS has diagonals 14 cm and 21 cm long.
Find the angle PQR to 3 significant figures.

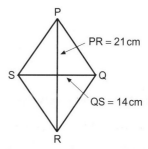

For Questions 3–5, find angle $a$.

**3**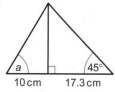

**4**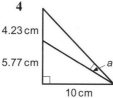

**5** Rectangle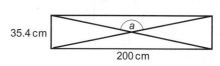

**6** A church C is 8.6 km North and 12.5 km West of a school S.
   **a** Calculate correct to 3 significant figures the bearing of S from C.
   **b** Calculate correct to 3 significant figures the bearing of C from S.

7 The grid represents a map on which villages X, Y and Z are shown. The sides of the grid squares represent 5 km.

Find, to 1 decimal place, the bearing of

a Y from X

b X from Y

c Z from X

d Z from Y

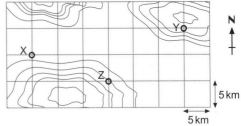

8 A 5 m flagpole is secured by two ropes PQ and PR. Point R is the mid-point of SQ. Find angle $a$ to 1 decimal place.

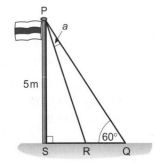

9 In a quadrilateral ABCD, AB = 3 cm, BC = 5.7 cm, CD = 4 cm and AD = 5 cm. Angle B = angle D = 90°. Calculate the angle at A to 1 decimal place.

10 Draw an equilateral triangle and bisect it. Prove that the exact values of the tangent ratios of 30°, 45° and 60° are $\dfrac{1}{\sqrt{3}}$, 1 and $\sqrt{3}$, respectively.

11 Given that $\tan 30° = \dfrac{1}{\sqrt{3}}$ and $\tan 60° = \sqrt{3}$, show that the exact value of the height of the tree in metres is given by $25\sqrt{3}$.

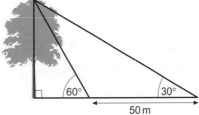

50 m

# Exercise 45 (Revision)

In this exercise, give answers correct to 3 significant figures.

For Questions 1–6, find the value of $x$.

1

35°
10
$x$

2

$x$
40°
8

3

15
25°
$x$

4

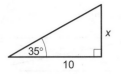

$x$
20°
4

5

$x$
35°
6

6

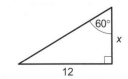

60°
$x$
12

For Questions 7–9, find angle $a$.

**7**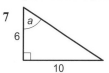

**8**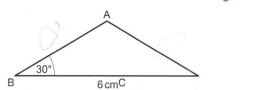

**9**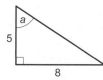

**10** Find the area of the isosceles triangle ABC, given that $\tan 30° = 0.577$.

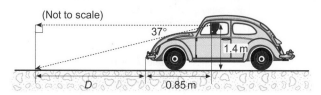

**11** Calculate the angle between the longest side and the diagonal of a 577 mm by 1000 mm rectangle.

# Exercise 45★ (Revision)

Give answers to 3 significant figures.

**1** The angle of elevation to the top of the CN Tower, Toronto, Canada, from a point on the ground 50 m away from the centre of the tower is 84.8°. Find the height of the CN Tower.

**2** A harbour H is 25 km due North of an airport A. A town T is 50 km due East of H.
  **a** Calculate the bearing of T from A.
  **b** Calculate the bearing of A from T.

**3** The diagram shows the lines of sight of a car-driver.

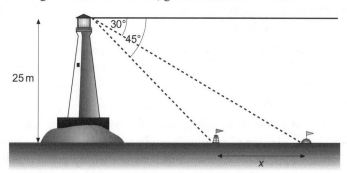

  **a** Calculate the driver's 'blind' distance $D$ in metres.
  **b** Why is this distance important in the car design?

**4** Show that the area of an equilateral triangle of side $2x$ is $x^2\sqrt{3}$, given that $\tan 60° = \sqrt{3}$.

**5** A lighthouse is 25 m high. From its top, the angles of depression of two buoys due North of it are 45° and 30°.

Given that $\tan 30° = \dfrac{1}{\sqrt{3}}$, show that the distance $x$, in metres, between the buoys is $25(\sqrt{3} - 1)$.

# Handling data 2

## Statistical investigation

Statistics is the science of collecting information (**data**) and analysing it. People do this to gather evidence, form conclusions (and make predictions), and then to take decisions.

A statistical investigation must have a clear purpose. This determines what data is needed, how it must be collected, what form it needs to be in, and how it should be displayed.

A school catering manager might survey pupils about their favourite foods before creating new menus. A government minister of transport must consider projected traffic flows when a new major road is being planned.

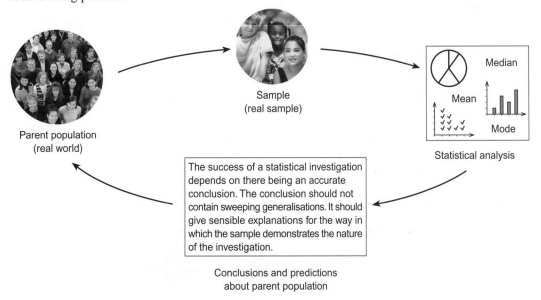

Parent population
(real world)

Sample
(real sample)

Median

Mean

Mode

Statistical analysis

The success of a statistical investigation depends on there being an accurate conclusion. The conclusion should not contain sweeping generalisations. It should give sensible explanations for the way in which the sample demonstrates the nature of the investigation.

Conclusions and predictions
about parent population

Data can be **qualitative** (such as opinions, colours, or clothes size) or **quantitative** (such as a length measurement, a mass, or a frequency).

**Primary data** is data that has been collected directly by a **researcher**. A researcher may also refer to **secondary data** – data that has been collected and made available by other organisations. Many **databases** are made available to the public by government departments and large companies, either by direct request or through the Internet.

In a perfect world, the researcher would take a **census** (in which everything or everybody is surveyed). However, this is rarely practical, and is often impossible. A **survey** costs money and takes time, and sometimes the process of 'testing' can destroy the 'objects'. For example, testing rope until it breaks destroys the rope.

So, instead, a **sample** from the population to be examined is tested. The size of the sample and the method of selection of the sample are important but no matter how carefully a sample is chosen, there is always a possibility of some **bias**. This is often allowed for by the inclusion of a **margin of error** in any conclusions. For example, the results of an election opinion poll might state that the Democrat vote will be 37%±3%.

# Collecting data

There are many ways of collecting data.

Questionnaires

Data logging

Databases

Simulation

It is important to be methodical and accurate when writing down the data, and when transcribing it into another form, for instance into a computer file, table, chart or diagram.

# Frequency tables

Before attempting to calculate any statistics, or to recognise any trends or patterns, it is sensible to sort the data and group it together.

These marks were obtained by a set of pupils in an IGCSE mock examination.

| | | | | | | | | | |
|---|---|---|---|---|---|---|---|---|---|
| 42 | 54 | 60 | 48 | 73 | 50 | 59 | 45 | 84 | 49 |
| 67 | 47 | 70 | 78 | 77 | 67 | 55 | 68 | 42 | 59 |
| 54 | 41 | 69 | 65 | 41 | 56 | 80 | 59 | 44 | 68 |
| 82 | 41 | 71 | 42 | 55 | 64 | 51 | 69 | 89 | 72 |
| 72 | 46 | 85 | 40 | 78 | 67 | 66 | 52 | 42 | 89 |
| 46 | 41 | 62 | 51 | 73 | 50 | 41 | 58 | 44 | 69 |

## Tally charts

Tally charts are used to count the numbers within each group. There are 60 numbers in the data set. The smallest is 40 and the largest is 89. It is most convenient to choose five groups of numbers: 40–49, 50–59, and so on.

Working through the figures in order, a *neat* tally mark is made alongside the appropriate group. Tally marks are grouped into fives for easier counting, and a 'slash' across |||| denotes '5'.

The frequencies can be added to check that the total is the same as the total number of items in the data set.

| Group | Tally | Frequency |
|---|---|---|
| 40–49 | ⵜ ⵜ ⵜ ||| | 18 |
| 50–59 | ⵜ ⵜ |||| | 14 |
| 60–69 | ⵜ ⵜ ||| | 13 |
| 70–79 | ⵜ |||| | 9 |
| 80–89 | ⵜ | | 6 |
| | | Total = 60 |

# Averages

The **mean** of the data is the $\dfrac{\text{total of all the values}}{\text{total number of values}}$.

The **mode** of the data is the value that occurs the most often.

The **median** of the data is the value in the middle when the data is arranged in ascending order. (For an even number of values, the median is the mean of the middle pair of values.)

Example set of data: 8, 9, 13, 19, 8, 15, 8     **Mean** $= \dfrac{8 + 9 + 13 + 16 + 8 + 15 + 8}{7} = \dfrac{77}{7} = 11$

Rearranging the data: 8, 8, 8, 9, 13, 15, 16:  **Mode** = 8 (3 times).  **Median** = 9 (middle value)

# Exercise 46

Use your calculator where appropriate.

**1**  A biased dice has six faces numbered 1, 2, 3, 4, 5 and 6. It has been thrown 40 times, and these are the scores.

| | | | | | | | | | |
|---|---|---|---|---|---|---|---|---|---|
| 1 | 6 | 3 | 2 | 3 | 2 | 1 | 2 | 4 | 1 |
| 4 | 2 | 4 | 1 | 6 | 1 | 3 | 5 | 2 | 6 |
| 3 | 2 | 3 | 1 | 2 | 5 | 4 | 2 | 5 | 1 |
| 2 | 1 | 6 | 2 | 4 | 1 | 3 | 6 | 3 | 5 |

Construct a tally chart for the data. Comment on the bias.

**2**  These are the number of appointments in the diary of a personnel manager for each working day in September.

| | | | | | | |
|---|---|---|---|---|---|---|
| 0 | 7 | 7 | 5 | 8 | 3 | 6 |
| 6 | 8 | 2 | 0 | 4 | 5 | 7 |
| 0 | 4 | 3 | 6 | 7 | 7 | 5 |

Construct a tally chart for the data.
Write down the median value and the modal value.

**3**  These are the weights, in kilograms, of 50 newborn babies in a hospital during August.

| | | | | | | | | | |
|---|---|---|---|---|---|---|---|---|---|
| 1.35 | 2.05 | 2.71 | 3.00 | 2.34 | 3.36 | 2.44 | 2.70 | 3.48 | 1.68 |
| 2.66 | 2.59 | 2.03 | 3.76 | 3.11 | 3.03 | 2.23 | 4.18 | 2.95 | 2.50 |
| 3.10 | 2.09 | 4.65 | 2.68 | 1.28 | 3.77 | 3.88 | 3.60 | 3.88 | 2.34 |
| 1.58 | 2.84 | 1.64 | 4.22 | 2.88 | 1.86 | 4.00 | 2.41 | 3.25 | 2.89 |
| 3.92 | 3.05 | 3.60 | 1.97 | 3.54 | 3.45 | 2.85 | 4.06 | 2.12 | 2.85 |

Construct a tally chart using the groups $1.0 \leqslant w < 1.5$, $1.5 \leqslant w < 2.0$, and so on.
What proportion of these babies weighs less than 3 kg?

**4** The weight printed on a packet of Skittles is 55 g.

The contents of 35 packets have been carefully weighed, and the weights have been recorded.

| | | | | |
|------|------|------|------|------|
| 54.7 | 58.0 | 57.0 | 56.7 | 58.1 |
| 56.6 | 56.9 | 57.5 | 55.8 | 56.4 |
| 56.1 | 57.2 | 55.3 | 57.8 | 57.3 |
| 57.8 | 56.5 | 55.9 | 57.5 | 57.7 |
| 55.5 | 56.2 | 55.3 | 58.1 | 57.6 |
| 55.6 | 57.0 | 54.9 | 56.4 | 55.0 |
| 57.1 | 56.8 | 56.7 | 56.9 | 55.3 |

Construct a tally chart using the groups $54.0 \leqslant w < 55.0$

$$55.0 \leqslant w < 56.0 \text{ and so on.}$$

Does the evidence suggest that 55 g is an average weight or a minimum weight?

For Questions 5–8, find the mean, the median and the mode for each data set.

**5** 7, 12, 14, 14, 3

**6** 1.4, 0.8, 2.4, 1.9, 0.8

**7** 21, 31, 11, 16, 18, 4, 4

**8** 1.2, 0.9, 1.1, 3.1, 0.7, 0.5.

**9** These are the times taken by a group of pupils to solve a puzzle.

48 sec    52 sec    88 sec    34 sec    37 sec    38 sec    45 sec

Calculate the mean and median times.

**10** The heights of the players in a netball team are listed as:

1.53 m    1.70 m    1.58 m    1.55 m    1.63 m    1.62 m    1.59 m

Calculate the mean and median heights of a player in this team.

**11** A keen golfer plays four rounds of golf. His scores are 88, 98, 91 and 91.

Which type of average (mean, median or mode) will he prefer to call his 'average'?

**12** Paula has received these marks out of 10 over the first three weeks of term: 3, 6, 8, 1, 5, 7, 8, 4, 2.

Which type of average will she prefer to call her 'average'?

**13** A group of 12 boys has a mean age of 10.75 years. One of them, aged 13.5, leaves the group.

What is the mean age of the remaining boys?

**14** Five pupils score these percentage marks in a German exam: 44, 56, 64, 35, 50.

**a** Calculate the mean mark.

**b** The teacher adds ten marks to each score. How does this affect the mean?

**15** Find six numbers, five of which are smaller than the mean of the group.

**16** Find eight numbers, seven of which are greater than the mean of the group.

# Displaying data

## Discrete data

Discrete data are measurements collected from a source that can be listed and counted. The number of absentees from a school is an example of discrete data.

The table and the **bar chart** show the numbers of driving tests taken by a group of students before they passed the test.

| Number of driving tests | Frequency |
|:---:|:---:|
| 1 | 23 |
| 2 | 35 |
| 3 | 20 |
| 4 | 12 |
| 5 | 6 |
| 6 | 3 |
| 7 | 0 |
| 8 | 1 |

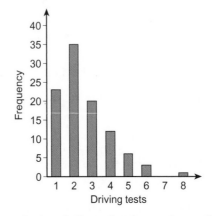

The labelling of the horizontal axis and the spaces between the bars indicate that the results are discrete.

## Continuous data

Continuous data are measurements collected from a source that *cannot* be listed and counted.

The table and bar chart show the lives of a sample of 75 Everglo batteries tested in continuous use in a model train until they were exhausted.

| Lifetime (hours) | Frequency |
|:---:|:---:|
| 6–7 | 8 |
| 7–8 | 12 |
| 8–9 | 14 |
| 9–10 | 22 |
| 10–11 | 16 |
| 11–12 | 3 |

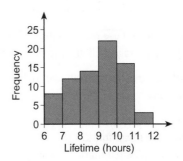

The labelling of the horizontal axis and the 'touching' bars indicate continuous data.

## Frequency polygon

A frequency polygon is made by joining the mid-points of the tops of the bars in a bar chart or histogram. It is useful for indicating trends, and for comparing two or more sets of results.

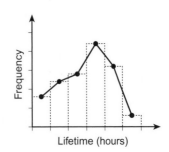

# Pie chart

These are the annual costs of running a typical family car.

| | Cost (£) | Cost as a proportion of the total | Angle of sector |
|---|---|---|---|
| **Petrol** | 1000 | $\frac{1000}{6000}$ | $\frac{1000}{6000} \times 360° = 60°$ |
| **Car tax** | 150 | $\frac{150}{6000}$ | $\frac{150}{6000} \times 360° = 9°$ |
| **Insurance** | 350 | $\frac{350}{6000}$ | $\frac{350}{6000} \times 360° = 21°$ |
| **Maintenance** | 900 | $\frac{900}{6000}$ | $\frac{900}{6000} \times 360° = 54°$ |
| **Depreciation** | 3600 | $\frac{3600}{6000}$ | $\frac{3600}{6000} \times 360° = 216°$ |
| Total = £6000 | | | Total = 360° |

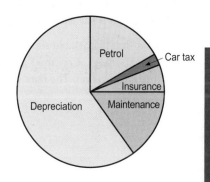

A pie chart is most suitable for displaying proportions.

# Pictogram

A pictogram is like a bar chart, except that the frequencies are shown by pictures instead of bars.

| Month | Car thefts (one car = 100 thefts) |
|---|---|
| Mar | 🚗 🚗 🚗 🚗 🚗 |
| Apr | 🚗 🚗 🚗 🚗 🚗 🚗 |
| May | 🚗 🚗 🚗 🚗 🚗 ◁ |
| Jun | 🚗 🚗 🚗 |
| Jul | 🚗 🚗 ◁ |
| Aug | 🚗 🚗 🚗 |

Operation Clampdown was introduced by a police authority on 1 June to combat car crime

# Exercise 47

Use your calculator where appropriate.

1　The following ingredients are required to make almond shortbread triangles:
8 ounces of plain flour,  5 ounces of ground almonds,  4 ounces of castor sugar,
4 ounces of fine semolina,  9 ounces of butter or margarine.
Display this information in a pie chart.

2　The table shows how 900 students travel to school.

| Mode of travel | Walk | Cycle | Car | Bus | Train |
|---|---|---|---|---|---|
| Number travelling | 510 | 80 | 50 | 140 | 120 |

Display the data as a pie chart.
What percentage of pupils travel to school by public transport?

**3** A dice is thrown 40 times, with these results.

| 1 | 6 | 3 | 2 | 3 | 2 | 1 | 2 | 4 | 1 |
|---|---|---|---|---|---|---|---|---|---|
| 4 | 2 | 4 | 1 | 6 | 1 | 3 | 5 | 2 | 6 |
| 3 | 2 | 3 | 1 | 2 | 5 | 4 | 2 | 5 | 1 |
| 2 | 1 | 6 | 2 | 4 | 1 | 3 | 6 | 3 | 5 |

Construct a bar chart of the scores.

**4** This frequency table shows the heights of a class of schoolboys. Construct a bar chart to represent this information. The bars should be 1 cm wide.

| Height (cm) | Frequency |
|---|---|
| $130 \leqslant h < 135$ | 4 |
| $135 \leqslant h < 140$ | 6 |
| $140 \leqslant h < 145$ | 9 |
| $145 \leqslant h < 150$ | 7 |
| $150 \leqslant h < 155$ | 3 |
| $155 \leqslant h < 160$ | 2 |

**5** This frequency table shows the heights of a class of schoolgirls of the same age as the boys in the question above.

**a** Construct a frequency polygon to represent this information. Choose a scale such that each group is 1 cm wide.

**b** Add the data for the boys from the question above to the diagram.

**c** Comment on the differences between the two classes, and guess their ages.

| Height (cm) | Frequency |
|---|---|
| $130 \leqslant h < 135$ | 3 |
| $135 \leqslant h < 140$ | 5 |
| $140 \leqslant h < 145$ | 8 |
| $145 \leqslant h < 150$ | 5 |
| $150 \leqslant h < 155$ | 4 |
| $155 \leqslant h < 160$ | 5 |

**6** A sports centre runs several evening clubs for junior members. The table shows the membership numbers.

**a** Display the details separately for the boys and the girls on one frequency polygon diagram.

**b** Investigate other ways of representing this information.

| | Girls | Boys |
|---|---|---|
| **Aerobics** | 12 | 6 |
| **Climbing** | 6 | 10 |
| **Gymnastics** | 16 | 4 |
| **Hockey** | 14 | 18 |
| **Judo** | 6 | 16 |
| **Tennis** | 12 | 8 |

**7** Calculate the mean of this data set: 5, 8, 3, 11, 9, 6, 8, 6.
The mean of another set of seven figures is 5. Calculate the mean of the combined set of 15 figures.

**8** The ages of the members of a fitness club are shown in the table.
Calculate the mean age of the group.

| 11 | 30 | 35 | 42 | 11 | 33 | 20 | 29 | 14 | 22 |
|---|---|---|---|---|---|---|---|---|---|
| 35 | 41 | 21 | 12 | 37 | 25 | 28 | 36 | 14 | 29 |
| 24 | 24 | 32 | 25 | 18 | 42 | 26 | 16 | 27 | 21 |

**9** These are the numbers of flights made each day over a 15-day period by the helicopter crew on an oil rig.

|    |    |    |    |    |
|----|----|----|----|----|
| 17 | 11 | 14 | 14 | 11 |
| 13 | 16 | 17 | 11 | 12 |
| 13 | 12 | 17 | 12 | 17 |

Find the mean, the median and the mode of the data. Comment on these averages.

**10** The bar chart shows the numbers of children in each of 30 families. *The information was gathered by asking a class of pupils to indicate their family details on a questionnaire.*

   **a** Calculate the mean and median number of children per family in this sample.

   **b** Comment on the bias in the sample and the effect that it will have on the statistics. Suggest how an unbiased sample could be gathered.

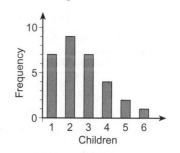

**11** The pie chart displays the distribution of mobile phones among the pupils at a school.

If 318 pupils have a mobile, how many pupils are there in the school?

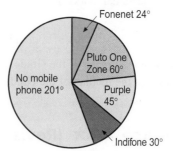

**12** In a class examination, the mean mark for the 18 boys was 66%, and the mean mark for the 12 girls was 71%. Calculate the mean mark for the whole class.

**13** In a cross-country race, these times were recorded by a team of eight runners.

| | | | | |
|---|---|---|---|---|
| 13 mins 53 secs | 14 mins 5 secs | 14 mins 28 secs | 14 mins 30 secs | 14 mins 56 secs |
| 15 mins 11 secs | 15 mins 12 secs | 15 mins 13 secs | | |

Calculate the mean time.

**14 a** Record the ages (in years and completed months) of every pupil in your class. Calculate the mean age and the median age.

   **b** Repeat **a**, but include your mother's or father's age in the data.

   **c** Is the mean or the median a better average for **b**?

**15** Explain briefly how the random number generator on your calculator can be used to simulate the experiment of throwing three coins and recording the number of heads.

   **a** Do this simulation and record the results for 40 throws in a tally chart.

   **b** Display the results in a bar chart.

   **c** Comment briefly on your results.

**16** Make sure that your calculator is in 'degree mode'. This operation INV tan RAN# (ignore the decimals) will then produce a *whole number* between 0 and 45.

   **a** Record 75 values in a tally chart.

   **b** Display the results on a bar chart with nine bars ($0 \leqslant n < 5$, $5 \leqslant n < 10$, and so on.)

   **c** Calculate the median and the mean.

   **d** Comment on whether you think that this process produces random numbers between 0 and 45.

# Exercise 48 (Revision)

Sixty pupils at a school took an arithmetic test. The marks, which are out of 50, are recorded in the tally chart below.

| Group | | Frequency |
|---|---|---|
| Grade E $0 \leq m < 10$ | 卌 | |
| Grade D $10 \leq m < 20$ | 卌 IIII | |
| Grade C $20 \leq m < 30$ | 卌 卌 I | |
| Grade B $30 \leq m < 40$ | 卌 卌 卌 卌 I | |
| Grade A $40 \leq m < 50$ | 卌 卌 IIII | |

1  Fill in the frequency column.

2  Display the data in a bar chart.

3  Draw a frequency polygon on your bar chart.

4  Display the data in a pie chart.

Here are the actual marks of the Grade A results.

| 42 | 45 | 40 | 44 | 46 | 41 | 45 |
|---|---|---|---|---|---|---|
| 49 | 46 | 42 | 41 | 43 | 43 | 42 |

5  Calculate the mean of the Grade A marks.

6  Calculate the median of the Grade A marks.

# Exercise 48★ (Revision)

The table shows the football results from the Premier League, from a fantasy league week in the 2003–2004 season. Goal scorers and the times of the goals are given below each team. (A football match lasts for 90 minutes.)

1  Find the total number of goals scored on this date.

2  Calculate the mean, median and mode for the number of goals per game.

3  Calculate the mean and median times for the first goal scored in a match.

4  Draw up a frequency table to show when the goals were scored. Use groups 0–15, 16–30, 31–45 etc.

5  Display these frequencies in a pie chart.

6  In which 15-minute period of these games are most goals scored?

| PREMIER LEAGUE | | | |
|---|---|---|---|
| **BOLTON** Freund (og) 4, Davies 85 | 2 | **LEICESTER** Ferdinand 16, Dickov 54 | 2 |
| **MAN CITY** Sun 13, Fowler (pen) 81 | 2 | **CHARLTON** Kishishev 66 | 1 |
| **MAN UTD** Scholes 5, Saha 24, Giggs 49 | 3 | **NEWCASTLE** | 0 |
| **FULHAM** Volz 17 | 1 | **ASTON VILLA** | 0 |
| **ARSENAL** Henry 51 | 1 | **PORTSMOUTH** | 0 |
| **TOTTENHAM** Keane 46 | 1 | **WOLVES** Ince 18 | 1 |
| **BLACKBURN** | 0 | **BIRMINGHAM** Savage (pen) 10 | 1 |
| **EVERTON** Jeffers 2 | 1 | **LIVERPOOL** Owen 31, 71 | 2 |
| **SOUTHAMPTON** Beattie 15 | 1 | **MIDDLESBROUGH** | 0 |
| **LEEDS** Smith 82, Viduka 90 | 2 | **CHELSEA** Veron 2, 17, Lampard 39, Crespo 48 | 4 |
| **BOLTON** | 0 | **MAN CITY** Anelka 34 | 1 |
| **FULHAM** Djetou 20, 87 | 2 | **BLACKBURN** Tugay 57, Emerton 67 | 2 |

# Summary 2

## Number

### Standard form

Numbers are written in standard form as $a \times 10^n$, where $1 \leqslant a < 10$ and $n$ is a positive or negative integer.

$$7500 = 7.5 \times 10^3$$
$$0.000\,075 = 7.5 \times 10^{-5}$$

### Ratio

Divide 105 hours in the ratio $3 : 4$. The sum of the ratios is $3 + 4 = 7$. One part $= \frac{105}{7} = 15$ hours, three parts $= 3 \times 15 = 45$ hours, and four parts $= 4 \times 15 = 60$ hours.

### Direct proportion

BT spent £159 308 000 on advertising in 2000. How much did BT spend on advertising per second in that year?

$$\text{Amount spent per second} = \frac{\text{total cost in year 2000}}{\text{seconds in that year}} = \frac{£159\,308\,000}{365 \times 24 \times 60 \times 60} = £5.05/\text{s.}$$

## Algebra

### Simplifying fractions

$$\frac{5a}{\cancel{12}^{6}} \times \frac{\cancel{2}^{1}b}{3} = \frac{5ab}{18}$$

$$\frac{5a}{12} \div \frac{2b}{3} = \frac{5a}{\cancel{12}_{4}} \times \frac{\cancel{3}^{1}}{2b} = \frac{5a}{8b}$$

$$\frac{5a}{12} + \frac{2b}{3} = \frac{5a + 8b}{12}$$

$$\frac{5a}{12} - \frac{2b}{3} = \frac{5a - 8b}{12}$$

### Solving equations

The way to solve equations is to isolate the unknown letter by systematically doing the same operation to both sides.

Solve this for $x$.

$3x^2 - 4 = 71$   (Add 4 to both sides)

$3x^2 = 75$   (Divide both sides by 3)

$x^2 = 25$   (Square root both sides)

$x = \pm 5$   (Note two answers)

Solve this for $y$.

$\dfrac{\sqrt{(y + 3)}}{4} - 2 = 1$   (Add 2 to both sides)

$\dfrac{\sqrt{(y + 3)}}{4} = 3$   (Multiply both sides by 4)

$\sqrt{(y + 3)} = 12$   (Square both sides)

$y + 3 = 144$   (Subtract 3 from both sides)

$y = 141$

## Using formulae

Formulae describe how items are related to each other. Simple substitution and the use of the BIDMAS (brackets, indices, divide, multiply, add, subtract) mnemonic will help you calculate their values.

The volume $V$ cubic centimetres of an object is given by $V = \pi(R^2 - r^2)h$.

Find the volume correct to 3 significant figures when $R = 10\,\text{cm}$, $r = 8\,\text{cm}$ and $h = 6\,\text{cm}$.

| | |
|---|---|
| **Facts** | $R = 10\,\text{cm}$, $r = 8\,\text{cm}$, $h = 6\,\text{cm}$, $V = ?\,\text{cm}^3$ |
| **Equation** | $V = \pi(R^2 - r^2)h$ |
| **Substitution** | $V = \pi \times (10^2 - 8^2) \times 6\,\text{cm}^3$ |
| **Working** | $V = 679\,\text{cm}^3$ (to 3 s.f.) |

## Indices

$$a^m \times a^n = a^{m+n} \qquad \text{(Add indices)}$$
$$a^m \div a^n = a^{m-n} \qquad \text{(Subtract indices)}$$
$$(a^m)^n = a^{mn} \qquad \text{(Multiply indices)}$$

## Inequalities

These are solved in a similar way to equations, except when both sides are multiplied or divided by a negative number. In that case, **the inequality is reversed**.

Solve the inequality showing the result on a number line.

$$2(x - 3) \leq 5(x - 3) \qquad \text{(Expand brackets)}$$
$$2x - 6 \leq 5x - 15 \qquad \text{(Add 15 to both sides)}$$
$$2x + 9 \leq 5x \qquad \text{(Subtract } 2x \text{ from both sides)}$$
$$9 \leq 3x \qquad \text{(Divide both sides by 3)}$$
$$3 \leq x$$

Therefore $\qquad x \geq 3$

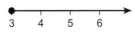

A solid circle means $\leq$ or $\geq$. An open circle means $<$ or $>$.

## Simultaneous equations

To solve simultaneous equations graphically:

- ♦ Draw the graphs for both equations on one set of axes.
- ♦ The solution is at the intersection point of the graphs.
- ♦ If the graphs do not intersect, there is no solution.
- ♦ If the graphs are the same, there are an infinite number of solutions.

Solving the simultaneous equations
$x + y = 6$, $2x - y = 0$

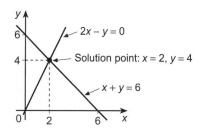

Solution is $x = 2$, $y = 4$.

## Inequalities

Inequalities can be shown graphically by shading unwanted regions to identify solutions in unshaded regions.

Solve the inequalities $x \geqslant 0$, $y \geqslant 0$, $x + y \leqslant 3$ and $y < 2$ by drawing suitable lines (solid or broken) and shading unwanted regions.

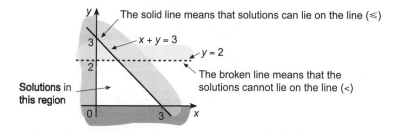

## Shape and space

### Trigonometry

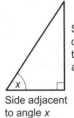

Side opposite to angle $x$

Side adjacent to angle $x$

$$\tan 56° = \frac{12\,\text{cm}}{a}$$

$$a = \frac{12\,\text{cm}}{\tan 56°}$$

$$= 8.09\,\text{cm (to 3 s.f.)}$$

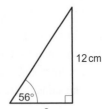

12 cm

56°

$a$

$$T \frac{\text{opp}}{\text{adj}}$$

## Handling data

**Discrete data** is data that can be listed and counted, for example the number of peas in a pod, or the number of pages in a book.

**Continuous data** is data that cannot be listed and counted, for example time, mass and length.

### Averages

Example set of data: 23, 5, 7, 8, 10, 7

$$\text{mean} = \frac{\text{total of all values}}{\text{total number of values}} = \frac{23 + 5 + 7 + 8 + 10 + 7}{6} = 10$$

**mode** = value that occurs most often = 7

**median** = value in the middle when the data is arranged in ascending order = 7.5
(The median of 5, 7, 7, 8, 10, 23 is 7.5, because it is in the middle of 7 and 8.)

### Displaying data

Coloured marbles in a bag are distributed as shown in the table.

| Colour | Frequency |
|--------|-----------|
| Red    | 1         |
| Yellow | 3         |
| Blue   | 2         |
| Green  | 4         |

**Pictogram**

| Marble colour | |
|---------------|--|
| Red    | 🔵 |
| Yellow | 🔵 🔵 🔵 |
| Blue   | 🔵 🔵 |
| Green  | 🔵 🔵 🔵 🔵 |

**Pie chart**

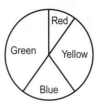

Red

Green

Yellow

Blue

**Bar chart**

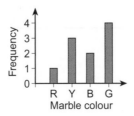

**Frequency polygon**

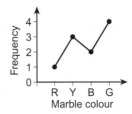

# Examination practice 2

1. A firm produces tomato soup in two sizes of can. Calculate the probable cost of the smaller can.

2. In 2000 £1 was worth approximately ¥175 (yen), FFr10 (French francs) and Dr550 (Greek drachmas). A mini-disc player cost ¥35 000 in Japan, FFr2250 in France, and Dr137 500 in Greece. Where would you advise a British tourist to have bought the player in 2000? Illustrate your answer with suitable calculations.

3. Simplify these.
   a $k^2 \times k^3$          b $p^6 \div p^2$
   c $6m^3 \div 2m^2$          d $(y^2)^3$

4. A regular polygon has twelve sides. Find the size of each interior and exterior angle.

5. Find angles $x$ and $y$, giving reasons for each step.

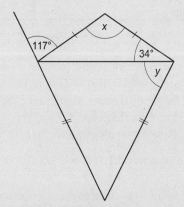

6. Construct an equilateral triangle of side 6 cm and draw the bisectors of each angle.

   Label the intersection point of the bisectors as P. Show construction marks clearly.

7.

   The Earth is approximately 150 million kilometres from the Sun, and Saturn is approximately 1430 million kilometres from the Sun.
   a When the three planets lie in a straight line, as shown, how far is Saturn from the Earth? (Answer in kilometres in standard form.)
   b The spacecraft Voyager 1 sent signals back to Earth from near Saturn that enabled us to see new and clear pictures. Radio waves travel at about 300 million metres per second. Approximately how long did it take the signals from Voyager 1 to reach Earth? (Answer in minutes.)

8. The time $t$ in minutes of a total eclipse of the Sun by the Moon is given by the approximate formula $t \simeq \dfrac{Dk - Kd}{60K}$ where $d$ and $D$ are the diameters in kilometres of the Moon and Sun, respectively, and $k$ and $K$ are the distances in kilometres from the Earth to the Moon and the Sun, respectively.

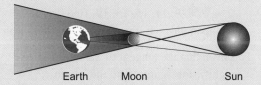

   Show that $t \simeq 2$ minutes 40 s if $d = 3.48 \times 10^3$, $D = 1.41 \times 10^6$, $k = 3.82 \times 10^5$ and $K = 1.48 \times 10^8$.

**9** Halley's Comet visits Earth every 76 years. At its furthest point, it is 35 astronomical units from Earth. An astronomical unit is $1.496 \times 10^{11}$ m.

**a** What is this maximum distance from Earth? Give your answer in metres, using standard index form.
**b** Light travels at $2.998 \times 10^8$ m/s. How far does it travel in one year? Give your answer in metres, using standard form.
**c** The distance in **b** is one *light year*. Give your answer to **a** in light years, using standard form.

*OCR*

**10 a** Draw a set of $x$ and $y$ axes from 0 to 10 units. On this plane, draw these graphs.
(i) $x = 4$ (ii) $y = 6$ (iii) $3x + 2y = 18$
**b** Use these lines to shade the regions not satisfied by the inequalities $x \leqslant 4$, $y \leqslant 6$, $3x + 2y \leqslant 18$, $x \geqslant 0$ and $y \geqslant 0$.

**11** Given the numbers:
2, 4, 7, 8, 9, 11, 15, 15, 19, find
**a** the mode **b** the median **c** the mean

**12** Given that the tangent of $32° = 0.625$ to 3 significant figures, find the length $x$.

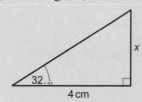

32.. 
4 cm

**13** The pie chart shows the numbers of overseas visitors to a certain mountain last year.

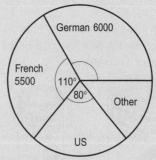

German 6000
French 5500
110°
80°
Other
US

**a** How many US visitors were there?
**b** What should the angle of the sector representing visitors from Germany be?

**14** In a triangle ABC, angle ABC and angle ACB are both 65° and BC = 8.4 cm. Find
**a** the perpendicular height of A above BC
**b** the area of the triangle

**15** A rectangle has sides of 6 cm and 8 cm. Find the smaller angle between the two diagonals.

**16** In the diagram, ABCD represents a high tower with a flagpole DE.

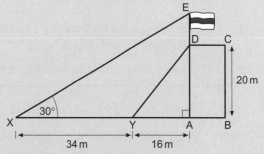

E
D C
20 m
30°
X Y A B
34 m 16 m

Calculate to 3 significant figures
**a** the angle of elevation of D from Y
**b** the height of E above A and hence the height of the flagpole
**c** the angle of depression from D to X

**17** For the numbers 1.2, 0.9, 1.1, 3.1, 0.9, 0.5, find
**a** the mean **b** the median **c** the mode

# Number 3

## Activity 11

♦ Show that if £120 **increases** in value by 8%, its new value will be £129.60. Calculate $120 \times 1.08$. Comment.

♦ Show that if £120 **decreases** in value by 8%, its new value will be £110.40. Calculate $120 \times 0.92$. Comment.

♦ Copy and complete these two tables.

| To increase by (%) | 100% → | Multiply by |
|---|---|---|
| 15 | | |
| 70 | | |
| | | 1.56 |
| | | 1.02 |
| | 108% | |
| | 180% | |

| To decrease by (%) | 100% → | Multiply by |
|---|---|---|
| 15 | | |
| 70 | | |
| | | 0.8 |
| | | 0.98 |
| | 92% | |
| | 20% | |

## Key Points

♦ To **increase** a number by $R\%$, multiply it by $\dfrac{100 + R}{100} = \left(1 + \dfrac{R}{100}\right)$.

♦ To **decrease** a number by $R\%$, multiply it by $\dfrac{100 - R}{100} = \left(1 - \dfrac{R}{100}\right)$.

## Compound percentages

Compound percentages are used when one percentage is followed by another in a calculation.

### Example 1

$120 is invested at 15% interest for 3 years. Find the value of the investment after 1 year, 2 years, and 3 years.

To increase by 15%, multiply by $\dfrac{100 + 15}{100} = (1 + \dfrac{15}{100}) = 1.15$.

*After 1 year*, the investment is $\$120 \times 1.15 = \$138$.

*After 2 years*, the investment is $\$138 \times 1.15 = \$158.70$.

(Notice that, after 2 years, you could write $(\$120 \times 1.15) \times 1.15 = \$120 \times 1.15^2 = \$158.70$.)

*After 3 years*, the investment is $= \$120 \times 1.15 \times 1.15 \times 1.15$
$$= \$120 \times (1.15)^3$$
$$= \$182.51 \text{ (to 2 d.p.)}.$$

In the second and third years, the interest earned by the investment has *not* been withdrawn. It has been added to the investment. This is called **compound interest**.

## Key Points

♦ To **increase** a number by $R\%$, for $n$ successive times, multiply by $\left(1 + \dfrac{R}{100}\right)^n$.

♦ To **decrease** a number by $R\%$, for $n$ successive times, multiply by $\left(1 - \dfrac{R}{100}\right)^n$.

## Activity 12

Inflation causes prices to increase, which in turn reduces the standard of living. Percentages are often used to measure inflation.

The cost of an article is $P$. Inflation makes the cost increase by $R\%$ in each of $n$ years, to $A$:

$$A = P\left(1 + \frac{R}{100}\right)^n$$

♦ In 2001, a loaf of bread costs $0.85. Copy and complete this table to show how its cost would be affected by different rates of inflation as shown. Enter the costs in dollars, correct to 3 significant figures.

| Time (years) | 10% | 20% | 30% | 40% |
|:---:|:---:|:---:|:---:|:---:|
| 3 | | | | |
| 6 | | | | |
| 9 | | | | |
| 12 | | | | |
| 15 | | | | |

Illustrate your data on a sheet of graph paper by drawing four curves.

Use your graph to estimate how long it might take for the cost of a loaf of bread to increase fivefold for each rate of inflation.

♦ In 2001, a car costs $85 000. Inflation is running at an average of 20% per annum (that is, per year).

Use your graph to estimate when the cost of a car would exceed $1 million. Use your calculator to find the answer, correct to 3 significant figures.

In some countries, inflation is so high that it is called 'hyperinflation'. This occurred in Peru in the 1980s, and by the late 1980s prices were increasing at 6000% a year. At one stage in Peru, a 50 000 inti note was worth 0.25p. Such rates of inflation devastate economies and affect everyone, including backpackers, as can be seen from this extract of a letter from Brazil.

> One thing that has greatly affected Penny and myself is inflation, because it is running at 25% a month. Prices seem to go up every 5 minutes. For example, the postage for one airmail letter has gone up from 42p to 91p in 3 months! Price increases are announced, on the radio and television, the night before they're due to come into effect. I was amazed to see a queue of 25 cars (I counted!) at a garage all waiting to fill up before the midnight deadline! Shop assistants descend on you the moment your big toe gets through the door because of the recession. To make things worse, nothing has a price tag on because they change so frequently, so you have to ask one of the assistants who then tries to sell you every imaginable accessory to make up a matching outfit.

Investigate the effect on prices, first using the inflation rate in Peru, and then using a current inflation rate from any country of your choice.

# Exercise 49

1 What would you multiply by to increase an amount by
   **a** 10%   **b** 20%   **c** 30%   **d** 1%   **e** 15%   **f** 25%

2 What would you multiply by to decrease an amount by
   **a** 10%   **b** 5%   **c** 15%   **d** 3%   **e** 13%   **f** 25%

For Questions 3–14, where appropriate, give your answers correct to 3 significant figures.

| Increase $400 by | **3** 10% | **4** 20% | **5** 30% |
| Increase 200 km by | **6** 1% | **7** 15% | **8** 25% |
| Decrease £300 by | **9** 10% | **10** 5% | **11** 15% |
| Decrease £500 by | **12** 3% | **13** 13% | **14** 25% |

15 A house increases in value by 3% in the first year and by 4% in the second year.
   Copy and complete 'to work out the answer, multiply the original value by $1.03 \times \ldots$'.

16 A caravan decreases in value by 8% in the first year and by 7% in the second year.
   Copy and complete 'to work out the answer, multiply the original value by $\ldots \times \ldots$'.

17 There are 500 pupils in a school. In the next year, the numbers increase by 10%, and, in the following year, they increase by 20%. Find the number of pupils at the end of two years.

**18** A rotting apple has a mass of 300 g. The next day its mass has decreased by 6%, and, on the following day, it has decreased by 7%. Find its mass after two days.

**19** A sum of money is invested at 4% compound interest for five years.
Copy and complete 'to calculate the answer, multiply by …'.

**20** A sum of money is invested at 9% compound interest for six years.
Copy and complete 'to calculate the answer, multiply by …'.

**21** €450 is invested at 7% compound interest. Find the value after six years.

**22** $630 is invested at 6% compound interest. Find the value after four years.

**23** In Snowland, inflation pushes up prices by 35% per year. In 2001, a car costs $40 000. Find the price of a similar car in 2004. How much might it cost in 2011?

**24** In Sandland, inflation pushes up prices by 75% per year. In 2001, a car costs $40 000. Find the price of a similar car in 2004. How much might it cost in 2011?

## Activity 13

◆ Using the formula $A = P\left(1 + \dfrac{R}{100}\right)^n$, find out what happens when $R = 0\%$, when $R = 100\%$, and when $R = 200\%$.

◆ In each family there are twice as many grandparents as parents, and twice as many greatgrandparents as grandparents.

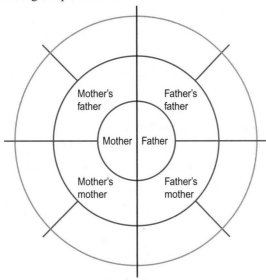

Use the formula to work out the number of great, great, … grandparents you had 30 generations ago, around 1000AD.

# Exercise 49★

Where appropriate, give your answers correct to 3 significant figures.

**1** What would you multiply by to increase an amount by
  **a** 12.5%   **b** 4%   **c** 5%

**2** What would you multiply by to decrease an amount by
  **a** 19.6%   **b** 10%   **c** 20%

**3** Increase 80 km by 12.5%.

**4** Decrease 1000 kg by 19.6%.

**5** Increase $400 by 4%, and then by another 5%.

**6** Decrease £400 by 10%, and then by a further 20%.

**7** What is the result of increasing a quantity by 20%, followed by decreasing it by 20%?
  Explain your answer.

**8** What is the result of decreasing a quantity by 20%, followed by increasing it by 20%?
  Explain your answer.

**9** €550 is invested at 8% compound interest. Find the value after five years.

**10** $730 is invested at 9% compound interest. Find the value after six years.

**11** An antique vase appreciates (gains) in value at the rate of 15% per year.
  If the vase is bought for €5000, how much will it be worth in 5 years, 10 years and 15 years?

**12** A car depreciates (loses) in value at the rate of 15% per year.
  If the car is bought for £20 000, how much will it be worth in 5 years, 10 years and 15 years?

**13** If Julius Caesar had invested the equivalent of 1 cent at 1% per annum compound interest, what would it have been worth in $ after 2000 years?

**14** Mr Scrooge inherits £20 000 and decides to put it under his mattress 'for a rainy day'.
  As a result, this money earns no interest, *and* it is reduced in value by inflation.
  If inflation is running at 14% each year, calculate the value of this money after 5 years, 10 years, 15 years and 20 years.

**15** If the average inflation rate is 20% per annum, estimate the cost of a SF2.50 box of biscuits in 50 years' time.

**16** If the average inflation rate is 30% per annum, estimate the cost of a 1 centime stamp in 50 years' time.

For Questions 17–22, copy and complete this table, using the equation $A = P\left(1 + \dfrac{R}{100}\right)^n$ to calculate the missing entries. Show all your working clearly.

| | $A(\$)$ | $P(\$)$ | $R(\%)$ | $n$ (years) |
|---|---|---|---|---|
| **17** | 1000 | | 5 | 10 |
| **18** | 1000 | | 10 | 5 |
| **19** | 10 000 | 500 | | 25 |
| **20** | 10 000 | 1000 | | 25 |
| **21** | 1000 | 500 | 5 | |
| **22** | 1000 | 500 | 10 | |

**23** A large sheet of paper is 0.1 mm thick. It is cut in half, and one piece is placed on top of the other. These two pieces are then cut in half a second time to make the pile four sheets thick.

Copy and complete this table.

| No. of times done | No. of sheets in pile | Height of pile (mm) |
|---|---|---|
| 2 | 4 | 0.4 |
| 3 | | |
| 5 | | |
| 10 | | |
| 50 | | |

How many times would you have to do this for the pile to reach the Moon, approximately $3.84 \times 10^5$ km away?

# Multiples, factors and primes

The **multiples** of 4 are 4, 8, 12, 16 …

The only numbers that divide exactly into 10 are called **factors of 10**, and these are 1, 2, 5 and 10.

If a number has no factors apart from 1 and itself, it is called a **prime number**.

---

**Remember**

**Prime numbers** are divisible only by 1 and themselves.

They are 2, 3, 5, 7, 11, 13, 17, 19, 23, 29, 31, 37, 41, 43, 47, 53, …, …, 67, 71, 73, …

Notice that 1 is *not* a prime number.

---

**Example 2**

Express 84 as the product of prime factors.

(Divide repeatedly by prime numbers)

$$84 = 2 \times 42$$
$$= 2 \times 2 \times 21$$
$$= 2 \times 2 \times 3 \times 7$$

| 2 | 84 |
|---|---|
| 2 | 42 |
| 3 | 21 |
| 7 | 7 |
| | 1 |

Therefore $84 = 2^2 \times 3 \times 7$

---

# Exercise 50

For Questions 1–4, list the first five multiples.

**1** 7        **2** 12        **3** 6        **4** 22

For Questions 5–8, list the factors.

**5** 12        **6** 18        **7** 30        **8** 32

For Questions 9–12, express the number as a product of prime factors.

**9** 28        **10** 70        **11** 60        **12** 96

**13** Is 7 a multiple of 161?

**14** Is 12 a factor of 516?

For Questions 15–18, express the number as a product of prime factors.

**15** 210        **16** 390        **17** 88        **18** 770

**19** Why is 511 not a prime?

**20** Why is 177 not a prime?

# Exercise 50★

**1** Is 13 a prime factor of 181?

**2** Is 13 a prime factor of 104?

For Questions 3–6, list the prime factors.

**3** 399        **4** 715        **5** 231        **6** 429

**7** What is the highest prime factor of 385?

**8** What is the highest prime factor of 273?

For Questions 9–12, write down, in numerical order, all the factors.

**9** 75        **10** 40        **11** 54        **12** 48

For Questions 13–16, express the number as a product of prime factors.

**13** 165        **14** 1155        **15** 399        **16** 715

**17** What are the two prime numbers between 53 and 67 in the sequence 43, 47, 53, …, …, 67, 71, 73?

**18** What number has only the prime factors $2^3$, 3, 7?

For Questions 19–22, express the number as a product of prime factors, using indices where necessary.

**19** 504        **20** 800        **21** 1008        **22** 1024

# Highest common factor (HCF) and lowest common multiple (LCM)

HCF: The highest factor that is common to a group of numbers.

LCM: The lowest multiple that is common to a group of numbers.

---

### Example 3

Find the HCF of 12 and 42.

Express 12 and 42 as the product of prime factors: $12 = 2 \times 2 \times 3$; $42 = 2 \times 3 \times 7$.

The *common* prime factors are 2, 3.

The common factors are 2, 3 and $2 \times 3 = 6$.

The *highest* common factor (HCF) is 6.

### Example 4

Find the LCM of 12 and 42.

Express 12 and 42 as the product of prime factors:   $12 = 2 \times 2 \times 3$; $42 = 2 \times 3 \times 7$.

Multiples of 12 must contain the factors 2, 2 and 3.

Multiples of 42 must contain the factors 2, 3 and 7.

Common multiples must contain the factors 2, 2, 3 and 7.

Common multiples are 84, 168, 252, …

The *lowest* common multiple (LCM) is 84.

N.B.  $12 \times 42 = 504$ is a common multiple, but not the lowest one.

---

The factors in Examples 3 and 4 can be illustrated by a Venn diagram.

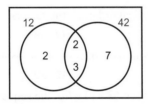

Can you see how this diagram can help you to calculate the highest common factor and lowest common multiple of the numbers 42 and 12?

---

## Key Points

♦ **HCFs** are used in cancelling fractions.

For example, divide the top and bottom of $\frac{12}{42}$ by 6 to get $\frac{2}{7}$.

They are also used in algebra when factorising, for example, $12x + 42y = 6(2x + 7y)$.

♦ **LCMs** are used in adding and subtracting fractions when the lowest common denominator has to be found.

For example, $\frac{1}{12} + \frac{5}{42} = \frac{7}{84} + \frac{10}{84} = \frac{17}{84}$.

# Exercise 51

You may find it helpful to draw a Venn diagram for these questions.

For Questions 1–6, find the highest common factor.

**1** 6 and 8      **2** 9 and 12      **3** 20 and 35      **4** 6 and 30

**5** 22 and 44      **6** 18 and 24

For Questions 7–12, find the lowest common multiple.

**7** 2 and 3      **8** 3 and 4      **9** 5 and 6      **10** 6 and 12

**11** 6 and 15      **12** 12 and 20

For Questions 13–15, find the highest common factor.

**13** $2x$ and $4xy$      **14** $6a^2$ and $9a$      **15** $12y^2$ and $8xy^2$

For Questions 16–18, find the lowest common multiple.

**16** $x$ and $y$      **17** $2a$ and $3b$      **18** $4y$ and $3x$

Simplify

**19** $\frac{6}{8}$      **20** $\frac{9}{12}$      **21** $\frac{20}{35}$      **22** $\frac{6}{30}$

**23** $\frac{22}{44}$      **24** $\frac{18}{24}$      **25** $\frac{6}{32}$      **26** $\frac{18}{33}$

**27** $\frac{15}{75}$      **28** $\frac{20}{28}$

Calculate

**29** $\frac{1}{6} + \frac{1}{8}$      **30** $\frac{1}{9} + \frac{1}{12}$      **31** $\frac{3}{20} - \frac{1}{35}$      **32** $\frac{5}{6} - \frac{1}{30}$

**33** $\frac{1}{5} + \frac{1}{6}$      **34** $\frac{1}{12} + \frac{5}{18}$      **35** $\frac{5}{6} - \frac{1}{4}$      **36** $\frac{7}{12} - \frac{11}{30}$

# Exercise 51★

For Questions 1–14, find the HCF and LCM.

**1** 12 and 18      **2** 45 and 75      **3** 30 and 105      **4** 70 and 42

**5** $3xy$ and $2yz$      **6** $5ab$ and $7bc$      **7** $4xy$ and $6xy$      **8** $10ab$ and $15ab$

**9** $x^2y$ and $xyz$      **10** $ab^2$ and $a^2b$      **11** $x^3y$ and $xy^4$

**12** $a^2b^3c$ and $ab^2c^2$      **13** $6x^2yz$ and $9xy^2z^2$      **14** $12ab^3c^2$ and $8a^2b^2c^3$

Calculate

**15** $\frac{1}{6} + \frac{3}{8}$      **16** $\frac{7}{12} + \frac{5}{18}$      **17** $\frac{7}{9} - \frac{5}{12}$      **18** $\frac{11}{45} - \frac{8}{75}$

**19** $\frac{11}{20} + \frac{4}{35}$      **20** $\frac{11}{30} + \frac{16}{105}$      **21** $\frac{11}{18} - \frac{5}{24}$      **22** $\frac{27}{70} - \frac{5}{42}$

# Calculator work

> ## Remember
>
> ◆ You should always make an **estimate** to check your calculator answer.
>   To make the estimate use the numbers corrected to 1 significant figure.
>
> ◆ **BIDMAS** will remind you of the correct order of operations.
>   First calculate the **B**rackets, and then the **I**ndices, followed by the **D**ivision and the **M**ultiplication, and, lastly, the **A**ddition and the **S**ubtraction.
>
> ◆ For long calculations you may have to add brackets. It helps to write out the sum before starting to use the calculator.
>
> ◆ Be familiar with your calculator. Read and keep the instruction booklet.

## Some calculator functions

◆ To calculate $4^5$, press   [4] [∧] [5]     (the answer is 1024)

◆ To enter $3.4 \times 10^{-2}$ (written in standard form), press   [3] [.] [4] [EXP] [(-)] [2]

◆ To enter $\frac{6}{7}$, press   [6] [aᵇ⁄꜀] [7]

◆ To enter $2\frac{4}{5}$, press   [2] [aᵇ⁄꜀] [4] [aᵇ⁄꜀] [5]

> ### Example 5
>
> Calculate these correct to 3 significant figures.
>
> **a** $\sqrt{23.8^2 - 18.4^2}$     (Estimate: $\sqrt{500 - 400} \simeq 10$)
>
> [√] [(] 23.8 [x²] [−] 18.4 [x²] [)] [=]    The answer is 15.1 (to 3 s.f.)
>
> **b** $2\frac{5}{7} - 3.8 \times 10^{-2}$     (Estimate: $3 - 0.04 \simeq 2.06$)
>
> 2 [aᵇ⁄꜀] 5 [aᵇ⁄꜀] 7 [−] 3.8 [EXP] [(-)] 2 [=]    The answer is 2.68 (to 3 s.f.)
>
> **c** $\sqrt{\dfrac{7.98^3}{5.91 + 1.09}}$     (Estimate: $\sqrt{500 \div 7} \simeq 8$)
>
> This is a long calculation, so write it out first:    $\sqrt{(7.98^3 \div (5.91 + 1.09))} =$
>
> [√] [(] 7.98 [∧] 3 [÷] [(] 5.91 [+] 1.09 [)] [)] [=]
>
> The answer is 8.52 (to 3 s.f.)

UNIT 3 ◆ Number

## Activity 14

♦ Key any three-digit number into your calculator.

♦ Multiply your number by 7, then multiply the answer by 11, then multiply this answer by 13.

♦ Repeat this with other three-digit numbers.

What do you notice? Can you explain this?

## Activity 15

Suppose that petrol costs $5 per litre. The petrol tank of a truck holds 142.15469 litres.
How much does a tank of petrol cost?
(Now turn your calculator upside-down.) Where should the driver buy his petrol?!
Where should he buy his petrol if he buys 236.6851 litres at $3 per litre?
Try and make up some puzzles yourself.

## Exercise 52

For each question, write an estimated answer and then give your final answer correct to 3 significant figures.

**1** $4.7 + 9.7 \div 4.3$

**2** $19 - 3.8 \div 2.3$

**3** $(4.7 + 9.7) \div 4.3$

**4** $(19 - 3.8) \div 2.3$

**5** $4.7 + \dfrac{9.7}{4.3}$

**6** $19 - \dfrac{3.8}{2.3}$

**7** $\dfrac{4.7 + 9.7}{4.3}$

**8** $\dfrac{19 - 3.8}{2.3}$

**9** $(7.6 + 3.8)^2 + 4.2$

**10** $\sqrt{4.8 + 9.6} + 5.8$

**11** $\dfrac{1}{9} + \dfrac{3}{11}$

**12** $\dfrac{4}{7} - \dfrac{1}{13}$

**13** $3\dfrac{2}{9} + 9.7$

**14** $2\dfrac{9}{11} - 0.831$

**15** $11.8 + 2.08^2$

**16** $0.971^2 - \dfrac{4}{7}$

**17** $\dfrac{9.73 - 1.08}{1.83^2}$

**18** $\dfrac{10.9 + 3.58}{\sqrt{4.5}}$

**19** $\sqrt{12.3^2 + 7.8^2}$

**20** $\sqrt{19.8^2 - 13.7^2}$

**21** $\sqrt{\dfrac{38}{12} + 4.08}$

**22** $\sqrt{\dfrac{38 + 4.08}{12}}$

**23** $\dfrac{3.8^2 + 4.9}{2.7}$

**24** $\dfrac{3.8^2}{2.7} + 4.9$

**25** $43.2 \div (4.7 - 0.87)$

**26** $19 \div (4.56 + 9.97)$

**27** $\dfrac{4.98}{2.09 \times 1.96}$

**28** $\dfrac{16.9}{3.94 \times 1.72}$

**29** $5^4$

**30** $4^5$

**31** $(2.4)^6$

**32** $(1.2)^7$

**33** $(2.4)^5 \times 2.4$

**34** $(1.2)^8 \div 1.2$

**35** $4680 + (2.4 \times 10^5)$

**36** $(5.6 \times 10^4) - 4560$

**37** $\dfrac{2.8 \times 10^8}{1.6 \times 10^{-2}}$

**38** $\dfrac{1.6 \times 10^{-2}}{2.8 \times 10^8}$

# Exercise 52★

For each question, make an estimate of the answer, and then calculate the answer correct to 3 significant figures.

**1** $\dfrac{45}{2.3 \times 5.7}$ 

**2** $\dfrac{98}{3.9 \times 7.2}$ 

**3** $\dfrac{0.059}{12.08 - 1.9}$

**4** $\dfrac{0.235}{23.9 + 98.4}$ 

**5** $\dfrac{12.8 - 55.9}{38.7 + 3.98}$ 

**6** $\dfrac{0.78 - 2.68}{0.065 + 0.123}$

**7** $13\frac{1}{7} - \frac{7}{9}$ 

**8** $13\frac{1}{7} \times \frac{7}{9}$ 

**9** $\dfrac{4.89}{1.8 \times 4.3} - \dfrac{3}{1.89}$

**10** $\dfrac{29.4}{34.9 + 67.1} + \dfrac{0.089}{12}$ 

**11** $\dfrac{1}{(0.678)^7}$ 

**12** $(1.2)^5$

**13** $(0.468)^3 + (0.0987)^5$ 

**14** $\dfrac{(0.468)^3}{(0.0987)^5}$ 

**15** $(3.4)^5 \div 10^4$

**16** $\dfrac{5.9 \times 10^{-7}}{9.5 \times 10^{-5}}$ 

**17** $\left(\dfrac{1}{0.95} + \dfrac{1}{9.98}\right)^9$ 

**18** $\left(\dfrac{1}{0.9^3} + \dfrac{1}{0.8^3}\right)^7$

**19** Does $(0.5)^2 + (0.6)^2 + (0.7)^2 + (0.8)^2$ equal $(0.5 + 0.6 + 0.7 + 0.8)^2$? Explain your answer.

**20** Find an easy way to calculate $\dfrac{1}{6^3} + \dfrac{1}{6^3} + \dfrac{1}{6^3} + \dfrac{1}{6^3} + \dfrac{1}{6^3} + \dfrac{1}{6^3}$.

# Exercise 53 (Revision)

**1** Write down the multiplying factor to increase a number by 15%.

**2** Write down the multiplying factor to decrease a number by 5%.

**3** Increase $40 by 10%, and then decrease your answer by 20%.

**4** A sum of money is invested at 7% compound interest for 5 years.
Copy and complete, 'to calculate the answer we multiply by …'.

**5** Write down the first 15 prime numbers, starting at 2.

For Questions 6–8, calculate the prime factors of the number.

**6** 55 

**7** 70 

**8** 30

**9** Why is 168 a multiple of 7? 

**10** Show that 7 is a factor of 175.

For Questions 11–13, write down the highest common factor.

**11** 14 and 30 

**12** 15 and 21 

**13** 30 and 50

For Questions 14–16, write down the lowest common multiple.

**14** 3 and 5 

**15** 4 and 6 

**16** 6 and 9

For Questions 17–22, use your calculator to work out the answer correct to 3 significant figures.

**17** $\dfrac{28.7 + 129}{3.7}$ 

**18** $15.9 + (5.87)^2$ 

**19** $\sqrt{5.8^2 + 7.6^2}$

**20** $(1.3)^5$         **21** $\dfrac{45.8}{12 \times 3.5}$         **22** $\sqrt{3.4 \times 10^5}$

**23** £650 is invested at 8% compound interest. Find the value of the investment, to the nearest £10, after 5 years, 10 years and 15 years.

**24** Mrs Scrooge inherits £10000, and decides to put it under her mattress 'for a rainy day'. Maybe her money is safe, but it is not earning any interest, and, because of inflation, it is actually losing value at the rate of 15% a year. Calculate the value of this money, correct to 3 significant figures, after 5 years, 10 years and 15 years.

**25** Calculate the percentage 'free' on this offer.

# Exercise 53★ (Revision)

**1** Increase 400 kg by 1.5%.

**2** Decrease 500 km by 0.5%.

**3** Write down the five prime numbers between 45 and 70.

For Questions 4–6, express each number as a product of prime factors, using indices where necessary.

**4** 340                **5** 112                **6** 216

**7** Which of these is a multiple of 3 and 7: 148, 196 or 777?

**8** Show that 18 is a factor of 666.

For Questions 9–11, write down the highest common factor.

**9** 84 and 140                **10** $3xy$ and $6x^2$                **11** $6a^3$ and $9a^2$

For Questions 12–14, write down the lowest common multiple.

**12** 14 and 30                **13** $4xy$ and $3x^2$                **14** $6a^2$ and $15ab$

For Questions 15–20, use your calculator to calculate the answer, correct to 3 significant figures.

**15** $\dfrac{45.9 - 1.069}{45.9 \times 1.069}$         **16** $16\dfrac{6}{7} \div \dfrac{1}{(0.591)^2}$         **17** $\dfrac{25.9}{34.8 - 9.17} - \dfrac{1}{(0.97)^3}$

**18** $(2.8 \times 10^{-2}) + (2.8 \div 10^{-2})$         **19** $(((1.001)^2)^3)^4$         **20** $1 + \dfrac{1}{1 + \dfrac{1}{(0.99)^5}}$

**21** £1500 was invested for 4 years at 7% compound interest, and then for another 7 years at 4% compound interest. What was the investment worth, correct to 3 significant figures, after 11 years?

**22** If you invest 1p at 1.5% per annum compound interest, what would it be worth, to 2 significant figures, after 2000 years?

# Algebra 3

## Simple factorising

Expanding $2a^2b(7a - 3b)$ gives $14a^3b - 6a^2b^2$. The reverse of this process is called **factorising**.

If the common factors are not obvious, first write out the expression to be factorised in full, writing numbers in prime-factor form. Identify each term that is common to all parts, and use these terms as common factors to be placed outside the bracket.

---

**Example 1**

Factorise $x^2 + 4x$.

$x^2 + 4x = x \times x + 4 \times x$
$\qquad\quad = x(x + 4)$

blue terms    black terms

**Example 2**

Factorise $6x^2 + 2x$.

$6x^2 + 2x = \mathbf{2} \times 3 \times x \times x + \mathbf{2} \times x$
$\qquad\quad = 2x(3x + 1)$

blue terms    black terms

**Example 3**

Factorise $14a^3b - 6a^2b^2$.

$14a^3b - 6a^2b^2 = \mathbf{2} \times 7 \times \mathbf{a} \times \mathbf{a} \times a \times \mathbf{b} - \mathbf{2} \times 3 \times \mathbf{a} \times \mathbf{a} \times \mathbf{b} \times b$
$\qquad\qquad\quad = 2a^2b(7a - 3b)$

blue terms    black terms

---

## Exercise 54

Factorise these completely.

**1** $x^2 + 3x$

**2** $x^2 + 5x$

**3** $x^2 - 4x$

**4** $x^2 - 7x$

**5** $5a - 10b$

**6** $21a - 7b$

**7** $xy - xz$

**8** $bc - 2cd$

**9** $2x^2 + 4x$

**10** $3x^2 + 6x$

**11** $3x^2 - 18x$

**12** $4x^2 - 8x$

**13** $ax^2 - a^2x$

**14** $a^2x^2 + a^3x^3$

**15** $6x^2y - 21xy$

**16** $28ab^2 - 21a^2b$

**17** $9p^2q + 6pq$

**18** $4ab^3 + 6a^2b$

**19** $ap + aq - ar$

**20** $ax + ay + ax^2$

# Exercise 54★

Factorise these completely.

**1** $5x^3 + 15x^4$

**2** $27a^4 + 9a^2$

**3** $3x^3 - 18x^2$

**4** $24p^4 - 6p^5$

**5** $9x^3y^2 - 12x^2y^4$

**6** $8xy^3 - 24x^3y$

**7** $x^3 - 3x^2 - 3x$

**8** $5a^2 + a^3 - 5a$

**9** $abc^2 - ab^2 + a^2bc$

**10** $a^3b^2c + a^2b^2c^2 - ab^2c^3$

**11** $4p^2q^2r^2 - 12pqr + 16pq^2$

**12** $9ab^2 + 15abc - 12a^2b$

**13** $30x^3 + 12xy - 21xz$

**14** $24ab^2 - 16a^2c + 20a^2b^2$

**15** $0.2h^2 + 0.1gh - 0.3g^2h^2$

**16** $0.7a^2b^3 - 2.1a^3b^2 + 1.4a^2b$

**17** $\frac{1}{8}x^3y - \frac{1}{4}xy^2 + \frac{1}{16}x^2y^2$

**18** $\frac{1}{3}pq^2 + \frac{1}{18}p^3q^3 - \frac{1}{9}p^2q^2$

**19** $\pi r^2 + 2\pi rh$

**20** $\frac{2}{3}\pi r^3 + \frac{1}{3}\pi r^2h$

**21** $16p^2qr^3 - 28pqr - 20p^3q^2r$

**22** $24ab^2c^3 + 32a^3b^2c - 48abc$

**23** $ax + bx + ay + by$

**24** $a(a + b)^2 + b(a + b)^2$

**25** $(x - y)^2 - (x - y)^3$

**26** $x(x + 1)(x + 3)(x + 5) - x(x + 3)(x + 5)$

# Further simplifying of fractions

To simplify $\frac{234}{195}$ it is easiest to factorise it first.

$$\frac{234}{195} = \frac{2 \times \cancel{3}^3\cancel{2} \times \cancel{13}^1}{\cancel{3}_1 \times 5 \times \cancel{13}_1} = \frac{6}{5}$$

Algebraic fractions are also best simplified by factorising first.

---

**Example 4**

Simplify $\frac{x^2 + 5x}{x}$.

$$\frac{x^2 + 5x}{x} = \frac{\cancel{x}(x + 5)}{\cancel{x}} = x + 5$$

**Example 5**

Simplify $\frac{2a^3 - 4a^2b}{2ab - 4b^2}$.

$$\frac{2a^3 - 4a^2b}{2ab - 4b^2} = \frac{\cancel{2}a^2(a \cancel{- 2b})}{\cancel{2}b(a \cancel{- 2b})} = \frac{a^2}{b}$$

---

# Exercise 55

Simplify these.

**1** $\frac{x^2 + x}{x}$

**2** $\frac{a - a^3}{a}$

**3** $\frac{2x + 2y}{2z}$

**4** $\frac{3a - 3b}{3c}$

**5** $\frac{2r + 2s}{r + s}$

**6** $\frac{5x - 5y}{x - y}$

**7** $\frac{a^2 - ab}{ab}$

**8** $\frac{x^2 + xy}{xy}$

**9** $\frac{at - bt}{ar - br}$

**10** $\frac{ax - ay}{xz - yz}$

**11** $\frac{x - xy}{z - zy}$

**12** $\frac{ab - a}{bc - c}$

# Exercise 55★

Simplify these.

**1** $\dfrac{ax + ay}{a}$

**2** $\dfrac{6a^2 + 2ab}{2a}$

**3** $\dfrac{z}{z^2 + z}$

**4** $\dfrac{2m}{m^2 - 2m}$

**5** $\dfrac{6x^2 + 9x^4}{3x^2}$

**6** $\dfrac{5a^3 - 15a^4}{5a^2}$

**7** $\dfrac{8x^3y^2 - 24x^2y^4}{12x^2y^2}$

**8** $\dfrac{15xy^3 + 5x^3y^2}{10xy^2}$

**9** $\dfrac{y^2 + y}{y + 1}$

**10** $\dfrac{ax + ay}{x + y}$

**11** $\dfrac{6x^2 - 12x^2y}{3xz - 6xyz}$

**12** $\dfrac{12x^2y^2 - 8xy^2}{6x^2y - 4xy}$

**13** $\dfrac{(a^2 + 2ac) - (ab^2 + 2ac)}{a^2 - ab^2}$

**14** $\dfrac{(xy + y) - (xy - z)}{y^2 + yz}$

**15** $\dfrac{2a^2 - ab}{3a^2 - ab} \times \dfrac{3ab - b^2}{2a^2 - ab}$

**16** $\dfrac{x^2 + 2xy}{xy^2 - 2y^3} \times \dfrac{3xy - 6y^2}{2x + 4y}$

**17** $\dfrac{1}{3x - y} \div \dfrac{1}{15x - 5y}$

**18** $\dfrac{1}{3a + 12b} \div \dfrac{1}{4b + a}$

**19** $\dfrac{5x^3 - 10x^2}{10x - 5x^2}$

**20** $\dfrac{6xy^2 - 3xy}{3xy^2 - 6xy^3}$

# Equations with fractions

Equations with fractions are easier to deal with than algebraic expressions because both sides of the equation can be multiplied by the lowest common denominator to clear the fractions.

## Key Point

Clear the fractions by multiplying **both** sides by the common denominator.

## Equations with numbers in the denominator

**Example 6**

$\dfrac{2x}{3} - 1 = \dfrac{x}{2}$  (Multiply both sides by 6)

$4x - 6 = 3x$

$x = 6$  Check: $4 - 1 = 3$

**Example 7**

$\dfrac{3}{4}(x - 1) = \dfrac{1}{3}(2x - 1)$  (Multiply both sides by 12)

$9x - 9 = 8x - 4$

$x = 5$  Check: $\dfrac{3}{4}(5 - 1) = \dfrac{1}{3}(10 - 1)$

# Exercise 56

Solve for $x$ the equations in Questions 1–22.

**1** $\dfrac{3x}{4} = 6$      **2** $\dfrac{5x}{3} = 10$      **3** $\dfrac{x}{5} = -2$      **4** $\dfrac{x}{7} = -3$

**5** $\dfrac{x}{4} = \dfrac{1}{2}$      **6** $\dfrac{x}{12} = \dfrac{5}{6}$      **7** $\dfrac{3x}{8} = 0$      **8** $\dfrac{2x}{5} = 0$

**9** $\dfrac{2x}{3} = -4$      **10** $\dfrac{4x}{5} = -8$      **11** $\dfrac{1}{3}(x + 7) = 4$      **12** $\dfrac{1}{5}(x - 2) = 1$

**13** $\dfrac{3(x - 10)}{7} = -6$      **14** $\dfrac{3(x - 7)}{4} = -9$      **15** $\dfrac{x}{2} - \dfrac{x}{3} = 1$      **16** $\dfrac{x}{3} - \dfrac{x}{5} = 2$

**17** $x - \dfrac{2x}{7} = 10$      **18** $\dfrac{3x}{5} + 2x = 26$      **19** $\dfrac{1}{4}(x + 1) = \dfrac{1}{5}(8 - x)$

**20** $\dfrac{1}{3}(x - 1) = \dfrac{1}{4}(x + 1)$      **21** $\dfrac{3 - x}{3} = \dfrac{2 + x}{2}$      **22** $\dfrac{x + 7}{7} = \dfrac{5 - x}{5}$

**23** $\left(\dfrac{x}{2} + 7\right)$ is three times $\left(\dfrac{x}{5} + 2\right)$. Find the value of $x$.

**24** $\left(\dfrac{x}{3} + 5\right)$ is five times $\left(\dfrac{x}{5} - 4\right)$. Find the value of $x$.

**25** Ryan does one-third of his journey to school by car, and one-half by bus. Then he walks the final kilometre. How long is his journey to school?

**26** Sam was boasting about a fish he had caught. When he was asked how long it was, he said he could not remember, but he did remember that the tail was 30 cm long, the head was one-sixth of the total length, and the body was equal to half of the tail plus two heads. How long was Sam's fish?

# Exercise 56★

Solve for $x$ the equations in Questions 1–18.

**1** $\dfrac{2x - 3}{5} = 3$      **2** $\dfrac{3x - 1}{7} = 2$

**3** $\dfrac{3}{8}\left(5x - 3\right) = 0$      **4** $\dfrac{4}{7}\left(6x + 5\right) = 0$

**5** $\dfrac{x + 1}{5} = \dfrac{x + 3}{6}$      **6** $\dfrac{x - 1}{3} = \dfrac{2x + 1}{7}$

**7** $\dfrac{x + 1}{7} - \dfrac{3(x - 2)}{14} = 1$      **8** $\dfrac{3(x - 2)}{2} - \dfrac{x - 5}{4} = 2$

**9** $\dfrac{6 - 3x}{3} - \dfrac{5x + 12}{4} = -1$      **10** $\dfrac{16 - 5x}{8} - \dfrac{4(2x + 5)}{5} = -2$

**11** $\dfrac{2(x + 1)}{5} - \dfrac{3(x + 1)}{10} = x$      **12** $\dfrac{3(x - 2)}{4} - \dfrac{x + 1}{6} = x$

**13** $\dfrac{2x - 3}{2} - \dfrac{x - 2}{3} = \dfrac{7}{6}$      **14** $\dfrac{x + 3}{2} + \dfrac{x + 2}{3} = \dfrac{4}{3}$

**15** $\dfrac{2x + 1}{4} - x = \dfrac{3x + 1}{8} + 1$      **16** $4 - \dfrac{x - 2}{2} = x + \dfrac{1 - 2x}{3}$

**17** $\dfrac{1}{2}\left(1 - x\right) - \dfrac{1}{3}\left(2 + x\right) + \dfrac{1}{4}\left(3 - x\right) = 1$      **18** $\dfrac{1}{5}\left(x + 1\right) - \dfrac{1}{15}\left(2x + 3\right) - \dfrac{1}{3}\left(1 - 3x\right) = -1$

**19** $\left(\dfrac{x}{14} + 3\dfrac{1}{2}\right)$ is twice $\left(\dfrac{x}{21} + 1\dfrac{2}{3}\right)$. Find the value of $x$.

**20** $\left(1 - \dfrac{2x}{5}\right)$ is 3 less than three times $\left(\dfrac{x}{2} - 5\right)$. Find the value of $x$.

**21** Diophantus was a famous ancient Greek mathematician. This was carved on his tomb.

> Here lie the remains of Diophantus. He was a child for one-sixth of his life. After one-twelfth more, he became a man. After one-seventh more, he married, and five years later his son was born. His son lived half as long as his father and died four years before his father.

How old was Diophantus when he died?

**22** Ethan goes on a charity journey. He walks one-tenth of the way at 6 km/hour, runs one-sixth at 12 km/hour, cycles one-fifth at 24 km/hour, and completes the remaining 32 km by car at 48 km/hour. How many kilometres does he travel, and how long does it take?

## Equations with *x* in the denominator

When the denominator contains $x$, the same principle of clearing fractions applies.

**Example 8**

$$\frac{3}{x} = \frac{1}{2}$$ (Multiply both sides by $2x$)

$$\frac{3}{x} \times 2x = \frac{1}{2} \times 2x$$

$$6 = x$$ Check: $\frac{3}{6} = \frac{1}{2}$

**Example 9**

$$\frac{4}{x} - x = 0$$ (Multiply both sides by $x$)

$$\frac{4}{x} \times x - x \times x = 0 \times x$$ (Remember to multiply everything by $x$)

$$4 - x^2 = 0$$

$$x^2 = 4$$

$$x = \pm 2$$ Check: $\frac{4}{2} - 2 = 0$ and $\frac{4}{-2} - (-2) = 0$

# Exercise 57

Solve these for $x$.

**1** $\dfrac{10}{x} = 5$      **2** $\dfrac{14}{x} = 2$      **3** $\dfrac{12}{x} = -4$      **4** $\dfrac{15}{x} = -3$

**5** $\dfrac{3}{x} = 5$      **6** $\dfrac{2}{x} = 7$      **7** $\dfrac{4}{x} = -\dfrac{1}{2}$      **8** $\dfrac{5}{x} = -\dfrac{1}{3}$

**9** $\dfrac{3}{5} = \dfrac{6}{x}$      **10** $\dfrac{3}{4} = \dfrac{12}{x}$      **11** $\dfrac{8}{x} = -\dfrac{10}{3}$      **12** $\dfrac{3}{x} = -\dfrac{2}{7}$

**13** $\dfrac{35}{x} = 0.7$      **14** $\dfrac{18}{x} = 0.6$      **15** $0.3 = -\dfrac{15}{2x}$      **16** $0.4 = -\dfrac{12}{5x}$

**17** $\dfrac{5}{3x} = 1$      **18** $\dfrac{6}{7x} = 1$      **19** $\dfrac{9}{x} - x = 0$      **20** $x - \dfrac{25}{x} = 0$

# Exercise 57★

Solve these for $x$.

**1** $\dfrac{52}{x} = 13$

**2** $\dfrac{80}{x} = 16$

**3** $2.5 = -\dfrac{20}{x}$

**4** $3.5 = -\dfrac{7}{x}$

**5** $\dfrac{15}{2x} = 45$

**6** $\dfrac{12}{5x} = 48$

**7** $\dfrac{8}{x} = -\dfrac{1}{8}$

**8** $\dfrac{1}{11} = -\dfrac{11}{x}$

**9** $\dfrac{2.8}{x} = 0.7$

**10** $\dfrac{5.4}{x} = 0.9$

**11** $\dfrac{16}{x} = \dfrac{x}{4}$

**12** $\dfrac{3}{x} = \dfrac{x}{27}$

**13** $\dfrac{12}{x} - 3x = 0$

**14** $\dfrac{45}{x} - 5x = 0$

**15** $\dfrac{3.2}{x} - 4.3 = 5.7$

**16** $\dfrac{1.4}{x} - 3.2 = 6.8$

**17** $\dfrac{1}{2x} + \dfrac{1}{3x} = 1$

**18** $\dfrac{1}{4x} + \dfrac{1}{3x} = 1$

**19** $\dfrac{1}{ax} + \dfrac{1}{bx} = 1$

**20** $\dfrac{1}{ax} - \dfrac{1}{bx} = 1$

# Simultaneous equations

## Elimination method

Solving simultaneous equations graphically is time-consuming. It can also be inaccurate, as the solutions are read from a graph. Algebraic methods are often preferable.

---

**Example 10**

Solve the simultaneous equations $2x - y = 35$, $x + y = 118$.

$$
\begin{array}{ll}
2x - y = 35 & (1) \\
\underline{x + y = 118} & (2) \\
3x \phantom{{}-y} = 153 & \text{(Adding equations (1) and (2))} \\
x \phantom{{}-y} = 51 &
\end{array}
$$

Substituting $x = 51$ into (1) gives $102 - y = 35 \implies y = 67$

The solution is $x = 51$, $y = 67$.

Check: Substituting $x = 51$, $y = 67$ into (2) gives $51 + 67 = 118$.

---

The method of Example 10 only works if the numbers before either $x$ or $y$ are of opposite sign and equal value. The equations may have to be multiplied by suitable numbers to achieve this.

**Example 11**

Solve the simultaneous equations $x + y = 5$, $6x - 3y = 3$.

$$x + y = 5 \qquad (1)$$
$$6x - 3y = 3 \qquad (2)$$

Multiply *both* sides of equation (1) by 3.

$$3x + 3y = 15 \qquad (3)$$
$$\underline{6x - 3y = 3} \qquad (2)$$
$$9x \quad = 18 \qquad \text{(Adding equations (3) and (2))}$$
$$x \quad = 2$$

Substituting $x = 2$ into (1) gives $2 + y = 5 \;\Rightarrow\; y = 3$

The solution is $x = 2$, $y = 3$.

Check: Substituting $x = 2$ and $y = 3$ into (2) gives $12 - 9 = 3$.

# Exercise 58

Solve these simultaneous equations.

1  $x + y = 11$, $x - y = 5$
2  $x + y = 1$, $2x - y = 5$
3  $2x - y = 3$, $x + y = 9$
4  $2x + 2y = 5$, $3x - 2y = 10$
5  $3x + y = 8$, $3x - y = -2$
6  $-x + 3y = 7$, $x - y = 3$
7  $-2x + y = -2$, $2x - 3y = 6$
8  $4x + 5y = -28$, $-4x - 7y = 28$
9  $x - y = -6$, $y + 2x = 3$
10  $3x + 4y = 7$, $3x - 4y = -1$

If the numbers in front of $x$ or $y$ are not of opposite sign, multiply by a negative number, as shown in Example 12.

**Example 12**

Solve the simultaneous equations $x + 2y = 8$, $2x + y = 7$.

$$x + 2y = 8 \qquad (1)$$
$$2x + y = 7 \qquad (2)$$

Multiply *both* sides of equation (2) by $-2$.

$$x + 2y = 8 \qquad (3)$$
$$\underline{-4x - 2y = -14} \qquad (2)$$
$$-3x \quad = -6 \qquad \text{(Adding equations (3) and (2))}$$
$$x \quad = 2$$

Substituting $x = 2$ into (1) gives $2 + 2y = 8 \;\Rightarrow\; y = 3$

The solution is $x = 2$, $y = 3$.

Check: Substituting $x = 2$ and $y = 3$ into equation (2) gives $4 + 3 = 7$.

Sometimes *both* equations have to be multiplied by suitable numbers, as in Example 13.

**Example 13**

Solve the simultaneous equations $2x + 3y = 5$, $5x - 2y = -16$.

$$2x + 3y = 5 \qquad (1)$$
$$5x - 2y = -16 \qquad (2)$$

Multiply (1) by 2. $\qquad 4x + 6y = 10 \qquad (3)$

Multiply (2) by 3. $\qquad \underline{15x - 6y = -48} \qquad (4)$

$$19x \qquad = -38 \qquad \text{(Adding equations (3) and (4))}$$
$$x \qquad = -2$$

Substituting $x = -2$ into (1) gives $-4 + 3y = 5 \quad \Rightarrow \quad y = 3$

The solution is $x = -2$, $y = 3$.

Check: Substituting $x = -2$ and $y = 3$ into equation (2) gives $-10 - 6 = -16$.

## Key Points

To solve two simultaneous equations by **elimination**:

♦ **Label** equations (1) and (2).

♦ **Multiply** one or both equations by suitable numbers so that the numbers in front of the terms to be eliminated are the same and the signs are different.

♦ **Eliminate** by adding the equations. Solve the resulting equation.

♦ **Substitute** your answer in one of the original equations to find the other answer.

♦ **Check** by substituting both answers into the other original equation.

## Exercise 59

Solve these simultaneous equations.

1  $3x + y = 11$, $x + y = 7$ $\qquad\qquad$ 2  $2x - y = 7$, $x - y = 3$

3  $x + 3y = 8$, $x - 2y = 3$ $\qquad\qquad$ 4  $x - 5y = 1$, $x + 3y = 25$

5  $2x + y = 5$, $3x - 2y = -3$ $\qquad\qquad$ 6  $3x + y = -5$, $5x - 3y = -13$

7  $2x + 3y = 7$, $3x + 2y = 13$ $\qquad\qquad$ 8  $3x + 2y = 5$, $6x + 5y = 8$

9  $2x + 5y = 9$, $3x + 4y = 10$ $\qquad\qquad$ 10  $3x + 4y = 11$, $5x + 6y = 17$

## Exercise 59★

Solve these simultaneous equations.

1  $2x + y = 5$, $3x - 2y = 11$ $\qquad\qquad$ 2  $3x + y = 10$, $2x - 3y = 14$

3  $3x + 2y = 7$, $2x - 3y = -4$ $\qquad\qquad$ 4  $4x + 7y = -5$, $3x - 2y = 18$

5  $3x + 2y = 4$, $2x + 3y = 7$ $\qquad\qquad$ 6  $2x + 3y = 5$, $3x + 4y = 7$

7  $7x - 4y = 37$, $5x + 3y = 44$ $\qquad\qquad$ 8  $5x - 7y = 27$, $3x - 4y = 16$

**9** $3x + 2y = 3$, $7x - 3y = 1.25$

**10** $4x - 3y = 2.6$, $10x + 5y = -1$

**11** $\dfrac{x}{2} + \dfrac{y}{3} = 4$, $\dfrac{y}{4} - \dfrac{x}{3} = \dfrac{1}{6}$

**12** $\dfrac{x}{5} + \dfrac{y}{2} = 1$, $\dfrac{x}{2} - \dfrac{y}{8} = -3$

**13** $\dfrac{a+1}{b+1} = 2$, $\dfrac{2a+1}{2b+1} = \dfrac{1}{3}$

**14** $\dfrac{c+d}{c-d} = \dfrac{1}{2}$, $\dfrac{c+1}{d+1} = 2$

Hint for Questions 15 and 16: let $p = \dfrac{1}{x}$, $q = \dfrac{1}{y}$.

**15** $\dfrac{2}{x} - \dfrac{1}{y} = 3$, $\dfrac{4}{x} + \dfrac{3}{y} = 16$

**16** $\dfrac{2}{x} - \dfrac{3}{y} = 1$, $\dfrac{8}{x} + \dfrac{9}{y} = \dfrac{1}{2}$

## Substitution method

The puzzle in Example 14 is often used as a brain-teaser. See if you can 'guess' the solution first.

### Example 14

A bottle and a cork together cost £1. The bottle costs 90p more than the cork. Find the cost of the bottle.

Let $b$ be the cost of the bottle in pence, and $c$ be the cost of the cork in pence. The total cost is 100p, and so

$$b + c = 100 \qquad (1)$$

The bottle costs 90p more than the cork, and so

$$b = c + 90 \qquad (2)$$

Substituting (2) into (1) gives

$$(c + 90) + c = 100$$
$$2c = 10 \quad \Rightarrow \quad c = 5$$

Substituting in (1) gives $b = 95$.

Therefore the bottle costs 95p, and the cork costs 5p.

Check: Equation (2) gives $95 = 5 + 90$.

## Exercise 60

Solve the following simultaneous equations by substitution.

**1** $x = y + 2$
$x + 4y = 7$

**2** $x = y + 1$
$x + 3y = 5$

**3** $y = x + 3$
$y + 2x = 6$

**4** $y = x + 2$
$y + 4x = 12$

**5** $y = 3x + 3$
$5x + y = 11$

**6** $y = 2x + 1$
$4x + y = 13$

**7** $x = y - 3$
$x + 3y = 5$

**8** $x = y - 4$
$x + 4y = 6$

**9** $y = 2x - 7$
$3x - y = 10$

**10** $y = 3x - 8$
$2x - y = 6$

# Exercise 60★

Solve the following simultaneous equations by substitution.

**1** $3x + 4y = 11$
    $x = 15 - 7y$

**2** $3x + 2y = 7$
    $x = 3 - y$

**3** $x = 7 - 3y$
    $2x - 2y = 6$

**4** $x = 10 - 7y$
    $3x - 4y = 5$

**5** $y = 5 - 2x$
    $3x - 2y = 4$

**6** $5x - 4y = 13$
    $y = 13 - 2x$

**7** $3x - 2y = 7$
    $4x + y = 2$

**8** $4x - 3y = 1$
    $3x + y = 17$

**9** $x - 2y + 4 = 0$
    $5x - 6y + 18 = 0$

**10** $x - 3y + 5 = 0$
     $3x + 2y + 4 = 0$

# Solving problems using simultaneous equations

**Example 15**

Tickets at a concert cost either £10 or £15. The total takings were £8750.
Twice as many £10 tickets were sold as £15 tickets. How many tickets were sold?

Let $x$ be the number of £10 tickets sold, and $y$ the number of £15 tickets sold.
The total takings were £8750, and so

$$10x + 15y = 8750$$

Divide by 5 to simplify.

$$2x + 3y = 1750 \qquad (1)$$

Twice as many £10 tickets were sold as £15 tickets, and so

$$x = 2y \qquad (2)$$

To check that equation (2) is correct, substitute simple numbers that obviously work, such as $x = 10$, $y = 5$.

Substituting (2) into (1) gives

$$4y + 3y = 1750$$
$$7y = 1750$$
$$y = 250$$

and so $x = 500$, from (2).

750 tickets were sold altogether.

Check: In (1), $1000 + 750 = 1750$.

**Example 16**

Akmed makes a camel journey of 20 km. The camel travels at 12 km/h for the first part of the journey, but then conditions worsen and the camel can only travel at 4 km/h for the second part of the journey. The journey takes 3 hours. Find the length of each part of the journey.

Let $x$ be the length in km of the first part of the journey, and $y$ be the length in km of the second part.

$$x + y = 20 \qquad (1) \qquad \text{(Total length is 20 km)}$$

Use the formula time $= \dfrac{\text{distance}}{\text{speed}}$.

$$\frac{x}{12} + \frac{y}{4} = 3 \qquad (2) \qquad \text{(Total time taken is 3 hours)}$$

Multiply equation (2) by 12.

$$x + 3y = 36 \qquad (3)$$
$$\underline{x + y \ = 20} \qquad (1) \qquad \text{(Subtract equation (1) from equation (3))}$$
$$2y = 16$$

$$y = 8$$

From equation (1), if $y = 8$ then $x = 12$, so the first part is 12 km and the second part is 8 km.

Check: These values work in equations (1) and (2).

## Key Points

◆ Define your variables.

◆ Turn each sentence from the problem into an equation.

◆ Solve the equations by the most appropriate method.

# Exercise 61

1 Find two numbers with a sum of 112 and a difference of 54.

2 Find two numbers with a sum of 127 and a difference of 41.

3 Find two numbers with a mean of 14 and a difference of 4.

4 Find two numbers with a mean of 24 and a difference of 8.

5 Two times one number added to four times another gives 34. The sum of the two numbers is 13. Find the numbers.

6 Three times one number added to two times another gives 29. The sum of the two numbers is 12. Find the numbers.

**7** For this rectangle, find $x$ and $y$ and the area.

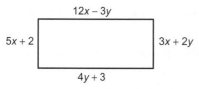

$12x - 3y$

$5x + 2$

$3x + 2y$

$4y + 3$

**8** For this equilateral triangle, find $x$ and $y$ and the perimeter.

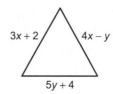

$3x + 2$

$4x - y$

$5y + 4$

**9** At McEaters, Pam bought two burgers and three colas, which cost her £3.45. Her friend Pete bought four burgers and two colas, and this cost him £4.94. How much did each item cost?

**10** Becky and her friend go to the fair. Becky has three rides on the dodgems and four rides on the big wheel at a total cost of £8.10. Her friend has four rides on the dodgems and two rides on the big wheel, at a total cost of £7.80. How much is each ride?

**11** A telephone box accepts only 20p coins or 50p coins. On one day, 39 coins were collected with a total value of £11.40. Find how many 50p coins were collected.

**12** At a concert, tickets cost either $40 or $60. 700 tickets were sold at a cost of $33 600. Find how many $40 tickets were sold.

**13** At a shooting range, each shot costs 20c. If you hit the target, you receive 30c. Emma has 20 shots and makes a loss of 70c. How many hits did she get?

**14** Emma is doing a multiple-choice test of 50 questions. She gets 2 marks for every question that is correct, but loses 1 mark for every question that is wrong. Emma answers every question and scores 67 marks. How many questions did she get right?

## Exercise 61★

**1** Find the intersection of the lines $y = x + 1$ and $3y + 2x = 13$ without drawing the graphs.

**2** Find the intersection of the lines $y = 2x - 1$ and $2y + 2x = 7$ without drawing the graphs.

**3** The line $y = mx + c$ passes through the points $(1, 1)$ and $(2, 3)$. Find $m$ and $c$.

**4** The line $y = mx + c$ passes through the points $(1, 2)$ and $(2, 1)$. Find $m$ and $c$.

**5** The denominator of a fraction is 5 more than the numerator. If both the denominator and numerator are increased by 3, the fraction becomes $\frac{3}{4}$. Find the original fraction.

**6** If 2 is subtracted from the numerator and denominator of a certain fraction, the answer is $\frac{4}{7}$. If 4 is added to the numerator and denominator, the answer is $\frac{2}{3}$. What is the fraction?

**7** Aidan can row at 3 m/s against the current and at 6 m/s with the current. Find the speed of the current.

**8** One year ago, Gill was five times as old as her horse. In one year's time, the sum of their ages will be 22. How old is Gill now?

**9** To cover a distance of 10 km, Jacob runs some of the way at 15 km/h, and walks the rest of the way at 5 km/h. His total journey time is 1 hour. How far did Jacob run?

**10** On a journey of 240 km, Archita travels some of the way on a motorway at 100 km/h and the rest of the way on minor roads at 60 km/h. The journey takes 3 hours. How far did she travel on the motorway?

**11** A 2-digit number is increased by 36 when the digits are reversed. The sum of the digits is 10. Find the original number.

**12** A 2-digit number is equal to seven times the sum of its digits. If the digits are reversed, the new number formed is 36 less than the original number. What is the number?

**13** To go to school in the morning, I first walk to the garden shed at 6 km/h to collect my bicycle. I then cycle to school at 15 km/h. The total journey normally takes 18.5 minutes.
One day, I am late and I run to the shed at 18 km/h and cycle at 27 km/h.
The journey takes me 10 min 10 s. How far is it from my house to the garden shed?

**14** When visiting his parents, Tyler drives at an average speed of 42 km/h through urban areas, and at an average speed of 105 km/h along the motorway. His journey usually takes him 2.5 hours. One day when there is fog, he sets off 1 hour early and only manages to average 28 km/h in the urban areas and 60 km/h on the motorway. He arrives 30 minutes late. How long was Tyler's journey?

## Exercise 62 (Revision)

Factorise

**1** $x^2 - 8x$

**2** $3x^2 + 12x$

**3** $6xy^2 - 30x^2y$

**4** $12x^3 + 9x^2 - 15x$

Simplify

**5** $\dfrac{x^2 - x}{x}$

**6** $\dfrac{x^2 + xy}{x^2 - xy}$

Solve these equations.

**7** $\dfrac{3x - 4}{4} = 2$

**8** $\dfrac{1}{4}(x - 2) = \dfrac{1}{7}(x + 1)$

**9** $\dfrac{2x + 7}{4} - \dfrac{x + 1}{3} = \dfrac{3}{4}$

**10** $\dfrac{4}{n} - 1 = \dfrac{2}{n}$

**11** Sarah shares out some sweets with her friends. She gives one-eighth of the sweets to Ann, one-sixth to Mary and one-third to Ruth. She then has nine sweets left over for herself. How many sweets did she have to start with?

For Questions 12–15, solve the pairs of simultaneous equations.

**12** $y - x = 4$ and $y + 2x = 1$

**13** $y + x = 3$ and $y - 2x = 3$

**14** $3x + 2y = 10$ and $5x - 4y = 2$

**15** $5x - 2y = -1$ and $10x - 3y = 1$

**16** At a sale, Andy buys two CDs and three tapes for £25.50.
His friend Charlie buys four CDs and five tapes for £47.50.
What is the cost of each item if all the CDs cost the same and all the tapes cost the same?

**17** Hailey is collecting 10p and 20p pieces. When she has 30 coins, the value of them is £4.10. How many of each coin does she have?

# Exercise 62★ (Revision)

Factorise

1 $3x^4 - 12x^3$

2 $\frac{4}{3}\pi r^3 + \frac{2}{3}\pi r^2$

3 $24x^3y^2 - 18x^2y$

4 $15a^2b^3c^2 - 9a^3b^2c^2 + 21a^2b^2c^3$

Simplify

5 $\dfrac{x^2 - xy}{xy - y^2}$

6 $\dfrac{ax - bx}{x^2 + xy} \div \dfrac{2a - 2b}{2x^2 + 2xy}$

For Questions 7–10, solve the equations.

7 $\frac{2}{7}(3x - 1) = 0$

8 $\dfrac{3x + 2}{5} - \dfrac{2x + 5}{3} = x + 3$

9 $1\frac{2}{3}\left(x + 1\right) = x + 5\frac{2}{3}$

10 $\dfrac{1}{x} + \dfrac{1}{2x} - \dfrac{1}{3x} = 2\frac{1}{3}$

11 Mrs Taylor has lived in many countries. She spent the first third of her life in England, the next sixth in France, one-quarter in Spain, $3\frac{1}{2}$ years in Italy, and one-fifth in Germany, where she is now living. How old is Mrs Taylor?

For Questions 12–15, solve the pairs of simultaneous equations.

12 $5x + 4y = 22$ and $3x + 5y = 21$

13 $5x + 3y = 23$ and $x + 2y = 6$

14 $3x + 8y = 24$ and $x - 2y = 1$

15 $6x + 5y = 30$ and $3x + 4y = 18$

16 The straight line $ax + by = 1$ passes through the points $(1, 4)$ and $(3, 1)$. Find the values of $a$ and $b$.

17 In ten years' time, Mike will be twice as old as his son Ben. Ten years ago, Mike was seven times as old as Ben. How old are Mike and Ben now?

# Graphs 3

## Travel graphs

This section covers distance–time graphs and speed–time graphs.

### Distance–time graphs

> **Example 1**
>
> A veteran car takes part in the annual London to Brighton motor rally, and then returns to London. Here is its distance–time graph.
>
>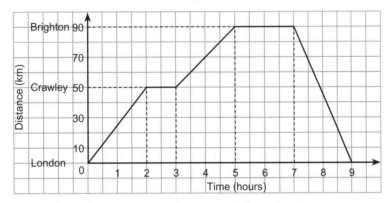
>
> **a** What is the speed of the car from London to Crawley?
>
> The speed from London to Crawley is $\frac{50\,\text{km}}{2\,\text{h}} = 25\,\text{km/h}$.
>
> **b** The car breaks down at Crawley. For how long does the car break down?
>
> The car is at Crawley for 1 hour.
>
> **c** What is the speed of the car from Crawley to Brighton?
>
> The speed from Crawley to Brighton is $\frac{40\,\text{km}}{2\,\text{h}} = 20\,\text{km/h}$.
>
> **d** The car is towed on a trailer back to London from Brighton. At what speed is the car towed?
>
> The speed from Brighton to London is $\frac{90\,\text{km}}{2\,\text{h}} = 45\,\text{km/h}$.

## Exercise 63

1 Ingar travels southbound on a motorway from Hamburg, while Franz travels northbound on the same road from Hannover. This distance–time graph (where the distance is from Hannover) shows the journeys of both travellers.

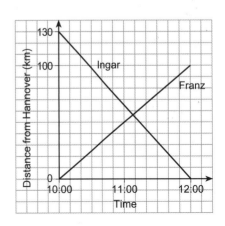

    **a** What is Ingar's speed in kilometres per hour?

    **b** What is Franz's speed in kilometres per hour?

    **c** At what time does Ingar reach Hannover?

    **d** How far apart are Ingar and Franz at 10:30?

    **e** At what time do Ingar and Franz pass each other?

**2** This distance–time graph shows the journeys
of a car and a motorcycle between
Manchester (M) and Birmingham (B).

**a** When did the car stop, and for how long?

**b** When did the car and the motorcycle pass
each other?

**c** How far apart were the car and the
motorcycle at 09:30?

**d** After the motorcycle's first stop, it
increased its speed until it arrived in
Birmingham. The speed limit on the road
was 70 miles/hour. Was the motorcyclist
breaking the speed limit?

**e** Over the whole journey (excluding stops),
what was the mean speed of the car?

**f** What was the mean speed of the
motorcycle (excluding stops)?

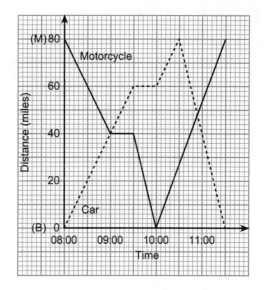

**3** Naseem leaves home at 09:00, and drives to Clare's house at a speed of 60 km/h for 1 hour.
Then she stops at a petrol station for 15 minutes. She continues on her journey at 40 km/h for
30 minutes, and then arrives at her destination. At 13:00, she starts her return journey, and
drives at a constant speed of 80 km/h without stopping.

**a** Draw a distance–time graph to illustrate Naseem's journey.

**b** Use this graph to estimate at what time Naseem returns home.

**4** Li and Jacki train for a triathlon by swimming 1 km along the coast, cycling 9 km in the
same direction along the straight coast road, and then running directly back to their starting
point via the same road. The times of this training session are shown in the table.

| Activity | Li's time (minutes) | Jacki's time (minutes) |
|---|---|---|
| Swimming | 20 | 15 |
| Rest | 5 | 5 |
| Cycling | 10 | 15 |
| Rest | 10 | 5 |
| Running | 35 | 50 |

**a** Draw a distance–time graph (in kilometres and hours) to illustrate this information, given
that Li and Jacki both start at 09:00. Let the time axis range from 09:00 to 10:30.

**b** Use your graph to estimate when Li and Jacki finish.

**c** Use your graph to estimate when Li and Jacki are level.

**d** Calculate the mean speed of both athletes over the whole session, excluding stops.

# Exercise 63★

1. A goat is tethered to a pole at A in the corner of a square field ABCD. The rope is the same length as the side of the field. The goat starts at B and trots at a constant speed to corner D keeping the rope taut. Sketch graphs for this journey:
   a Distance from A against time
   b Distance from B against time
   c Distance from C against time
   d Distance from D against time

   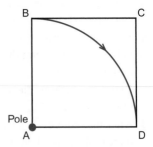

2. Harry and Jack are two footballers who are put through an extra training session of running at *identical constant speeds*. For all three exercise drills, Harry and Jack always start simultaneously from A and C, respectively.

   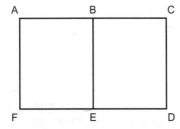

   **Drill 1** is that Harry and Jack both run one clockwise circuit.
   **Drill 2** is that Harry runs a circuit clockwise, and Jack runs a circuit anticlockwise.
   **Drill 3** is that Harry and Jack run directly towards D and F, respectively.
   Sketch three graphs of the distance of Harry from Jack against time, one for each drill.

3. Three motorcyclists A, B and C set out on a journey along the same road. Their journey is shown in the travel graph.
   a Place the riders in order (first, second and third) after
      (i) 0 s    (ii) 15 s    (iii) 30 s
   b When are all the riders the same distance along the road?
   c Which rider travels at a constant speed?
   d Which rider's speed is gradually
      (i) increasing?    (ii) decreasing?

   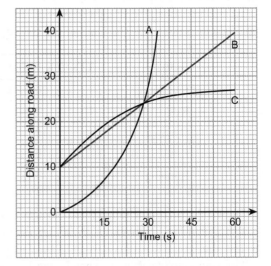

UNIT 3 ◆ Graphs

**4** The diagram shows the distances, in kilometres, between some junctions on a motorway.
The junctions are numbered as **1**, **2** ... and Ⓢ is the service area.
Driver A (northbound) joins **5** at 08:00, arrives at Ⓢ at 09:00, rests for
half an hour, and then continues his journey, passing **1** at 12:00.
Driver B (southbound) joins **1** at 08:00, arrives at Ⓢ at 10:00, rests for
1 hour, and then continues her journey, passing **5** at 12:00.

**a** Draw a graph of the distance in kilometres from **1** against the time in
hours to show both journeys.

**b** When does driver A pass driver B?

**c** What are A and B's final speeds?

**d** Find their mean speeds, excluding stops.

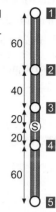

## Speed–time graphs

Travel graphs of speed against time can be used to find out more about speed changes and distances travelled.

---

**Example 2**

A train changes speed as shown in the speed–time graph.

The train's speed is *increasing* between
A and B, so it is *accelerating*.

The train's speed is *decreasing* between
C and D, so it is *decelerating (retarding)*.

The train's speed is constant at 20 m/s (and
therefore the acceleration is zero) between
B and C for 30 s. It has travelled 600 m
(20 × 30 m). This is the area under the
graph between B and C.

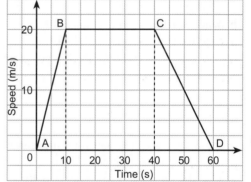

**a** Find the *total* distance travelled by the
train, and thus find the mean speed for the whole journey.

Total distance travelled = area under graph

$$= \left(\frac{1}{2} \times 10 \times 20\right) + (30 \times 20) + \left(\frac{1}{2} \times 20 \times 20\right) = 900$$

Therefore, mean speed $= \dfrac{900\,\text{m}}{60\,\text{s}} = 15\,\text{m/s}$

**b** Find the train's acceleration between A and B, B and C, and C and D.

Acceleration between A and B = gradient of line AB

$$= \frac{20\,\text{m/s}}{10\,\text{s}} = 2\,\text{m/s}^2$$

Between B and C the speed is constant, so the acceleration is zero.

Acceleration between C and D = gradient of line CD

$$= \frac{-20\,\text{m/s}}{20\,\text{s}} = -1\,\text{m/s}^2 \qquad \text{(The − sign indicates retardation.)}$$

> **Remember**
>
> In a speed–time graph,
>
> **acceleration** = gradient of line = $\dfrac{\text{change in speed}}{\text{time}}$
>
> **distance travelled** = area under the graph

## Exercise 64

**1** A speed–time graph for a journey of 15 s is shown.
   **a** Find the acceleration over the first 10 s.
   **b** Find the retardation over the last 5 s.
   **c** Find the total distance travelled.
   **d** Find the mean speed for the journey.

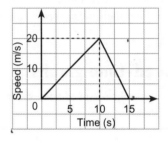

**2** A speed–time graph for a journey of 3 hours is shown.
   **a** Find the acceleration over the first 2 hours.
   **b** Find the retardation over the last hour.
   **c** Find the total distance travelled.
   **d** Find the mean speed for the journey in kilometres
      per hour.

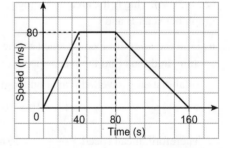

**3** A speed–time graph is shown for the journey of
   a train between two stations.
   **a** Find the acceleration over the first 40 s.
   **b** Find the retardation over the final 80 s.
   **c** Find the total distance travelled.
   **d** Find the mean speed for the journey in metres
      per second.

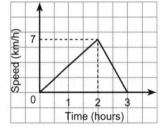

**4** The speed–time graph for a Sparker firework is shown.
   **a** Find the Sparker's maximum speed.
   **b** When does the firework have zero acceleration?
   **c** Estimate the total distance travelled by the firework.
   **d** By calculation, estimate the mean speed of the firework
      during its journey.

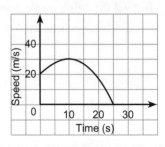

# Exercise 64★

1 A cycle-taxi accelerates from rest to 6 m/s in 10 s, remains
  at that speed for 20 s, and then slows steadily to rest in 12 s.
  **a** Draw the graph of speed, in metres per second, against time
     in seconds for this journey.
  **b** Use your graph to find the cycle-taxi's initial acceleration.
  **c** What was the final acceleration?
  **d** What was the mean speed over the 42 s journey?

2 The acceleration of the first part of the journey shown is 3 m/s².
  **a** Find the maximum speed $S$ metres per second.
  **b** Find the total distance travelled.
  **c** Find the mean speed of the whole journey.

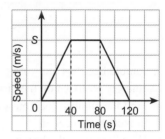

3 The speed–time graph shows an initial constant
  retardation of 2 m/s² for $t$ seconds.
  **a** Find the total distance travelled.
  **b** Find the deceleration at $3t$ seconds.
  **c** Find the mean speed of the whole journey.

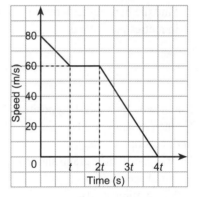

4 Sasha and Kim race over $d$ metres. Sasha accelerates from rest for 6 s to a speed of 8 m/s, which
  she maintains for the next 40 s before she tires and uniformly decelerates at $\frac{4}{7}$ m/s² until she stops.
  Kim accelerates from rest for 4 s to a speed of 8 m/s, which she maintains until 44 s have elapsed
  before she also tires and uniformly decelerates to a stop at $\frac{1}{2}$ m/s².
  **a** Draw the speed–time graph in metres per second and seconds for both girls on the same axes.
  **b** Use your graph to find who wins the race.
  **c** What was the mean speed for each runner?
  **d** Over what distance, in m, do the girls race?
  **e** Who is in the lead after (i) 100 m?  (ii) 300 m?

5 Explain carefully why this graph could never
  represent the distance–time or speed–time graph for
  a bumble-bee's journey over the course of a day.

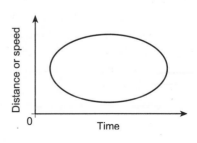

# Exercise 65 (Revision)

**1** The graph shows the journeys of Cheri and Felix, who went on a cycling trip.
  **a** How long did Cheri stop?
  **b** At what time did Felix start?
  **c** Find Cheri's mean speed.
  **d** How far apart were they at 10:20?

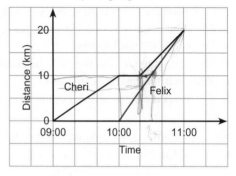

**2** A bumble-bee flies out from its nest to some flowers, and returns to its nest some time later. Its journey is shown on the distance–time graph.
  **a** Find the bumble-bee's outward journey speed in metres per second.
  **b** How long does the bee stay at the flowers?
  **c** Find the bumble-bee's return journey speed in metres per second.

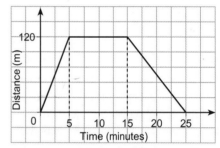

**3** Robin sails from a resting position in his boat to a uniform speed of 4 m/s in 30 s.
He then remains at this speed for a further 60 s, before he slows down at a constant retardation until he stops 15 s later.
Draw a speed–time graph showing this journey, and use it to find Robin's
  **a** initial acceleration
  **b** acceleration at 60 s
  **c** retardation
  **d** mean speed for the whole journey

**4** The speed–time graph illustrates the journey of a cyclist.
  **a** Find the distance travelled in the first 50 s.
  **b** Find the total distance travelled.
  **c** Find the mean speed of the cyclist.
  **d** Find the acceleration when $t = 80$ s.

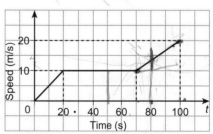

# Exercise 65★ (Revision)

**1** This distance–time graph shows the journeys of Elisa and Albert, who cycle from their houses to meet at a lake. After staying at the lake for 90 minutes, Elisa cycles back home at 8 km/hour and Albert returns home at 12 km/hour.

   **a** Copy the graph and represent these facts on your graph.

   **b** When did each person arrive home?

   **c** What is the mean speed in m/s for each cyclist? (Exclude the stop.)

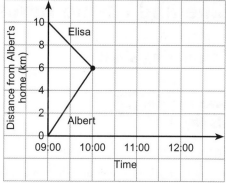

**2** A squash ball is hit against a wall by Ramesh. He remains stationary throughout the ball's flight, as shown in the distance–time graph.

   **a** Find the speed with which the ball approaches the wall.

   **b** When does the ball pass Ramesh, and at what speed?

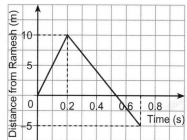

**3** This speed–time graph illustrates the speed of a lorry in metres per second.

Which of these statements are true?

Which are false?

Show working to justify your answers.

   **a** The initial acceleration over the first 30 s changes.

   **b** The braking distance is 150 m.

   **c** The mean speed for the whole journey is 12.5 m/s.

   **d** The greatest speed is 60 km/hour.

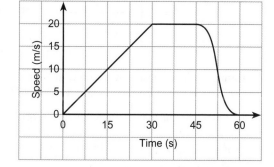

**4** This speed–time graph is for a toy racing car. The initial retardation is $2 \, \text{m/s}^2$.

   **a** Find the total distance travelled.

   **b** Find the deceleration at $6t$ seconds.

   **c** Find the mean speed of the whole journey.

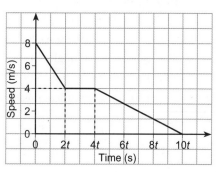

# Shape and space 3

## Sine and cosine ratios

### Calculating sides

**Activity 16**

Triangles P, Q and R are all similar.

♦ For each triangle, measure the three sides in millimetres, and then copy and complete the table.

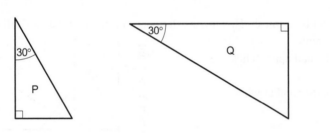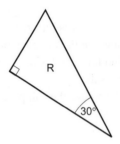

| Triangle | Side length (mm) | | |
|---|---|---|---|
| | *o* Opposite to 30° | *a* Adjacent to 30° | *h* Hypotenuse |
| P | | | |
| Q | | | |
| R | | | |

♦ Calculate the *o* : *h* ratio for each of the three triangles P, Q and R. This ratio is the same for *any* similar right-angled triangle. In this case, for 30°, it is 0.5. It is called the **sine ratio** of 30°, or **sin 30°**.

♦ Use a calculator to complete this table correct to 3 significant figures.

| *d* | 0° | 15° | 30° | 45° | 60° | 75° | 90° |
|---|---|---|---|---|---|---|---|
| **sin *d*** | | | 0.500 | | 0.866 | | |

♦ Calculate the *a* : *h* ratio for each of the three triangles P, Q and R. This ratio is the same for any similar right-angled triangle. In this case, for 30°, it is equal to 0.866 (to 3 decimal places) and is called the **cosine ratio** of 30°, or **cos 30°**.

♦ Use a calculator to complete this table correct to 3 significant figures.

| *d* | 0° | 15° | 30° | 45° | 60° | 75° | 90° |
|---|---|---|---|---|---|---|---|
| **cos *d*** | | | 0.866 | | 0.500 | | |

# Key Points

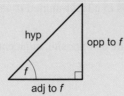

$$\sin f = \frac{\textbf{opposite side}}{\textbf{hypotenuse}} = \frac{o}{h}$$

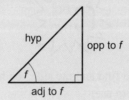

hyp

opp to f

f

adj to f

 **S** $\frac{\text{opp}}{\text{hyp}}$

$$\cos f = \frac{\textbf{adjacent side}}{\textbf{hypotenuse}} = \frac{a}{h}$$

hyp

opp to f

f

adj to f

**C** $\frac{\text{adj}}{\text{hyp}}$

---

## Example 1

Find the value of $p$ correct to 3 significant figures.

$$\sin 32° = \frac{p}{10}$$

$$10 \times \sin 32° = p$$

$$p = 5.30 \ \text{(to 3 s.f.)}$$

 **S** $\frac{\text{opp}}{\text{hyp}}$

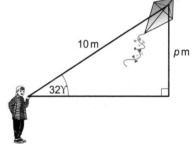

10 m

$p$ m

32°

$\boxed{1}\ \boxed{0}\ \boxed{\times}\ \boxed{\sin}\ \boxed{3}\ \boxed{2}\ \boxed{=}\ \boxed{5.29919}$ (to 6 s.f.)

So the height of the kite is 5.30 metres.

## Example 2

Find the length of side $q$ correct to 3 significant figures.

$$\cos 26° = \frac{q}{35}$$

$$35 \times \cos 26° = q$$

$$q = 31.5 \ \text{(to 3 s.f.)}$$

**C** $\frac{\text{adj}}{\text{hyp}}$

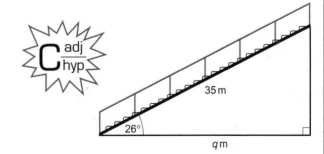

35 m

26°

$q$ m

$\boxed{3}\ \boxed{5}\ \boxed{\times}\ \boxed{\cos}\ \boxed{2}\ \boxed{6}\ \boxed{=}\ \boxed{31.4578}$ (to 6 s.f.)

So the length of the side is 31.5 m.

### Remember

When using trigonometrical ratios in a right-angled triangle, it is important to choose the correct one.

- ◆ Identify the sides of the triangle as opposite, adjacent or hypotenuse to the angle you are interested in.

- ◆ Write down these ratios.

- ◆ Mark off the side you have to find and the side you have been given. The ratio with the two marks is the correct one to use.

### Example 3

Find the length $y$ cm of the minute hand correct to 3 significant figures.

$$\cos 43° = \frac{75}{y}$$

$$y \times \cos 43° = 75$$

$$y = \frac{75}{\cos 43°}$$

$$y = 103 \text{ (to 3 s.f.)}$$

```
7  5  ÷  cos  4  3  =   102.550
```
(to 6 s.f.)

So the length of the minute hand is 103 cm.

## Exercise 66

Give your answers to the questions in this exercise correct to 3 significant figures.

For Questions 1–6, find $x$.

**1**

**2**

**3**

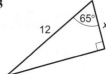

**4**

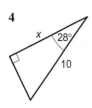

**5**

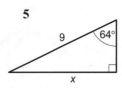

**6**

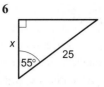

For Questions 7–12, find the lengths of the sides marked $y$.

**7**

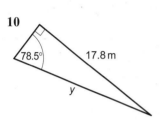

**8**

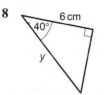

**9**

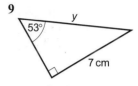

**10**

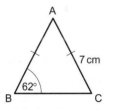

**11**

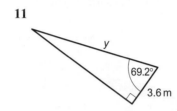

**12**

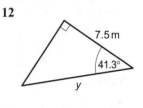

**13** A 3.8 m ladder making a 65° angle with the ground rests against a vertical wall.
Find the distance of the foot of the ladder from the wall.

**14** A kite is at the end of a 70 m string. The other end of the string is attached to a point on level ground. The string makes an angle of 75° with the ground.
At what height above the ground is the kite flying?

# Exercise 66★

Give your answers to the questions in this exercise correct to 3 significant figures.

**1** Find BC in this isosceles triangle.

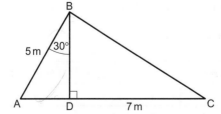

**2** Find LM in this isosceles triangle.

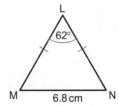

**3** Calculate
  **a** AD
  **b** BD
  **c** the area ABC
  **d** the angle C

**4** Find the area of an equilateral triangle of perimeter 12 m.

5 The cable car climbs at 48° to the horizontal
   up the mountainside. BC = 52 m and DE = 37 m.
   Calculate
   **a** the total length of the cable AC
   **b** the vertical height gained from C to A

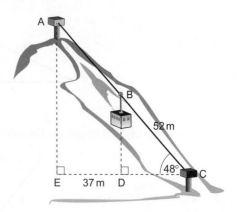

6 A submarine dives at a constant speed of 10 m/s at a diving angle measured from the vertical of
   75°. If the submarine starts its dive from the surface, how deep is the front end of the submarine at
   the end of a 1-minute dive?

7 A regular cyclical pentagon is inscribed inside a circle of radius 5 cm. Find the length of one side.

8 The Petronas Tower II in Kuala Lumpur, Malaysia, is one of the tallest buildings in the world.
   From a position X on level ground, the angle of elevation to the top is 80°.
   Position Y lies 84.81 m further back from X in a direct line with the building.
   The angle of elevation from Y to the top is 70°. How high is the Petronas Tower II?

## Calculating angles

**Activity 17**

If you are given the **opposite** side to angle $f$ and the **hypotenuse** in a triangle, you can find **sin $f$**.
You can therefore use the [INV] [sin] buttons on a calculator to find the angle $f$.

If you are given the **adjacent** side to angle $f$ and the **hypotenuse**, you can find **cos $f$**. You can use the
[INV] [cos] buttons on a calculator to find the angle $f$.

♦ Check that the calculator is in degree mode.

♦ Copy and complete the table
  for $d$ and $b$ to the nearest
  degree.

| **sin $d$** | 0 | 0.259 | 0.500 | 0.866 | 0.966 | 1 |
|---|---|---|---|---|---|---|
| **$d$** | | | 30° | | | |
| **cos $b$** | 1 | 0.966 | 0.866 | 0.500 | 0.259 | 0 |
| **$b$** | | | | 60° | | |

## Example 4

Find angle $d$ correct to 3 significant figures.

$$\cos d = \frac{1.5}{2}$$

$$d = 41.4° \text{ (to 3 s.f.)}$$

$\boxed{\text{INV}}$ $\boxed{\text{cos}}$ $\boxed{(}$ $\boxed{1}$ $\boxed{.}$ $\boxed{5}$ $\boxed{÷}$ $\boxed{2}$ $\boxed{)}$ $\boxed{=}$ $\boxed{\text{4I.4096}}$ (to 6 s.f.)

## Example 5

Find angle $b$ correct to 3 significant figures.

$$\sin b = \frac{3.5}{4.3}$$

$$b = 54.5° \text{ (to 3 s.f.)}$$

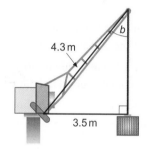

$\boxed{\text{INV}}$ $\boxed{\text{sin}}$ $\boxed{(}$ $\boxed{3}$ $\boxed{.}$ $\boxed{5}$ $\boxed{÷}$ $\boxed{4}$ $\boxed{.}$ $\boxed{3}$ $\boxed{)}$ $\boxed{=}$ $\boxed{\text{54.4840}}$ (to 6 s.f.)

## Remember

To find an angle in a right-angled triangle.

♦ Write down these ratios.

$$S\frac{\text{opp}}{\text{hyp}} \qquad C\frac{\text{adj}}{\text{hyp}} \qquad T\frac{\text{opp}}{\text{adj}}$$

♦ Mark off the sides of the triangle you have been given.

♦ The correct ratio to use is the one with two marks.

♦ Use the $\boxed{\text{INV}}$ and $\boxed{\text{sin}}$ , $\boxed{\text{cos}}$ or $\boxed{\text{tan}}$ buttons on a calculator to find the angle, having made sure that the calculator is in degree mode.

# Exercise 67

Give your answers to the questions in this exercise correct to 3 significant figures.

For Questions 1–6, find each angle marked *d*.

**1**

**2**

**3**

**4**
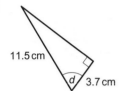

**5**
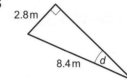

**6**
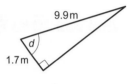

**7** A small coin is thrown off the Eiffel Tower in Paris. It lands 62.5 m away from the centre of the base of the 320 m-high structure. Find the angle of elevation from the coin to the top of the tower.

**8** A rectangle has a side of 10 m and diagonals of 25 m.
Find the angle between the longer side and the diagonal.

**9** A steam train travels at 20 km/h for 15 minutes along a straight track inclined at *f*° to the horizontal. During this time it has risen vertically by 150 m. Calculate the angle *f*°.

**10** A control-line toy aircraft at the end of a 15 m wire is flying in a horizontal circle of radius 5 m.
What angle does the wire make with the ground?

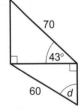

# Exercise 67★

Give your answers to the questions in this exercise correct to 3 significant figures.

For Questions 1–6, find each angle marked *d*.

**1**
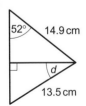

**2**
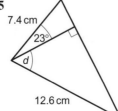

**3**
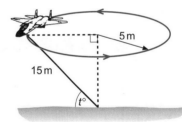

**4**
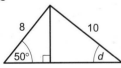

**5**

**6**
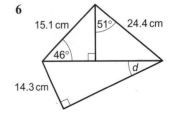

**7** Martina starts from point M. She cycles 15 km North, and then 4 km East. She finally stops at point P. Find the bearing of P from M, and then the bearing of M from P.

**8** A camera tripod has three equally spaced legs, each of length 1.75 m, around a circle of radius 52 cm on horizontal ground. Find the angle that the legs make with the horizontal.

**9** An isosceles triangle of sides 100 cm, 60 cm and 60 cm has a base angle $f$. Find $f$.

**10** ABCD is a quadrilateral. Angles BAD and DCB are both 90°. AB = 8 m, DB = 9 m and DC = 4.5 m. Calculate angle ABC.

## Investigate

- Investigate the value of $(\sin x)^2 + (\cos x)^2$ for all angles $x$. Show this result algebraically.

- Compare the value of $\dfrac{\sin x}{\cos x}$ for all angles $x$. Show this result algebraically.

- For an equilateral triangle with sides of 2 units, express the sine, cosine and tangent ratios of angles of 30° and 60° as exact fractions.

# Exercise 68

Give your answers to the questions in this exercise correct to 3 significant figures.
For Questions 1–6, find each side marked $x$.

**1**
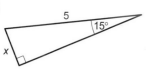

**2**
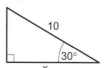

**3**
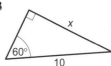

**4**
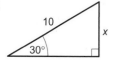

**5**
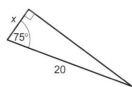

**6**
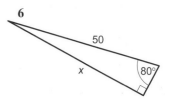

For Questions 7–10, find each angle marked $a$.

**7**

**8**
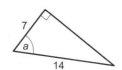

**9**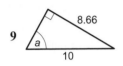

**10**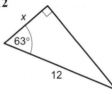

For Questions 11–22, find each marked side or angle.

**11**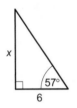

**12**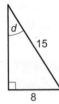

**13**

**14**

**15**

**16**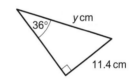

**17**

**18**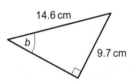

**19**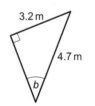

**20**

**21**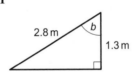

**22**

**23** Jake runs from position A on a constant bearing of 060° for 500 m.
How far North of A is he at the end of his run?

**24** A straight 20 m wheelchair ramp rises 348 cm.
Find the angle that the slope makes with the horizontal.

In Questions 25 and 26, the triangle ABC has a right angle at B.

**25** In triangle ABC, C = 23° and AC = 14 km. Calculate AB.

**26** In triangle ABC, AC = 9 m and AB = 4 m. Calculate angle BAC.

# Exercise 68★

Give your answers to Questions 1–10 correct to 2 significant figures.

For Questions 1–6, find $x$.

**1**

9, $x$, 30°

**2**
11, 60°, $x$

**3**
86.6, 60°, $x$

**4**
10, 60°, $x$, 30°

**5**
20, 60°, $x$, 30°

**6** Regular hexagon
$x$, 100

For Questions 7 and 8, find the angles marked $a$.

**7**
$a$, 70.7, 100

**8**
20°, $a$, 25, 50

**9** The area of the triangle is $1500\,\text{cm}^2$. Find the angle $a$.

60 cm, $a$, 100 cm

**10** Find the angle $a$ and sides $x$ and $y$.

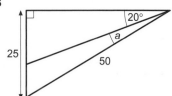
10, 5, $a$, $a$, $x$, $y$

Give your answers to Questions 11–15 correct to 3 significant figures.

**11** A hot-air balloon drifts for 90 minutes at a constant height on a bearing of 285° at a steady speed of 12 km/h. How far is it then from its starting position
  **a** North or South?
  **b** West or East?

**12** For this cross-section of a railway bridge, calculate the depth of the valley, and the length of the bridge.

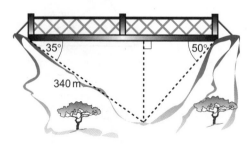

**13** Calculate the height $H$ of these stairs.

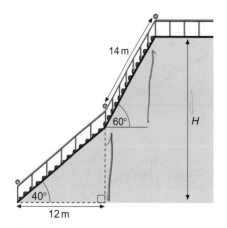

**14** A helicopter hovers in a fixed position 150 m above the ground. The angle of depression from the helicopter to church A (due West of the helicopter) is 32°. The angle of elevation from church B (due East of the helicopter) to the helicopter is 22°. Calculate the distance $d$ between the two churches.

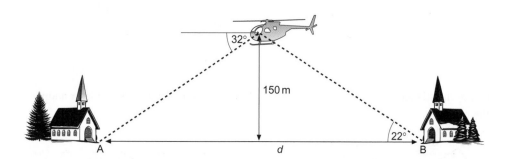

**15** A motorboat is 10 km South of a lighthouse and is on a course of 053°.
What is the shortest distance between the motorboat and the lighthouse?

**16** Gita sits on her garden swing, and swings. At the highest point, the angle that the 3 m chain makes with the vertical is 60°. Find the difference in height between the highest and lowest points of her ride. Give your answer correct to 2 s.f.

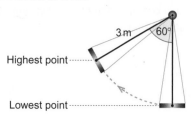

## Investigate

Draw on the same axes, the graphs of $y = \tan a$, $y = \sin a$ and $y = \cos a$ for $0° \leqslant a \leqslant 90°$.
Comment.

## Investigate

A swing is on a chain of length $x$, measured in metres. Show that the height (in metres) travelled by the swing above its lowest position, when the chain makes an angle of $a$ (degrees) to the vertical, is given by $h = x(1 - \cos a)$.

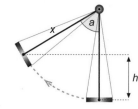

Investigate how $h$ changes with $a$ for swings of length $2\,$m, $3\,$m and $10\,$m.
What limits should you impose on angle $a$?

## Exercise 69 (Revision)

Give your answers to the questions in this exercise correct to 3 significant figures.

For Questions 1–6, find the length of the side marked $x$ and the size of angle $d$.

**1**

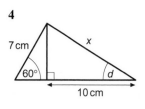

**2**

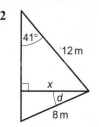

**3**

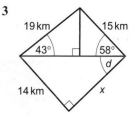

**4**

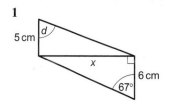

**5**

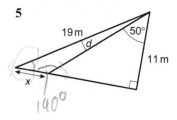

**6**

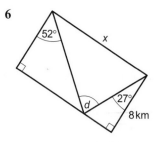

**7**  Calculate the area of an equilateral triangle of sides $10\,$cm.

**8**  The coordinates of triangle ABC are A(1, 1), B(7, 1) and C(7, 5). Calculate the value of angle CAB.

**9** The area of triangle ABC is 50 cm².
  **a** Find the value of sin *f*.
  **b** What is the size of angle *f*?

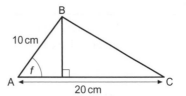

**10** The centre of the clock face is in a tower 20 m above the ground.
The hour hand is 1 m long.
  **a** How far above the ground is the end of the hour hand at 2am
  (*h* in the diagram)?
  **b** At 7am?
  **c** At 10.30am?

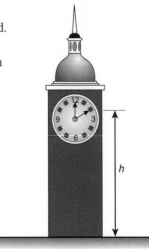

# Exercise 69★ (Revision)

**1** A scuba diver dives directly from A to B, and then to C, and then to the seabed floor at D. She then realises that she has only 4 minutes' worth of air left in her tank. She can ascend vertically at 10 cm/s. AB = 8 m, BC = 12 m and CD = 16 m. Can she reach the surface before her air supply runs out?

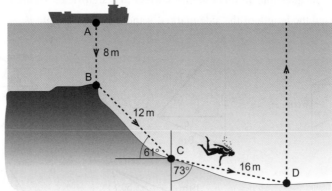

**2** A hiker walks from her base camp for 10 km on a bearing of 050°, and then walks a further 14 km on a new bearing of 140°. There is then a thunderstorm, and she decides to return directly back to base camp.
  **a** Find the distance and bearing of the return journey.
  **b** The hiker's speed is a constant 1.5 m/s. If her return journey starts at 15:00, at what time will she arrive back?

**3** A rectangular packing case ABCD is leaning against a vertical wall XY. AB = 2 m, AD = 3 m and angle DCZ = 25°. Calculate the height of A above the floor YZ.

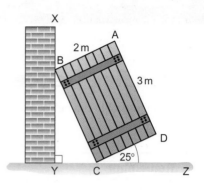

**4** A hare wants to cross a busy road. He measures the angle of elevation from the edge of the road to the top of a lamp-post directly on the other side as 25°. From a position 12 m further back from the road, the angle is 15°.

  **a** Calculate the width of the road.

  **b** The hare can scamper across the road at 1 m/s.
    Find the time he could take to cross the road.

  **c** The traffic on the road travels at 60 miles/hour.
    Find how far apart the vehicles must be for him to cross safely. (1 mile $\simeq$ 1600 m.)

**5** Given that $\sin 60° = \dfrac{\sqrt{3}}{2}$, show that side $x$ of the triangle is given by $p\sqrt{3}$, where $p$ is a whole number. Find the value of $p$.

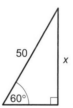

**6** Given that $\sin 60° = \dfrac{\sqrt{3}}{2}$ and $\cos 45° = \dfrac{1}{\sqrt{2}}$, show that side $x$ of the triangle is given by $q\sqrt{6}$ where $q$ is a whole number. Find the value of $q$.

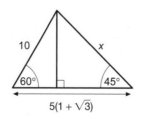

# Handling data 3

## Extending frequency tables into calculation tables

A calculation table can be used to organise large sets of data and this reduces the chance of errors being made.

> **Remember**
>
> - **Discrete data** are measurements, collected from a source, that can be listed and counted. For example, the number of absentees from school will be 1 or 2 or 3, and so on.
>
> - **Continuous data** are measurements, collected from a source, that cannot be listed and counted. For example, the weight of an average portion of cake could be anything from 100 g to 150 g.
>
> - The **mean** = $\dfrac{\text{total of all values}}{\text{total number of values}}$.
>
> - The **mode** is the value with the highest frequency.
>
> - The **median** is the value (or mean of the pair of values) in the middle, after the results have been sorted into ascending order.
>
> - The **range** = largest value − smallest value.
>
> - **Sigma** ($\Sigma$) means 'the sum of'.
>
> - The **mid-point** is the mean of the *exact* group boundaries.

## Averages from discrete data

> **Example 1**
>
> These tables show the results of rolling a dice 40 times.
>
> **Data**
>
> | | | | | |
> |---|---|---|---|---|
> | 1 | 3 | 5 | 3 | 5 |
> | 1 | 6 | 3 | 2 | 3 |
> | 4 | 2 | 4 | 1 | 6 |
> | 3 | 2 | 3 | 1 | 2 |
> | 2 | 1 | 6 | 2 | 4 |
> | 2 | 1 | 2 | 4 | 1 |
> | 1 | 3 | 5 | 2 | 3 |
> | 5 | 4 | 2 | 5 | 1 |
>
> $\rightarrow$
>
> **Frequency table**
>
> | Score $x$ | Frequency $f$ |
> |---|---|
> | 1 | 9 |
> | 2 | 10 |
> | 3 | 8 |
> | 4 | 5 |
> | 5 | 5 |
> | 6 | 3 |
> | | Total = 40 |
>
> $\rightarrow$
>
> **Calculation table**
>
> | Score $x$ | Frequency $f$ | $x \times f$ |
> |---|---|---|
> | 1 | 9 | $1 \times 9 = 9$ |
> | 2 | 10 | $2 \times 10 = 20$ |
> | 3 | 8 | $3 \times 8 = 24$ |
> | 4 | 5 | $4 \times 5 = 20$ |
> | 5 | 5 | $5 \times 5 = 25$ |
> | 6 | 3 | $6 \times 3 = 18$ |
> | | Total = 40 | Total = 116 |

The multiplication in the third column, $x \times f$, of the calculation table simply adds all the 1s together, then the 2s, and so on.

For this data, find the mean, the median and the mode.

$$\text{Mean} = \frac{\text{total of all values}}{\text{total number of values}} = \frac{116}{40} = 2.9.$$

The median (or middlemost value) is 3, as the twentieth and twenty-first values are both 3.

The mode (the score with the highest frequency) is 2.

## Mean from grouped data (discrete or continuous)

### Example 2

This frequency table gives the times of the first 60 runners in a cross-country race.

**Frequency table**

| Time (minutes) | Frequency |
|---|---|
| 17–18 | 4 |
| 18–19 | 7 |
| 19–20 | 8 |
| 20–21 | 13 |
| 21–22 | 12 |
| 22–23 | 9 |
| 23–24 | 7 |
| | $\Sigma = 60$ |

$\rightarrow$

**Calculation table**

| Time (minutes) | Mid-point $t$ | Frequency $f$ | $t \times f$ |
|---|---|---|---|
| $17 \leqslant t < 18$ | 17.5 | 4 | $17.5 \times 4 = 70$ |
| $18 \leqslant t < 19$ | 18.5 | 7 | $18.5 \times 7 = 129.5$ |
| $19 \leqslant t < 20$ | 19.5 | 8 | $19.5 \times 8 = 156$ |
| $20 \leqslant t < 21$ | 20.5 | 13 | $20.5 \times 13 = 266.5$ |
| $21 \leqslant t < 22$ | 21.5 | 12 | $21.5 \times 12 = 258$ |
| $22 \leqslant t < 23$ | 22.5 | 9 | $22.5 \times 9 = 202.5$ |
| $23 \leqslant t < 24$ | 23.5 | 7 | $23.5 \times 7 = 164.5$ |
| | | $\Sigma = 60$ | $\Sigma = 1247$ |

The exact values of the times have not been recorded, but the *boundaries* of the groups are defined *exactly* in the calculation table.

For example, the group '17–18' includes all the times from 17.00 to 17.99. (17.00 is in this group, 18.00 is in the next group.) So the group is defined exactly as $17 \leqslant t < 18$.

The mid-points of the groups are used to *estimate* the sum of the figures in each group. For example, we estimate each of the twelve times in the group '21–22' as 21.5. So the sum of these twelve times is estimated as $12 \times 21.5 = 258$. (In fact, some will be less than 21.5, and some will be more then 21.5, but 'on average' 258 will be a good estimate of the total.)

$$\textit{Estimate of the mean} = \frac{1247}{60} = 20.8 \text{ (to 3 s.f.)}.$$

## Group boundaries

Take care to find the *exact* boundaries. Data tables can be misleading.

♦ Is the data discrete or continuous?

♦ Has the data been 'rounded' from continuous to discrete as it was collected?

♦ How is age rounded differently from other quantities?

## Activity 18

Define the exact group boundaries, and find the class widths and mid-points for these sets of data.

♦ The times taken by 25 swimmers to breaststroke 50 m

| Time (s) | 55–60 | 60–65 | 65–70 | 70–75 |
|---|---|---|---|---|
| Frequency | 3 | 8 | 12 | 2 |

♦ The lengths of 50 earthworms to the nearest centimetre

| Length (cm) | 3–5 | 6–8 | 9–11 | 12–14 |
|---|---|---|---|---|
| Frequency | 16 | 17 | 12 | 5 |

♦ The noon temperature in degrees Celsius in Brighton over 20 days in May

| Temperature (°C) | 0–8 | 8–16 | 16–24 | 24–32 |
|---|---|---|---|---|
| Frequency | 3 | 5 | 11 | 1 |

♦ The weights of 20 babies born in a hospital ward on 1 January

| Weight (kg) | 1–2 | 2–3 | 3–4 | 4–5 |
|---|---|---|---|---|
| Frequency | 4 | 6 | 7 | 3 |

♦ The ages of children at a nursery/ primary school

| Age (years) | 2–3 | 4–5 | 6–7 | 8–9 |
|---|---|---|---|---|
| Frequency | 8 | 12 | 11 | 9 |

## Exercise 70

1 A dice has six faces numbered 1, 2, 3, 4, 5 and 6. It is thrown 30 times and the scores are recorded here.

| 1 | 6 | 3 | 2 | 3 | 2 | 1 | 2 | 4 | 1 |
|---|---|---|---|---|---|---|---|---|---|
| 4 | 2 | 4 | 1 | 6 | 1 | 3 | 5 | 2 | 6 |
| 3 | 2 | 3 | 1 | 2 | 5 | 4 | 2 | 5 | 1 |

Construct a calculation table with three columns.

**a** Work out the frequencies of each score and add them to your table.

**b** Calculate the mean score.

**c** Write down the median score.

**2** The number of advertisements was recorded during thirty commercial breaks on a TV channel one evening, and are shown here.

| 4 | 3 | 8 | 4 | 7 | 5 | 7 | 4 | 6 | 5 |
| 7 | 5 | 5 | 3 | 7 | 4 | 6 | 6 | 3 | 6 |
| 4 | 5 | 4 | 6 | 4 | 8 | 3 | 4 | 5 | 8 |

Construct a calculation table with three columns.

**a** Work out the frequencies of each result and add them to your table.

**b** Calculate the mean number of advertisements per break.

**c** Write down the median number of advertisements per break.

**3** The sizes of each family living in a block of flats are shown in the table.

**a** How many families are there?

**b** Construct a calculation table to work out the mean number of children per family.

**c** What is the median number of children per family?

| No. of children $x$ | Frequency $f$ |
|---|---|
| 0 | 12 |
| 1 | 14 |
| 2 | 16 |
| 3 | 6 |
| 4 | 2 |

**4** A traffic survey recorded the number of people in cars at a busy junction during the morning rush hour.

**a** How many cars were photographed?

**b** Construct a calculation table to work out the mean number of occupants per car.

**c** Find the median number of occupants per car.

| Occupants of car $x$ | Frequency $f$ |
|---|---|
| 1 | 41 |
| 2 | 23 |
| 3 | 9 |
| 4 | 5 |
| 5 | 2 |

**5** A schoolmaster recorded the number of times each pupil was late for his mathematics class for half a term.

**a** How many pupils are there in the class?

**b** Construct a calculation table to work out the mean number of times late per pupil.

**c** How many times late do you think his pupils were allowed before they were punished?

| No. of times late $L$ | No. of pupils $f$ |
|---|---|
| 0 | 11 |
| 1 | 8 |
| 2 | 6 |
| 3 | 2 |
| 4 | 2 |
| 5 | 1 |

6 A questionnaire filled in by all the students at a college who had passed their driving test yielded these results.

a How many students completed the questionnaire?

b Construct a calculation table to work out the mean number of attempts required to pass the test.

c What percentage of this group needed more than two attempts to pass?

| No. of driving tests $t$ | No. of students $f$ |
|---|---|
| 1 | 19 |
| 2 | 29 |
| 3 | 18 |
| 4 | 9 |
| 5 | 5 |

7 The heights of all the players in a basketball club are shown in the table.

a How many players are there in the club?

b Write down the mid-point of the smallest group.

c Construct a calculation table and estimate the mean height of a player at this club.

| Height $h$ (cm) | No. of players $f$ |
|---|---|
| $165 \leqslant h < 170$ | 2 |
| $170 \leqslant h < 175$ | 4 |
| $175 \leqslant h < 180$ | 5 |
| $180 \leqslant h < 185$ | 6 |
| $185 \leqslant h < 190$ | 7 |
| $190 \leqslant h < 195$ | 7 |
| $195 \leqslant h < 200$ | 4 |

8 The delay times of all the charter flights leaving an airport were recorded on one day at the height of the holiday season.

a How many flights were there?

b Write down the mid-point of the group $0 \leqslant d < 30$.

c Construct a calculation table and estimate the mean flight delay time on this day.

| Delay $d$ (mins) | No. of flights $f$ |
|---|---|
| $0 \leqslant d < 30$ | 6 |
| $30 \leqslant d < 60$ | 12 |
| $60 \leqslant d < 90$ | 16 |
| $90 \leqslant d < 120$ | 13 |
| $120 \leqslant d < 150$ | 8 |
| $150 \leqslant d < 180$ | 5 |

## Exercise 70★

1 The weights of 20 babies born in a hospital ward on 1 January are recorded in the table.

Construct a calculation table and estimate the mean weight.

| Weight $w$ (kg) | Frequency $f$ |
|---|---|
| $1 \leqslant w < 2$ | 4 |
| $2 \leqslant w < 3$ | 6 |
| $3 \leqslant w < 4$ | 8 |
| $4 \leqslant w < 5$ | 2 |

2   Conservationists monitor crocodiles in their natural habitat using aerial photographs. The lengths of 30 crocodiles are recorded in the table.

Construct a calculation table and estimate the mean length of the crocodiles.

| Length $l$ (cm) | Frequency $f$ |
|---|---|
| $170 \leqslant l < 180$ | 4 |
| $180 \leqslant l < 190$ | 3 |
| $190 \leqslant l < 200$ | 11 |
| $200 \leqslant l < 210$ | 7 |
| $210 \leqslant l < 220$ | 5 |

3   The waiting times, to the nearest minute, of patients at a morning surgery are shown in the table.

Construct a calculation table giving the exact boundaries of each group and estimate the mean waiting time.

| Waiting time $t$ (mins) | No. of patients $n$ |
|---|---|
| 5–9 | 7 |
| 10–14 | 8 |
| 15–19 | 5 |
| 20–24 | 5 |
| 25–29 | 4 |
| 30–34 | 1 |

4   A class of girls were timed over 400 m as part of a fitness test at school. The times were measured to the nearest second.

Construct a calculation table giving the exact boundaries of each group and estimate the mean time for the class.

| Time $t$ (s) | No. of pupils $n$ |
|---|---|
| 60–69 | 1 |
| 70–79 | 5 |
| 80–89 | 9 |
| 90–99 | 7 |
| 100–109 | 4 |
| 110–119 | 3 |
| 120–129 | 1 |

5   The ages of women giving birth in a hospital during January 2000 are given in the table.

| Age (years) | Frequency |
|---|---|
| 16–21 | 19 |
| 22–27 | 64 |
| 28–33 | 51 |
| 34–39 | 26 |

Construct a calculation table giving the exact boundaries of each group, and estimate the mean age of the mothers.

6   The times of cross-country runners in a race were distributed as shown in the table.

| Time $t$ (mins) | Frequency $f$ |
|---|---|
| 11.5–14.5 | 3 |
| 14.5–17.5 | 7 |
| 17.5–20.5 | 11 |
| 20.5–23.5 | 4 |

Construct a calculation table giving the exact boundaries of each group, and estimate the mean time, correct to the nearest second.

7   The UK National Lottery draws six numbered
    balls and a bonus ball. Balls are coloured white,
    blue, pink, green or yellow. Note that the white
    group has nine, and not ten, balls.
    The data in this table shows the bonus ball
    numbers drawn in the first 415 UK National
    Lottery draws.
    Draw up a calculation table, and calculate the
    mean value of the bonus ball number.

| Colour | Number $x$ | Frequency $f$ |
|--------|-----------|---------------|
| White  | 1–9       | 73            |
| Blue   | 10–19     | 70            |
| Pink   | 20–29     | 92            |
| Green  | 30–39     | 80            |
| Yellow | 40–49     | 100           |
|        |           | Total = 415   |

8   This table shows the numbers drawn (excluding
    the bonus ball) in the first 415 UK National
    Lottery draws.
    Draw up a calculation table, and calculate the
    mean value of the numbers.

| Colour | Number $x$ | Frequency $f$ |
|--------|-----------|---------------|
| White  | 1–9       | 471           |
| Blue   | 10–19     | 480           |
| Pink   | 20–29     | 520           |
| Green  | 30–39     | 493           |
| Yellow | 40–49     | 526           |

## Exercise 71 (Revision)

1   The share price of Cyplex over a 28-day period is recorded here. Construct a calculation table with
    three columns.

    $13   $14   $12   $11   $10   $13   $13
    $13   $14   $12   $10   $9    $9    $13
    $10   $11   $13   $13   $10   $11   $9
    $12   $13   $13   $11   $12   $13   $12

    a   Work out the frequencies of the prices and add them to your table.
    b   Calculate the mean price over this period.
    c   Write down the median price.

2   The numbers of errors that a proofreader found in the proofs
    of a new novel are given in the table.
    a   Draw up a calculation table. Work out the mean
        number of errors per page.
    b   If the reader is paid 20p for each error that he finds,
        how much does he earn from the book?

| No. of errors $x$ | No. of pages $n$ |
|-------------------|------------------|
| 0                 | 1                |
| 1–3               | 15               |
| 4–6               | 22               |
| 7–9               | 30               |
| 10–12             | 60               |
| 13–15             | 60               |
| 16–18             | 40               |
| 19–21             | 22               |

# Exercise 71★ (Revision)

**1** A government department, worried about fuel costs, surveyed the engine size of its fleet of cars. The results are given here.

Construct a calculation table and calculate an estimate of the mean engine size of the fleet.

| Engine size | No. of cars $f$ |
|---|---|
| 750–1000 cc | 4 |
| 1000–1300 cc | 13 |
| 1300–1600 cc | 9 |
| 1600–2000 cc | 6 |
| 2–3 litres | 5 |
| 3–5 litres | 3 |

**2** This table summarises Joseph's calls from the itemised telephone bill for July. His father allows him $5 worth of calls, but charges him for any excess.

**a** Construct a calculation table that will allow you to answer the rest of the question.

**b** Calculate an estimate for the total time that Joseph spent on the telephone.

**c** Calculate an estimate for the mean length, to the nearest second, of a call.

**d** Estimate the mean time, to the nearest second, that Joseph spent on the telephone each day.

**e** If calls cost 0.8 cents/minute, estimate how much Joseph owes his father.

| Duration of calls $t$ (mins) | No. of calls $f$ |
|---|---|
| $0 \leqslant t < 10$ | 20 |
| $10 \leqslant t < 20$ | 5 |
| $20 \leqslant t < 30$ | 4 |
| $30 \leqslant t < 40$ | 12 |
| $40 \leqslant t < 50$ | 15 |
| $50 \leqslant t < 60$ | 7 |
| $60 \leqslant t < 70$ | 4 |
| $70 \leqslant t < 80$ | 2 |
| | Total = 69 |

## Number

### Compound percentages

- To increase a number by $R\%$, multiply by $\left(1 + \dfrac{R}{100}\right)$.

  £50 increased by 14% = £50 × (1.14) = £57.

- To decrease a number by $R\%$, multiply by $\left(1 - \dfrac{R}{100}\right)$.

  £50 decreased by 14% = £50 × (0.86) = £43.

- To increase a number by $R\%$, $n$ successive times, multiply by $\left(1 + \dfrac{R}{100}\right)^n$.

  £50 increased by 14% per year for 3 successive years = £50 × (1.14)$^3$ = £74.08.

- To decrease a number by $R\%$, $n$ successive times, multiply by $\left(1 - \dfrac{R}{100}\right)^n$.

  £50 decreased by 14% per year for 3 successive years = £50 × (0.86)$^3$ = £31.80.

### Multiples, factors and primes

**Multiples** of 6 are 6, 12, 18, 24, 30 …

**Factors** of 18 are 1, 2, 3, 6, 9 and 18. These are the only numbers that divide exactly into 18.

**Prime numbers** are 2, 3, 5, 7, 11, …. Prime numbers have no factors apart from 1 and themselves.

**Prime factors** of 18 are 2 and 3. These are the only primes that divide exactly into 18.

### Highest common factor (HCF) and lowest common multiple (LCM)

Find the HCF of 18 and 48.

Prime factors of 18 are 2 × 3 × 3. Prime factors of 48 are 2 × 2 × 2 × 2 × 3. Factors common to both numbers are 2 and 3. Thus the highest common factor of 18 and 48 is 2 × 3 = 6.

Find the LCM of 18 and 48.

The lowest number that both numbers divide into must contain 2, 3 and 3 and 2, 2, 2, 2 and 3. Thus the lowest common multiple of 18 and 48 is 2 × 2 × 2 × 2 × 3 × 3 = 144.

### Calculator work

$$\left(\frac{34.2 - 12.6}{3.1 - 2.9}\right)^5 = 1.47 \times 10^{10} \quad \text{(to 3 s.f.)}$$

$$\sqrt{\frac{4.35 \times 10^5}{6.47 \times 10^{-7}}} = 8.20 \times 10^5 \quad \text{(to 3 s.f.)}$$

# Algebra

## Simple factorising

$$2x^3 + 6x^5 = 2x^3(1 + 3x^2)$$

$$9xy^3 + 15x^3yz - 12x^2y^2 = 3xy(3y^2 + 5x^2z - 4xy)$$

## Equations with fractions

**1** Solve for $x$.

$$\frac{2(x-3)}{3} - \frac{x-1}{2} = 4 \qquad \text{(Multiply both sides by } 3 \times 2 = 6)$$

$$4(x-3) - 3(x-1) = 24 \qquad \text{(Expand out brackets)}$$

$$4x - 12 - 3x + 3 = 24 \qquad \text{(Add 9 to both sides)}$$

$$x - 9 = 24$$

$$x = 33 \qquad \text{(Check: } 20 - 16 = 4)$$

**2** Solve for $x$.

$$\frac{2}{x} = \frac{1}{3} \qquad \text{(Multiply both sides by } 3x)$$

$$\frac{2}{x} \times 3x = \frac{1}{3} \times 3x$$

$$6 = x \qquad \left(\text{Check: } \frac{2}{6} = \frac{1}{3}\right)$$

## Simultaneous equations

$$x + 2y = 5 \qquad (1) \qquad \text{(Multiply (1) by 3 to give equation (3))}$$

$$3x - 4y = 25 \qquad (2)$$

$$\underline{3x + 6y = 15} \qquad (3) \qquad \text{(Subtract equation (2) from equation (3))}$$

$$10y = -10$$

$$\Rightarrow \quad y = -1 \qquad \text{(Substitute into equation (1))}$$

$$x - 2 = 5 \qquad (1)$$

$$\Rightarrow \quad x = 7 \qquad \text{(Check by substitution into equation (2))}$$

$$21 - (-4) = 25 \qquad (2) \qquad \text{(Correct)}$$

# Graphs

## Distance–time graphs

Gradient of slope = speed

**Velocity**

Gradient OA $= \dfrac{10\,\text{m}}{2\,\text{s}} = 5\,\text{m/s}$

Gradient AB $= 0 = 0\,\text{m/s}$

Gradient BC $= \dfrac{-10\,\text{m}}{5\,\text{s}} = -2\,\text{m/s}$

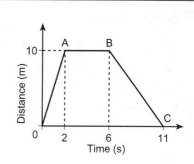

## Speed–time graphs

Gradient of slope = acceleration

Area under graph = distance travelled

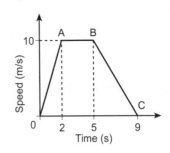

**Acceleration**

Gradient OA $= \dfrac{10\,\text{m/s}}{2\,\text{s}} = 5\,\text{m/s}^2$ (speeding up)

Gradient AB $= 0\,\text{m/s}^2$ (constant speed)

Gradient BC $= \dfrac{-10\,\text{m/s}}{4\,\text{s}} = -2.5\,\text{m/s}^2$ (slowing down)

**Distance travelled**

Average speed $= \dfrac{\text{distance travelled}}{\text{time}} = \dfrac{\frac{1}{2} \times (3 + 9) \times 10}{9} = 6\frac{2}{3}\,\text{m/s}$

# Shape and space

## Trigonometry

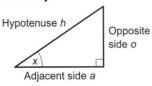

- $\sin x = \dfrac{o}{h}$

- $\cos x = \dfrac{a}{h}$

- $\tan x = \dfrac{o}{a}$

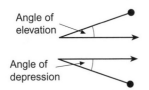

Bearings are measured from North and clockwise.

The top of a 30 m electric pylon is observed from a position X, 50 m away from the centre of its base. Find the angle of elevation of the top of the pylon from X.

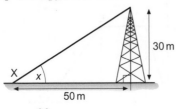

$\tan x = \dfrac{30}{50}$

$x = 31.0°$ (to 3 s.f.)

A ship travels on a bearing of 060° for 12 km. How far North is the ship from its starting point?

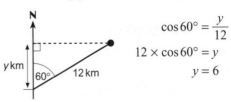

$\cos 60° = \dfrac{y}{12}$

$12 \times \cos 60° = y$

$y = 6$

So the ship is 6 km North of its starting point.

# Handling data

Grouped data: Estimate of mean $= \dfrac{\Sigma f \times t}{\Sigma f}$

$$= \frac{\text{sum of (frequency} \times \text{mid-point)}}{\text{sum of frequencies}}$$

| Time (s) | Mid-point $t$ | Frequency $f$ | $f \times t$ |
|----------|---------------|---------------|--------------|
| 20–25 | 22.5 | 5 | $5 \times 22.5 = 112.5$ |
| 25–35 | 30.0 | 15 | $15 \times 30.0 = 450$ |
| 35–40 | 37.5 | 10 | $10 \times 37.5 = 375$ |
| | | Sum = 30 | Total = 937.5 |

The mean is therefore $937.5 \div 30 = 31.25$ s.

**1** Calculate these.

**a** $15^3$

**b** $\dfrac{5.0 \times 10^7}{2.5 \times 10^2}$

**c** $\dfrac{2.5 + 3 \times 1.5}{2 \times 3.1 - 2.7}$

**d** $\sqrt{\dfrac{7.02 + 3.46}{3.51 - 0.89}}$

**c** $\sqrt{1.2 \times 10^5 \times \sin 60°}$

**d** $\sqrt{\dfrac{3.2 \times 10^{-2} \times 2.3 \times 10^2}{1.3 \times 10^{-3}}}$

**2 a** Find the highest common factors of
  (i) 15 and 21     (ii) 36 and 48
 **b** Find the lowest common multiples of
  (i) 6 and 8      (ii) 36 and 16

**3 a** Solve for $x$.
  (i) $3(x - 6) = 12 - 2x$   (ii) $\dfrac{3x - 1}{7} = 5$
 **b** Factorise
  (i) $x^2 + 7x$     (ii) $x^2y + 2xy^2 + 3xy$

**4** A motor boat starts from rest and accelerates uniformly for 10 s before it reaches a speed of 15 m/s. It maintains this speed for 30 s before it slows down uniformly, becoming stationary in a further 20 s.
 **a** Draw the speed–time graph.
 **b** Use this graph to find
  (i)  the distance travelled
  (ii) the acceleration over the first 10 s
  (iii) the deceleration over the last 20 s
  (iv) the mean speed for the journey in kilometres per hour

**5** The heights of 1000 army recruits were measured, with the results shown in the table.

| Height (cm) | 165–170 | 170–175 | 175–180 | 180–185 | 185–190 |
|---|---|---|---|---|---|
| Frequency | 100 | 250 | 400 | 200 | 50 |

Calculate, to 3 significant figures, an estimate of the mean height of the recruits.

**6** Calculate these, and give your answers to 3 significant figures.
 **a** $\dfrac{6.3}{4.2 - 1.8} + \dfrac{3.5}{1.7}$
 **b** $\left(\dfrac{1.2}{2.1} + \dfrac{2.1}{1.2}\right)^5$

**7** The costs for a school production of *Les Miserables* were as shown in the table.

| Script hire | 50 scripts at £4.50 each |
|---|---|
| Set design | £325 |
| Costumes | £250 |
| Programmes | £120 |

Tickets for the play cost £2 each, and the programmes were 50p each. The number of tickets sold for the three performances were as shown in the table.

| Day | Tickets | Programmes |
|---|---|---|
| Thursday | 156 | 90 |
| Friday | 165 | 110 |
| Saturday | 189 | 150 |

 **a** Calculate the total gain or loss made by the production.
 **b** Express this figure as a percentage of the production cost.
 **c** What price should the tickets have been if the school wanted to make a 50% profit?

**8 a** Solve these for $x$.
  (i) $\dfrac{5}{2x - 1} = \dfrac{1}{2}$   (ii) $\dfrac{2}{3x - 1} = \dfrac{5}{3x + 1}$
 **b** Factorise these.
  (i) $10x^3y^2 - 15x^2y^3$
  (ii) $12xy^2z^3 + 16x^3y^2z - 24xyz$
 **c** Find the dimensions of the two shapes given that they have the same area.

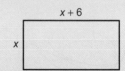

**9** Scottish mountains over 3000 feet are called Monroes. The frequency distribution of these Monroes is given in the table. 15

| Height (feet) | 3000 –3300 | 3300 –3600 | 3600 –3900 | 3900 –4200 | 4200 –4500 |
|---|---|---|---|---|---|
| Frequency | 300 | 135 | 80 | 20 | 5 |

  **a** Calculate an estimate of the mean height of a Monroe, being careful to show all your working.

  **b** Estimate the percentage of Monroes which are at least 4000 feet high.

**10** In the College Games, Michael Jackson won the 200 metres race in a time of 20.32 seconds.

  **a** Calculate his average speed in metres per second. Give your answer correct to 1 decimal place.

  **b** Change your answer to part **a** to kilometres per hour. Give your answer correct to 1 decimal place.

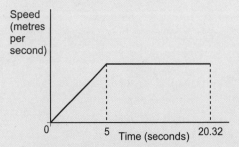

The diagram shows a sketch of the speed/time graph for Michael Jackson's race.

  **c** Calculate his maximum speed in metres per second. Give your answer correct to 1 decimal place.

  **d** Calculate his acceleration over the first 5 seconds. State the units in your answer. Give your answer correct to 2 significant figures.

**11** Solve the following simultaneous equations for $x$ and $y$.

  **a** $x + 2y = 12$
     $x - 3y = -13$

  **b** $2x + 3y = 3$
     $x - 5y = 8$

**12** A window cleaner uses a 5 m ladder. The ladder leans against the vertical wall of a building with the base of the ladder 1 m from the wall on level ground.
$\cos f° = 0.980$

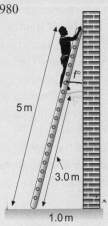

  **a** Calculate the height of the top of the ladder above the ground.

  **b** The window cleaner climbs 3 m up the ladder (see diagram). How far is his lower foot from the wall?

**13**

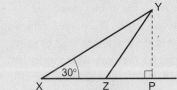

XYZ is an obtuse-angled triangle. Point P is the foot of the perpendicular from Y onto XZ produced. XY = 10 cm, XZ = 5 cm and angle ZXY = 30°. Calculate these, giving your answers correct to 2 significant figures:

  **a** The length XP

  **b** The length ZP

  **c** The length YP

  **d** The area of triangle XYZ.

**14** The diagram shows the positions of three telephone masts $A$, $B$ and $C$.

Mast $C$ is 5 kilometres due East of Mast $B$.

Mast $A$ is due North of Mast $B$, and 8 kilometres from Mast $C$.

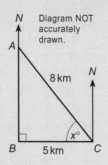

Diagram NOT accurately drawn.

**a** Calculate the distance of $A$ from $B$. Give your answer in kilometres, correct to three significant figures.

**b** (i) Calculate the size of the angle marked $x°$. Give your angle correct to one decimal place.

(ii) Calculate the bearing of $A$ from $C$. Give your bearing correct to one decimal place.

(iii) Calculate the bearing of $C$ from $A$. Give your bearing correct to one decimal place.

**a** Calculate, correct to 3 significant figures or to the nearest degree,
(i) $WN$, (ii) $ON$, (iii) $NS$, (iv) angle $ODS$.

**b** Explain briefly why, if angle $ODS$ were less than $90°$, a signal from the satellite would not be able to reach Delhi.

**c** A radio signal is sent from $W$ to $D$ via $S$. Calculate the distance it will have to travel.

**d** Radio signals travel at $300\,000$ km/s. How long does the signal take to travel from $W$ to $D$?

*MEG*

**15**

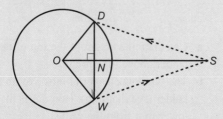

In the diagram, the circle, centre $O$, represents the Earth. The point $S$ represents a communications satellite used to pass radio messages between Washington ($W$) and Delhi ($D$). The line $OS$ intersects the line $WD$ at right angles at $N$.

$OW = 6370$ km.   $OS = 42\,300$ km.

Angle $WOD = 100°$.

# Number 4

## Inverse percentages

In Unit 3 you calculated simple percentages, starting with the original quantity. Now you will work back to find the original quantity.

> **Remember**
>
> To **increase** a number by $R\%$, multiply it by $(1 + \frac{R}{100})$.
>
> To **decrease** a number by $R\%$, multiply it by $(1 - \frac{R}{100})$.

---

**Example 1**

A house in Italy is sold for €138 000, giving a profit of 15%.

Find the original price that the owner paid for the house.

Let €$x$ be the original price.

$$x \times 1.15 = 138\,000 \quad \text{Selling price after a 15\% increase}$$
$$x = \frac{138\,000}{1.15}$$
$$x = €120\,000$$

**Example 2**

An ancient Japanese book is sold for ¥34 000 (yen), giving a loss of 15%.

Find the original price that the owner paid for the book.

Let ¥$x$ be the original price.

$$x \times 0.85 = 34\,000 \quad \text{Selling price after a 15\% increase}$$
$$x = \frac{34\,000}{0.85}$$
$$x = ¥40\,000$$

# Exercise 72

Give all your answers correct to 3 significant figures.

**1** Decrease $456 by 3%.

**2** Increase $876 by 5%.

**3** A price doubles. What is this, as a percentage increase?

**4** A rare stamp triples in value. What is this, as a percentage increase?

**5** A pair of shoes is sold for $48 at a profit of 20%. Find the original cost to the shop.

**6** A shirt is sold for $33 at a profit of 10%.
Find the original cost to the shop.

**7** Find the price of this chest of drawers
*before* the reduction.

**8** A picture is sold for $32, making a loss of 20%. Find the cost price of the picture.

**9** The height of a tree increases from 12 m to 13.6 m. Find the percentage increase.

**10** Shares drop in value from $651 to $542. Find the reduction as a percentage of the original price.

**11** A garden chair is bought for €68 and sold at a 12% profit. Find its selling price.

**12** A table is bought for €480 and sold at a loss of 15%. Find its selling price.

**13** The value of a carpet increases by 6% to €78.56. Find the original value.

**14** A house is sold for €234 000, which gives a profit of 12%. Find the actual profit.

**15** A computer loses 32% of its value in the first year. It is worth $1650 after one year.
How much was it bought for, to the nearest $?

**16** Find, as a percentage, how much smaller 52.8 is than 60.

**17** Find, as a percentage, how much larger 23.54 is than 22.

**18** 1 mile = 1.609 km (to 4 s.f.). Find the percentage error if 1 mile is taken as 1.6 km.

# Exercise 72★

Give all your answers correct to 3 significant figures.

**1** A portable digital radio is sold for $52.80, giving the shop a profit of 20%.
Find the cost to the shop of the radio.

**2** A garden chair is sold for $25.30 at a profit of 10%. Find the cost to the shop.

**3** The price of a house is reduced by 15% to $153 000. What was the original price of the house?

**4** A vase is bought for $23.50, which includes a profit of 17.5%. Calculate the amount of profit.

**5** A rare stamp is worth $81 after an increase in value of 35%. What was its value before the increase?

**6** A CD player is valued at \$220 after a decrease in value of 45%. What was its value before the decrease?

**7** In a chemistry laboratory, there are two beakers, each containing 500 ml of a liquid. 5% of beaker A is water and 4.5% of beaker B is water. As a percentage, how much more water is in beaker A?

**8** A seed is planted. It grows to a height of 45 cm, which is 72% of its maximum height. Find its maximum height.

**9** A farm is sold for €457 000, which gives a profit of 19%. Find the profit.

**10** Dr Peisner's salary increases by €450 to €45 600. Find his percentage salary increase.

**11** Miss Meg's salary increases by €450 to €1760. Find her percentage salary increase.

**12** Curry powder is sold at one store for 36 c/g (cents per gram) and at another for 45 c/g. Find the percentage saving with the better buy.

**13** A DVD recorder is reduced by 20% and then by a further 10% of the original price. What was the original price of the DVD recorder if it was sold for €324 after the second reduction?

**14** A price increases threefold. What is this as a percentage increase?

**15** A price decreases threefold. What is this as a percentage decrease?

**16** A shopkeeper makes 14% profit on all the items sold.
  **a** He sells five items, at \$3.40, \$4.60, \$5.98, \$6.07, and \$7.89, respectively.
  Find the cost to the nearest cent of each before the profit was added.
  **b** Explain why a quick method of calculating the answers is to multiply by 0.877 (3 d.p.).

**17** Find, as a percentage, how much smaller $\frac{7}{9}$ is than 91%.

**18** Find, as a percentage, how much larger $1.9 \times 10^{-3}$ is than $8.8 \times 10^{-4}$.

**19** The usual profit made by a jeweller on a gold watch is 35%. During a sale, 20% is taken off the marked price. What is the cost to the shop of an article that sells for €3672 during the sale?

**20** Calculate the percentage difference due to gender and comment.

| Events | 1965 | 1975 | 1985 | 1995 |
|---|---|---|---|---|
| Men's 100 m | 10.06 s | 9.95 s | 9.93 s | 9.84 s |
| Women's 100 m | 11.23 s | 11.07 s | 10.76 s | 10.49 s |
| Men's 1500 m | 3 minutes 35.6 s | 3 minutes 32.2 s | 3 minutes 29.7 s | 3 minutes 28.82 s |
| Women's 1500 m | 4 minutes 19.0 s | 4 minutes 01.4 s | 3 minutes 52.4 s | 3 minutes 50.46 s |

# Rounding

Rounding introduces errors. It is important not to round too much during the intermediate steps of a calculation, otherwise the final answer will be inaccurate.

---

**Example 3**

The full display on a calculator shows

$$\sqrt{2} = 1.4142136$$

Approximating to 1 decimal place and squaring gives

$$1.4 \times 1.4 = 1.96$$

Using 2 decimal places gives

$$1.41 \times 1.41 = 1.9881$$

Using the memory function produces the exact value of 2 for $\sqrt{2} \times \sqrt{2}$.

---

**Remember**

Always write down your working to at least one more significant figure (or decimal place) than you are asked for in the answer. Do not round your answers until all the working has been done.

---

# Upper and lower bounds

We are often given information that is not exact.

For example, a scientific journal may state that 'the Earth is moving around the Sun at 66 000 miles/hour'. If this figure has been rounded to the nearest 1000 miles/hour, the exact speed will lie between 65 500 miles/hour and 66 500 miles/hour.

This range can be shown on a number line.

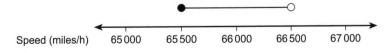

The range can also be written as 66 000 ± 500 miles/hour.

## Example 4

The populations rounded, to the *nearest thousand*, of Cardiff and Nottingham are given as 274 000 and 271 000 respectively. Find the population difference.

The maximum population and the minimum population are shown in this table

| City | Maximum population | Minimum population |
|------|--------------------|--------------------|
| **Cardiff** | 274 500 | 273 500 |
| **Nottingham** | 271 500 | 270 500 |

and on this diagram.

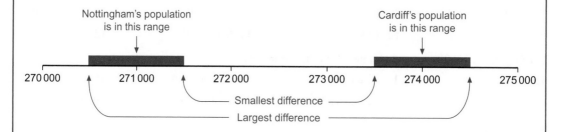

The smallest population difference is 273 500 − 271 500 = 2000.

The largest population difference is 274 500 − 270 500 = 4000.

The *exact* population difference lies between 2000 and 4000.

## Example 5

If $c = \dfrac{a}{b}$, and $a = 2.3 \pm 0.1\,\text{cm}$ and $b = 4.5 \pm 0.5\,\text{cm}$, find the maximum and minimum values of $c$.

For the maximum value of $c$, $a$ must be as large as possible, and $b$ must be as small as possible. Therefore $c$ (maximum) $= \dfrac{2.4}{4} = 0.6$.

For the minimum value of $c$, $a$ must be as small as possible, and $b$ must be as large as possible. Therefore $c$ (minimum) $= \dfrac{2.2}{5} = 0.44$.

$c$ can be expressed as $0.52 \pm 0.08$.

UNIT 4 ◆ Number

# Exercise 73

For Questions 1–4, copy and complete this table.

| | Given dimension (to nearest) | Maximum dimension | Minimum dimension | Dimension written as $a \pm b$ |
|---|---|---|---|---|
| 1 | 230 m (to 10 m) | | | |
| 2 | 70 kg (to 10 kg) | | | |
| 3 | 74 °F (to 1 °F) | | | |
| 4 | 19 m² (to 1 m²) | | | |

For Questions 5–12, write each interval as 'between … and … '.

**5** $3.5 \pm 0.1$        **6** $12.6 \pm 0.5$        **7** $5.6 \pm 0.01$        **8** $10.2 \pm 0.05$

**9** $4.9 \pm 0.2$        **10** $12.9 \pm 0.3$        **11** $6.5 \pm 0.01$        **12** $13.2 \pm 0.05$

For Questions 13–20, write each interval in the form $a \pm b$.

**13** 2.5 to 2.7        **14** 12.2 to 12.3        **15** 21.3 to 21.4        **16** 8.15 to 8.25

**17** 2.1 to 2.3        **18** 12.7 to 12.9        **19** 46.3 to 46.4        **20** 9.35 to 9.45

# Exercise 73★

For Questions 1–6, write down the minimum and maximum values for each measurement, to the given degree of accuracy.

**1** 6, 17, 123, to the nearest unit.

**2** 7, 40, 700, to 1 significant figure.

**3** 2.5, 14.5, 146.0, to the nearest 0.5.

**4** 50, 230, 4560, to the nearest 10.

**5** 0.2, 7.6, 12.4, to the nearest 0.2 of a unit.

**6** 0.34, 7.23, 12.89, to 2 decimal places.

**7** A sheep is weighed by a farmer as 43 kg to the nearest kg.
What is the sheep's greatest and least possible weight?

**8** An oil Sheikh estimates his wealth to be $2.2 \times 10^7$ to the nearest million dollars.
What is his greatest and least possible worth?

**9** If these two metal strips are placed end to end, find the maximum and minimum lengths.

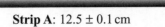

**Strip A**: $12.5 \pm 0.1$ cm        **Strip B**: $5.6 \pm 0.1$ cm

If, instead, the strips are placed side by side, as shown, find the minimum difference in length.

UNIT 4 ◆ Number

10 A rectangular Indian carpet's dimensions are given, to the nearest metre, as 8 m by 10 m. Find the maximum and minimum values for the carpet's perimeter and its area.

11 If $p = \dfrac{x}{yz}$, and $x = 23.1 \pm 0.5$, $y = 12.1 \pm 0.3$ and $z = 1.2 \pm 0.1$, calculate correct to 3 significant figures the maximum and minimum values of $p$.

12 If $w = \dfrac{a + b}{c}$, and $a = 1.2 \pm 0.05$, $b = 3.7 \pm 0.03$ and $c = 1.1 \pm 0.1$, calculate correct to 3 significant figures the maximum and minimum values of $w$.

13 A circle has an area of $7.4\,\text{cm}^2$. Find, correct to 3 significant figures, its radius and its circumference.

14 A rectangle has a side length of 5.79 cm and an area of $39\,\text{cm}^2$. Find, correct to 3 significant figures, the other side length and the length of the diagonal.

# Estimating

On many occasions it is acceptable and desirable not to calculate an exact answer, but to make a sensible estimate.

You have already been approximating numbers to a certain number of decimal places (2.47 = 2.5 to 1 d.p.), significant figures (34.683 = 34.7 to 3 s.f.) and rounding to a fixed suitable value (12 752 = 13 000 to the nearest 1000). Estimation is usually done without a calculator and is performed in a way that makes the working simple!

## Activity 19

Write down the following estimates to the stated degree of accuracy.

| | |
|---|---|
| ♦ The length of your pen. | (nearest cm) |
| ♦ The area of this page. | (nearest cm$^2$) |
| ♦ The time for you to walk a mile. | (nearest 30 s) |
| ♦ The weight of a dog. | (nearest kg) |
| ♦ The population of your town. | (nearest 1000) |
| ♦ The acute angle between these lines. | (nearest degree) |

Check your estimates.

**Example 6**

Estimate the answers to these calculations.

**a**  $19.7 \times 3.1$

$$19.7 \times 3.1 \qquad \approx 20 \times 3$$
$$= 60 \qquad \text{(exact answer is 61.07)}$$

**b**  $121.3 \times 98.6$

$$121.3 \times 98.6 \qquad \approx 120 \times 100$$
$$= 12\,000 \qquad \text{(exact answer is 11\,960.18)}$$

**c**  $252.03 \div 81.3$

$$252.03 \div 81.3 \qquad \approx 240 \div 80$$
$$= 3 \qquad \text{(exact answer is 3.1)}$$

**d**  $(11.1 \times (7.8 - 5.1))^2$

$$(11.1 \times (7.8 - 5.1))^2 \qquad \approx (10 \times (8{-}5))^2$$
$$= (30)^2$$
$$= 900 \qquad \text{(exact answer is 898.2009)}$$

# Estimating using standard form

It is often useful to use standard form to work out an estimate. Make sure you can write a number in standard form. You can refer to the rules of indices on page 263.

**Example 7**

Change to 1 s.f.  Write in standard form  Use rules of indices

$$0.067\,68 \times 38\,750 \approx 0.07 \times 40\,000 = 7 \times 10^{-2} \times 4 \times 10^4 \approx 30 \times 10^{-2+4} = 30 \times 10^2 = 3000$$

$$0.0753 \div 0.003\,68 \approx \frac{0.08}{0.004} = \frac{8 \times 10^{-2}}{4 \times 10^{-3}} = 2 \times 10^{(-2)-(-3)} = 2 \times 10^1 = 20$$

**Example 8**

Use standard form to calculate an estimate of $\sqrt{3.3 \times 10^7}$.

$$\sqrt{3.3 \times 10^7} = \sqrt{33 \times 10^6} \qquad \text{(Change 33 to the nearest square whole number)}$$
$$\approx \sqrt{36 \times 10^6}$$
$$= \sqrt{36} \times \sqrt{10^6} = 6 \times 1000 = 6000$$

**Example 9**

Use standard form to work out an estimate for $(4.5 \times 10^7) + (4.5 \times 10^6)$.

Write your answer, correct to 1 significant figure, in standard form.

$$(4.5 \times 10^7) + (4.5 \times 10^6) = (45 \times 10^6) + (4.5 \times 10^6) \quad \text{(Change the index numbers to be the same)}$$
$$= 49.5 \times 10^6 \qquad \text{(Add like terms)}$$
$$\approx 5 \times 10^7$$

# Exercise 74

Estimate the answers to these.

**1** $3.1 \times 47.9$      **2** $5.1 \times 19.6$      **3** $23.2 \div 7.8$

**4** $394.82 \div 78.1$      **5** $(7.3 + 12.1) \times 15.9$      **6** $24.1 \times (17.9 - 8.3)$

**7** $79.868 \times 0.101$      **8** $1239.32 \times 0.24$      **9** $20.92 \div 0.11$

**10** $315.71 \div 0.53$

Use standard form to calculate these, giving each answer in standard form.

**11** $(2 \times 10^4) \times (3 \times 10^3)$      **12** $(4 \times 10^5) \times (3 \times 10^3)$      **13** $(8 \times 10^6) \div (2 \times 10^3)$

**14** $(6 \times 10^7) \div (3 \times 10^2)$      **15** $(2 \times 10^4) + (3 \times 10^3)$      **16** $(7 \times 10^4) + (2 \times 10^3)$

**17** $(9 \times 10^4) - (3 \times 10^3)$      **18** $(8 \times 10^6) - (3 \times 10^5)$      **19** $\sqrt{2.5 \times 10^5}$

**20** $\sqrt{8.1 \times 10^5}$

Use standard form to calculate an estimate for these, giving each answer in standard form correct to 1 significant figure.

**21** $2670 \times 760$      **22** $880 \times 3420$      **23** $8490 \div 56.9$

**24** $6830 \div 29.5$      **25** $(6.8 \times 10^6) + (2.3 \times 10^5)$      **26** $(4.1 \times 10^4) + (2.1 \times 10^5)$

**27** $(4.8 \times 10^6) - (3.2 \times 10^5)$      **28** $(9.1 \times 10^7) - (3 \times 10^6)$      **29** $\sqrt{6.3 \times 10^5}$

**30** $\sqrt{7.8 \times 10^7}$

# Exercise 74★

Estimate the answers to these.

**1** $\dfrac{3.1 \times 19.7}{14.8}$      **2** $\dfrac{52.1 \times 94.3}{4.1}$      **3** $1.98^3$      **4** $(4.1^3 \div 2.1^2) \times \sqrt{25.6}$

**5** Estimate the volume and surface area of a closed box of dimensions 7.9 cm × 5.1 cm × 14.8 cm.

**6** A square has sides of length 7.23 cm. Estimate the length of its diagonal.

**7** Estimate the area of the rectangle.

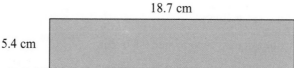

18.7 cm

5.4 cm

**8** Estimate the shaded area.

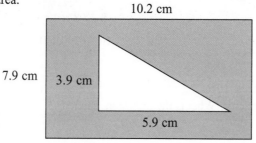

10.2 cm

7.9 cm

3.9 cm

5.9 cm

Use standard form to calculate these, giving each answer in standard form.

**9** $(3 \times 10^3) \times (4 \times 10^5)$      **10** $(6 \times 10^4) \times (4 \times 10^{-2})$      **11** $(6 \times 10^8) \div (3 \times 10^5)$

**12** $(3.5 \times 10^{-4}) \div (7 \times 10^{-2})$      **13** $(7 \times 10^8) + (6 \times 10^6)$      **14** $(2.8 \times 10^{-3}) - (7 \times 10^{-5})$

Use standard form to calculate an estimate for these, giving each answer as an ordinary number correct to 1 significant figure.

**15** $(1.2 \times 10^2) \times (4.5 \times 10^2)$      **16** $(6.8 \times 10^6) \times (3.1 \times 10^{-4})$      **17** $(7.98 \times 10^{-4}) \div (3.79 \times 10^{-3})$

**18** $(2.56 \times 10^{-3}) \div (6.28 \times 10^{-1})$      **19** $\sqrt{3.2 \times 10^{-3}}$      **20** $\sqrt{6.3 \times 10^7}$

**21** A warehouse sells 8.7 million cups at an average price of \$0.65 each. Use standard form to calculate an estimate of the total sales. Give your answer correct to 1 significant figure.

**22** A human kidney has a volume of $120\,cm^3$. It contains 1.35 million nephrons (small tubes). Use standard form to calculate an estimate of the average number of nephrons per cubic centimetre. Give your answer correct to 1 significant figure.

Use standard form to calculate an estimate for these, giving each answer in standard form, correct to 1 significant figure.

**23** $5003 \times 393$      **24** $0.041 \times 5700$      **25** $47.8 \times 0.0059$

**26** $0.00057 \times 0.00287$      **27** $\dfrac{80\,920}{0.004\,18}$      **28** $\dfrac{530}{0.000\,802}$

**29** $\dfrac{0.6597}{729.8}$      **30** $0.0678 \div 0.00321$

## Exercise 75 (Revision)

**1** Write down the multiplying factor to increase a number by 15%.

**2** Write down the multiplying factor to decrease a number by 5%.

**3** Abu sells her donkey for \$186, giving her a profit of 24%. Find the original price she paid for the creature.

**4** Khalid sells a diamond ring at a loss of 24% for \$912. Find the original price he paid for the item.

**5** Estimate the value of $\dfrac{2.1 \times 5.8 \times 3.1}{2.9 \times (11.9 - 8.8)}$ .

**6** Estimate the surface area of a square cube of side 9.876 cm.

For Questions 7–10, use standard form to calculate an estimate, giving each answer in standard form correct to 1 significant figure.

**7** $(1.4 \times 10^6) \times (3.7 \times 10^4)$      **8** $(7.8 \times 10^8) \div (3.7 \times 10^4)$

**9** $(3.6 \times 10^5) + (2.8 \times 10^4)$      **10** $\sqrt{790\,000}$

**11** Write $34 \pm 0.5$ as between '… and …'.

**12** $80\,ml$ is written to the nearest $10\,ml$.
What is the maximum capacity? What is the minimum capacity?

**13** 17.2 is written correct to 1 decimal place.
What is the maximum value? What is the minimum value?

**14** A square field is measured to be of side 135 m to the nearest 5 metres. Find its greatest and least possible area.

# Exercise 75★ (Revision)

**1** Increase 400 kg by 1.5%.

**2** Decrease 500 km by 0.5%.

**3** Mr Yeoh sells an antique rug for $1800, giving him a profit of 60%. Find the original price he paid for the rug.

**4** Mrs Wong sells a set of golf clubs for $704 at a loss of 12%. Find the original price she paid for the clubs.

**5** $125 is increased by 25%. By what percentage, correct to 3 significant figures, must $80 be increased to give the same result?

**6** A doctor and a nurse each get an 8% increase in salary. The doctor's new salary is $119 556. Find the nurse's old salary if her new salary is the same as the doctor's increase.

**7** The result of reducing $2y$ by $x$% is the same as increasing $y$ by $x$%. Find $x$.

**8** Estimate the volume of a cube if the diagonal of one of its faces is 10 cm.

For Questions 9–12, use standard form to calculate an estimate, giving each answer in standard form correct to 1 significant figure.

**9** $0.005\,89 \times 876\,000$

**10** $0.000\,287 \div 785.9$

**11** $\sqrt{0.000\,456}$

**12** $(3.56 \times 10^{-6}) - (8.34 \times 10^{-7})$

**13** Write the interval 3.25 to 3.35 in the form $a \pm b$.

**14** 6.83 is written to an accuracy of 2 decimal places.
What is its maximum value? What is its minimum value?

**15** The area of a circle is given as $45.5 \pm 0.5$ cm. Find, correct to 3 significant figures, the maximum possible radius and the minimum possible circumference.

**16** A surveyor measures the distance along the ground to the base of a tree to be 12 m to the nearest metre. If the angle of elevation to the top of the tree is 40° to the nearest degree, find
  **a** the greatest possible tree height
  **b** the least possible tree height

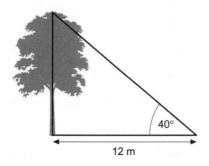

12 m

# Algebra 4

## Change of subject

It is sometimes helpful to write an equation, or formula, in a different way. For example, to draw the graph of the equation $2y - 4 = 3x$ it is easier to compile a table of values if $y$ is the subject, that is, on its own and on one side. Notice that the method used in Example 1 is the same as that used to solve a similar equation, as shown in Example 2.

---

**Example 1**

Make $y$ the subject in $2y - 4 = 3x$.

$2y - 4 = 3x$     (Add 4 to both sides)

$2y = 3x + 4$     (Divide both sides by 2)

$y = \dfrac{3x + 4}{2}$

$y = \dfrac{3}{2}x + 2$

**Example 2**

Solve $2y - 4 = 2$.

$2y - 4 = 2$     (Add 4 to both sides)

$2y = 6$     (Divide both sides by 2)

$y = 3$

---

The rearranged equation in Example 1 also allows you to state the gradient of the line, that is, $\frac{3}{2}$, and the $y$ intercept of $+2$. Both of these are required to *sketch* the graph of the line.

---

## Key Point

To rearrange an equation or formula, apply the same rule that is used to solve equations.
**Do the same operation to both sides.**

---

## Exercise 76

Make $x$ the subject of the equations.

**1** $x + 2 = a$      **2** $x + 4 = b$      **3** $x - p = 5$      **4** $x - a = 3$

**5** $c = x + a$      **6** $p = x - q$      **7** $5x = b$      **8** $q = 7x$

**9** $3x + a = b$      **10** $4x - p = q$      **11** $t - 2x = s$      **12** $-3x + d = f$

**13** $ax + b = 4$      **14** $px - q = 3$      **15** $f = ex - g$      **16** $a = bx + c$

**17** $x(a + b) = c$      **18** $q = (r - s)x$      **19** $8b + cx = d$      **20** $5q - rx = s$

**21** $3(x + b) = a$      **22** $d = 2(f + x)$      **23** $a(x + b) = c$      **24** $h(x - k) = f$

**25** $a = \dfrac{x}{b}$      **26** $\dfrac{x}{q} = r$      **27** $\dfrac{px}{q} = r$      **28** $c = \dfrac{bx}{d}$

**29** $p + q = \dfrac{x}{r}$      **30** $\dfrac{x}{a} = c - b$      **31** $d = \dfrac{x - b}{c}$      **32** $\dfrac{x + p}{q} = r$

# Exercise 76★

In Questions 1–22, make $x$ the subject of the equations.

**1** $ax + b = c$

**2** $ax - \pi = d$

**3** $d = \dfrac{x - b}{c}$

**4** $\dfrac{x + b}{a} = Q$

**5** $c = \dfrac{bx}{d}$

**6** $\dfrac{bx}{d} = P$

**7** $a(x + c) = e$

**8** $f = e(x - s)$

**9** $P = \pi x + b^2$

**10** $e = \dfrac{x}{a} - b^2$

**11** $\dfrac{bx}{d^2} = T$

**12** $\dfrac{x - b^2}{c} = a$

**13** $\pi - x = b$

**14** $TP = Q - x$

**15** $ab - dx = c$

**16** $\dfrac{b}{d} - x = P$

**17** $\dfrac{a}{x} = b$

**18** $t = \dfrac{s}{x}$

**19** $\dfrac{a + b}{x} = c$

**20** $r = \dfrac{p - q}{x}$

**21** $p = q + \dfrac{s}{x}$

**22** $\dfrac{v}{x} + s = t$

**23** Make $r$ the subject of the formula $A = 2\pi r$.

**24** Make $h$ the subject of the formula $V = \pi r^2 h$.

**25** Make $h$ the subject of the formula $V = \dfrac{1}{3}\pi r^2 h$.

**26** Make $h$ the subject of the formula $mgh = \dfrac{1}{2}mv^2$.

**27** Make $x$ the subject of the formula $y = mx + c$.

**28** Make $t$ the subject of the formula $v = u + at$.

**29** Make $s$ the subject of the formula $v^2 = u^2 + 2as$.

**30** Make $y$ the subject of the formula $a^2 x + b^2 y = d$.

**31** Make $a$ the subject of the formula $m = \dfrac{1}{2}(a + b)$.

**32** Make $h$ the subject of the formula $A = 2\pi r\,(r + h)$.

**33** Make $a$ the subject of the formula $S = \dfrac{a(1 - r^n)}{1 - r}$.

**34** Make $a$ the subject of the formula $s = ut + \dfrac{1}{2}at^2$.

**35** Make $a$ the subject of the formula $S = \dfrac{n}{2}\{2a + (n - 1)d\}$.

---

**Example 3**

Make $x$ the subject of the equation $ax^2 + b = c$.

$ax^2 + b = c$      (Subtract $b$ from both sides)

$ax^2 = c - b$      (Divide both sides by $a$)

$x^2 = \dfrac{c - b}{a}$      (Square root both sides)

$x = \sqrt{\dfrac{c - b}{a}}$

## Exercise 77

Make $x$ the subject of these equations.

**1** $ax^2 = b$

**2** $bx^2 = c$

**3** $\dfrac{x^2}{a} = b$

**4** $\dfrac{x^2}{y} = w$

**5** $x^2 + C = 2D$

**6** $x^2 - B = 3C$

**7** $\dfrac{x^2}{a} + b = c$

**8** $\dfrac{x^2}{y} - z = a$

**9** $ax^2 + 2b = c$

**10** $px^2 - q = r$

**11** $ax + dx = t$

**12** $bx - cx = D$

**13** $a(x - b) = x$

**14** $p(x - q) = x$

**15** $a(x + 1) = b(x + 2)$

**16** $d(x - e) = c(b - x)$

**17** Make $r$ the subject of the formula $A = 4\pi r^2$.

**18** Make $c$ the subject of the formula $E = mc^2$.

**19** Make $v$ the subject of the formula $a = \dfrac{v^2}{r}$.

**20** Make $x$ the subject of the formula $E = \dfrac{\lambda x^2}{2l}$.

**21** Make $r$ the subject of the formula $V = \dfrac{4}{3}\pi r^3$.

**22** Make $d$ the subject of the formula $F = \dfrac{k}{3\pi}d^3$.

**23** Make $l$ the subject of the formula $T = 2\pi\sqrt{l}$.

**24** Make $x$ the subject of the formula $\dfrac{F}{\pi} = 4\sqrt{x}$.

## Exercise 77★

Make $x$ the subject of these equations.

**1** $Rx^2 = S$

**2** $b = cx^2$

**3** $g = cx^2 + a$

**4** $\dfrac{x^2}{a} + c = b$

**5** $ax = bx - c$

**6** $px + q = tx$

**7** $c - dx = ex + f$

**8** $a(x - b) = c(d - x)$

**9** $\tan b + a(x + c) = x$

**10** $\cos b + ax = cx$

**11** $p = \sqrt{s + \dfrac{x}{t}}$

**12** $P = \dfrac{\sqrt{x + Q}}{R}$

**13** $\dfrac{Ab - x^2}{D} = a$

**14** $y = z - \dfrac{x^2}{a}$

**15** Make $r$ the subject of the formula $V = \frac{1}{3}\pi r^2 h$.

**16** Make $v$ the subject of the formula $E = \frac{1}{2}mv^2$.

**17** Make $v$ the subject of the formula $mgh = \frac{1}{2}mv^2$.

**18** Make $r$ the subject of the formula $F = G\frac{mM}{r^2}$.

**19** Make $x$ the subject of the formula $y = \frac{1}{a^2 + x^2}$.

**20** Make $x$ the subject of the formula $\frac{x^2}{a^2} + \frac{y^2}{b^2} = 1$.

**21** Make $a$ the subject of the formula $s = \frac{1}{12}(b - a)^2$.

**22** Make $d$ the subject of the formula $F = \frac{k}{d^3}$.

**23** Make $Q$ the subject of the formula $r = \frac{S}{\sqrt{PQ}}$

**24** Make $l$ the subject of the formula $T = 2\pi\sqrt{\frac{l}{g}}$.

**25** Make $d$ the subject of the formula $F = \frac{k}{\sqrt[3]{d}}$

**26** Make $x$ the subject of the formula $D = \frac{q^{-1}}{\sqrt[3]{x - 1}}$.

**27** Make $x$ the subject of the formula $y = \frac{1}{\sqrt{1 - x^2}}$.

**28** Make $c$ the subject of the formula $x = \frac{-b + \sqrt{b^2 - 4ac}}{2a}$.

**29** Make $x$ the subject of the formula $y = \frac{x + p}{x - p}$.

**30** Make $t$ the subject of the formula $m = \frac{1 + at}{1 - at}$.

# Using formulae

## Activity 20

The period $T$ seconds taken for a pendulum of length $L$ metres to swing to and fro is given by the formula

$$T = 2\pi\sqrt{\frac{L}{g}}$$

where $g$ is the acceleration due to gravity $= 9.81\,\text{m/s}^2$.

♦ What is the period of a pendulum 1 km long? (Give your answer correct to 1 decimal place.)

♦ Show that, when $L$ is made the subject, the formula becomes

$$L = g\left(\frac{T}{2\pi}\right)^2$$

♦ Use the rearranged formula to complete this table, giving your answers correct to 2 significant figures.

| Period | 1 s | 10 s | 1 minute | 1 hour | 1 day |
|---|---|---|---|---|---|
| Length | | | | | |

Here are some useful formula that you can use to calculate some more areas.

## Circles

The area of a circle with radius $r$ is $\pi r^2$.

Therefore the **area of a semicircle** with radius $r$ is $\dfrac{\pi r^2}{2}$.

The **perimeter of a shape** is the distance all the way around the shape.
For a circle, the perimeter is called its circumference.
The circumference of a circle with radius $r$ is $2\pi r$.

## Activity 21

Take a rectangular piece of paper and make it into a hollow cylinder.

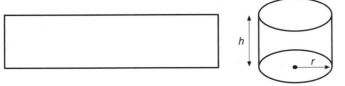

Use your paper to explain why the **curved surface area of your cylinder = $2\pi rh$**.

### Example 5

The volume $V$ m³ of a pyramid with a square base of length $a$ m and a height of $h$ m is given by $V = \dfrac{1}{3}a^2 h$.

**a** Find the volume of the Great Pyramid of Cheops, where $a = 232$ m and $h = 147$ m.

Substituting $a = 232$ m and $h = 147$ m gives
$V = \dfrac{1}{3} \times 232^2 \times 147 = 2\,640\,000$ m³ to 3 s.f.

**b** Another square-based pyramid has a volume of $853\,000$ m³ and a height of $100$ m. Find the length of the side of the base.

First make $a$ the subject of the formula: $a^2 = \dfrac{3V}{h} \Rightarrow a = \sqrt{\dfrac{3V}{h}}$

Substituting $V = 853\,000$ and $h = 100$ gives $a = \sqrt{\dfrac{3 \times 853\,000}{100}} \Rightarrow a = 160$ m to 3 s.f.

# Exercise 78

**1** The time $T$ minutes, to cook a joint of meat weighing $w$ kg is $T = 45w + 20$.
  **a** Find the time to cook a joint weighing 3 kg.
  **b** Find the weight of a joint of meat that took 110 minutes to cook.

**2** The cost $C$ in dollars of hiring a car for $d$ days and covering $k$ kilometres is $C = 50d + 0.2k$.
  **a** Find the cost of hiring a car for 7 days and covering 750 km.
  **b** If $d = 5$ and $C = 600$ find $k$.

**3** The area $A$ cm$^2$ of a triangle with base $b$ cm and height $h$ cm is given by $A = \frac{bh}{2}$.
  **a** Find $A$ when $b = 12$ and $h = 3$.
  **b** Find $h$ when $A = 25$ and $b = 5$.

**4** The distance $s$ m travelled by an object with acceleration $a$ m/s$^2$ for $t$ seconds is $s = \frac{1}{2}at^2$.
  **a** Find $s$ when $a = 10$ and $t = 4$.
  **b** Find $a$ when $t = 6$ and $s = 144$.
  **c** Find $t$ when $s = 27$ and $a = 6$.

**5** The increase in length $e$ cm of an aluminium rod of length $l$ cm when heated by $T$ degrees Celsius is $e = 0.00003lT$.
  **a** Find $e$ when $l = 100$ and $T = 50$.
  **b** Find $T$ if $l = 150$ and $e = 0.9$.

**6** The number $K$ kilowatts used by an electrical appliance with current $I$ amps and voltage $V$ volts is $K = \frac{IV}{1000}$.
  **a** Find $K$ if $I = 3$ and $V = 240$.
  **b** Find $I$ if $V = 110$ and $K = 1.65$.

**7** A formula to calculate income tax $I$ in \$ for a salary of \$$S$ is $I = 0.2(S - 8250)$.
  **a** Find $I$ if $S = 20\ 000$.
  **b** Find $S$ if $I = 3500$.

**8** The cost $C$ dollars of a pizza $d$ cm in diameter is $C = 0.03r^2$.
  **a** Find the cost of a pizza that is 22 cm in diameter.
  **b** What is the diameter of a pizza that costs \$20?

**9** Find the formulae for the area and the perimeter of the quarter circle shown.
  **a** Find the area and perimeter when $r = 5$ cm.
  **b** Find $r$ when $A = 50$.
  **c** Find $r$ when $P = 50$.
  **d** Find the value of $r$ that makes $A = P$.

**10** Find the formulae for the area and the perimeter of the semicircle shown.
  **a** Find the area and perimeter when $r = 10$ cm.
  **b** Find $r$ when $A = 75$.
  **c** Find $r$ when $P = 75$.
  **d** Find the value of $r$ that makes $A = P$.

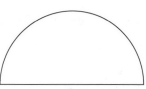

UNIT 4 ◆ Algebra

# Exercise 78★

**1** The cost $C$, in \$, of using a phone is $C = 0.02n + 18$ where $n$ is the number of units used.
  **a** Find $C$ when 200 units are used.
  **b** Find the number of units used if the cost is \$26.

**2** The area $A$ cm$^2$ of a ring with outer radius $R$ cm and inner radius $r$ cm is $A = \pi(R^2 - r^2)$.
  **a** Find $A$ if $R = 12$ and $r = 6$.
  **b** Find $R$ if $A = 37.7$ and $r = 2$.

**3** A formula to find the volume $V$ cm$^3$ of a cylindrical can with radius $r$ cm and height $h$ cm is $V = \pi r^2 h$.
  **a** Find the volume of a cola can with radius 3.2 cm and height 11 cm.
  **b** Find the radius of another can with volume of 440 cm$^3$ and height 14 cm.

**4** The stopping distance $d$ m of a car travelling at $v$ km/h is given by $d = 0.0065v^2 + 0.75v$.
  **a** Find the stopping distance of a car travelling at 50 km/h.
  **b** Estimate the speed of a car which takes 100 m to stop.

**5** The volume $V$ cm$^3$ of a cone with base radius $r$ cm and height $h$ cm is $V = \frac{1}{3}\pi r^2 h$.
  **a** Find $V$ when $r = 5$ and $h = 8$.
  **b** Find $h$ when $V = 113$ and $r = 6$.
  **c** Find $r$ when $V = 670$ and $h = 10$.

**6** The distance $d$ km that a person can see when at a height of $h$ metres above the surface of the sea is $d = \sqrt{12.8h}$.
  **a** Find how far a person can see when $h = 20$ m.
  **b** How high do you have to be to see 50 km?

**7** A formula used in Mechanics is $v = \sqrt{u^2 + 20s}$ where $u$ and $v$ are speeds in m/s and $s$ is distance in metres.
  **a** Find $v$ when $u = 12$ and $s = 30$.
  **b** Find $s$ when $v = 16$ and $u = 4$.
  **c** Find $u$ when $v = 12$ and $s = 5$.

**8** $\frac{1}{f} = \frac{1}{u} + \frac{1}{v}$ is a formula used in optics, with all the units measured in cm.
  **a** Find $f$ when $u = 10$ and $v = 20$.
  **b** Find $u$ when $v = 30$ and $f = 12$.

**9** Find the formulae for the area and the perimeter of the shape shown.
  **a** Find the area and perimeter when $r = 4$ cm.
  **b** Find $r$ when $A = 30$.
  **c** Find $r$ when $P = 70$.
  **d** Find the value of $r$ that makes $A = P$.

**10** Find the formulae for the area and the perimeter of the shape shown.
  **a** Find the area and perimeter when $r = 8$ cm.
  **b** Find $r$ when $A = 100$.
  **c** Find $r$ when $P = 40$.
  **d** Find the value of $r$ that makes $A = P$.

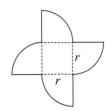

UNIT 4 ◆ Algebra

# Exercise 79 (Revision)

Make $x$ the subject of these equations.

**1** $ax = b$
**2** $\dfrac{x}{c} = a$
**3** $bx + c = a$

Make $y$ the subject of these equations.

**4** $by^2 = d$
**5** $\sqrt{ay} = b$
**6** $ay - cy = d$
**7** $c(y - b) = y$

**8** From a height $h$ metres above sea level, the horizon appears to be $1.6\sqrt{4.9h}$ kilometres away.

   **a** What is the distance, correct to 2 significant figures, to the horizon from the top of a 100 m high cliff?

   **b** How high must you be to see 50 km to the horizon?
   Give your answer correct to 2 significant figures.

   **c** You are sitting on the beach with your eye 1.8 m above sea level, and you can just see the top of a liner. The liner is 40 m high.
   How far out to sea is the liner? Give your answer to the nearest kilometre.

**9** Elise is organising a leavers' prom. The caterer tells her that the cost $C$ dollars of each meal when $n$ meals are supplied is given by $C = 20 + \dfrac{300}{n}$.

   **a** Find the cost of each meal if 50 people come to the prom.

   **b** How many people must come if the cost of each meal is to be less than \$23?

   **c** What must Elise pay the caterer if 75 meals are supplied?

**10** The shape shown is two quarter-circles of radius $r$ cm.

   **a** Find the formulae for the area $A$ and the perimeter $P$ of the shape.

   **b** Find the area and perimeter when $r = 5$ cm.

   **c** Find $r$ when $A = 20$.

   **d** Find $r$ when $P = 60$.

   **e** Find the value of $r$ that makes $A = P$.

# Exercise 79★ (Revision)

Make $x$ the subject of these equations.

**1** $c - ax = b$
**2** $\dfrac{b}{x} + d = a$
**3** $a(b - x) = \tan c$

Make $y$ the subject of these equations.

**4** $\dfrac{a}{y^2} + c = b$
**5** $a(y - c) + d = by$
**6** $c = a + \sqrt{\dfrac{b - y}{d}}$

**7** A rubber ball is dropped onto the floor from a height of $h$ metres. The time taken for it to stop bouncing is $t$ seconds. This is given by the formula

$$t = \left(\frac{1 + e}{1 - e}\right)\sqrt{\frac{h}{5}}$$

where $e$ is a number that measures the 'bounciness' of the rubber ball.
When dropped from a height of 1.25 m, it bounces for 2.5 s.

   **a** Find the value of $e$.

   **b** For how long would the rubber ball bounce if it was dropped from a height of 1.8 m?

   **c** The same rubber ball is dropped again, and this time it bounces for 3.5 s.
   Find the height from which it was dropped.

**8** The shape shown is a right-angled triangle together with two quarter-circles of radius $r$ cm.

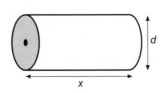

  **a** Find the formulae for the area $A$ and the perimeter $P$ of the shape.

  **b** Find the area and perimeter when $r = 3$ cm.

  **c** Find $r$ when $A = 12$.

  **d** Find $r$ when $P = 34$.

  **e** Find the value of $r$ that makes $A = P$.

**9** These two 3D figures have the same volume. Find the ratio $x : y$.

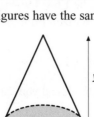

Hint: Volume of cone is $\frac{1}{3}$ shaded area × height.

      Volume of cylinder is shaded area × length.

**10** The tangent $t$ of the angle between two straight lines with gradient $m_1$ and $m_2$ is given by

$$t = \frac{m_1 - m_2}{1 + m_1 m_2}.$$

  **a** Find $t$ when $m_1 = 2$ and $m_2 = \frac{1}{2}$.

  **b** Make $m_2$ the subject of the formula.

  **c** If $t = 0.5$ and $m_1 = 3$ find $m_2$.

# Graphs 4

## Quadratic graphs $y = ax^2 + bx + c$

You have seen how to plot straight lines of type $y = mx + c$; but, in reality, many graphs are curved.

**Quadratic curves** are those in which the highest power of $x$ is $x^2$, and they produce curves called **parabolas**. The three graphs below show parts of some quadratic curves.

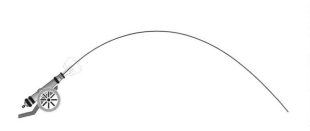

## Activity 22

Mathematicians and scientists often try to find a formula to connect two quantities. The first step is usually to plot the graph. For these three graphs, suggest two quantities from real life that might be plotted as $x$ and $y$ to produce the shapes of graphs shown here.

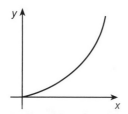

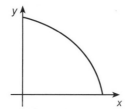

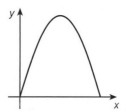

---

**Example 1**

Plot the curve $y = 2x^2 - 3x - 2$ in the range $-2 \leqslant x \leqslant 4$.

Construct a table and plot a graph from it.

| $x$ | $-2$ | $-1$ | 0 | 1 | 2 | 3 | 4 |
|---|---|---|---|---|---|---|---|
| $2x^2$ | 8 | 2 | 0 | 2 | 8 | 18 | 32 |
| $-3x$ | 6 | 3 | 0 | $-3$ | $-6$ | $-9$ | $-12$ |
| $-2$ | $-2$ | $-2$ | $-2$ | $-2$ | $-2$ | $-2$ | $-2$ |
| $y$ | 12 | 3 | $-2$ | $-3$ | 0 | 7 | 18 |

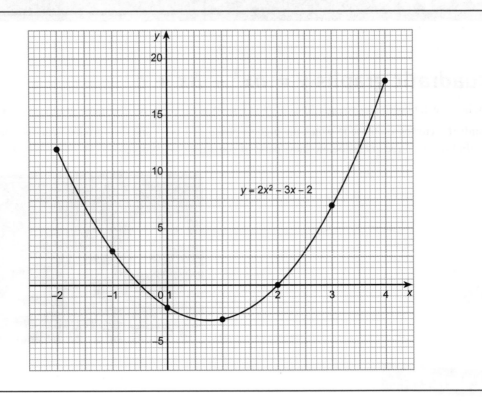

$y = 2x^2 - 3x - 2$

## Remember

♦ Plot enough points to enable a smooth curve to be drawn, especially where the curve turns.

♦ Do not join up the points with straight lines. Plotting intermediate points will show you that this is incorrect.

## Key Points

♦ Expressions of the type $y = ax^2 + bx + c$ are called **quadratics**. When they are plotted, they produce **parabolas**.

♦ If $a > 0$, the curve is U-shaped.

♦ If $a < 0$, the curve is an inverted U shape.

## Example 2

Plot the curve $y = -3x^2 + 3x + 6$ in the range $-2 \leqslant x \leqslant 3$.

Construct a table and plot a graph from it.

| $x$ | $-2$ | $-1$ | 0 | 1 | 2 | 3 |
|---|---|---|---|---|---|---|
| $-3x^2$ | $-12$ | $-3$ | 0 | $-3$ | $-12$ | $-27$ |
| $+3x$ | $-6$ | $-3$ | 0 | 3 | 6 | 9 |
| $+6$ | 6 | 6 | 6 | 6 | 6 | 6 |
| $y$ | $-12$ | 0 | 6 | 6 | 0 | $-12$ |

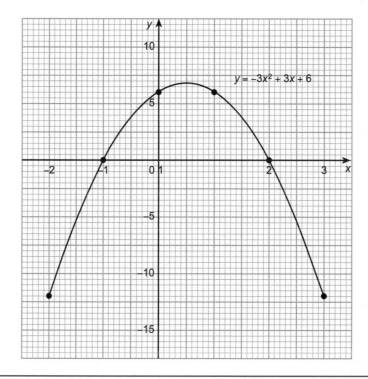

$y = -3x^2 + 3x + 6$

## Example 3

Adiel keeps goats, and she wants to use a piece of land beside a straight stone wall for grazing. This grazing land must be rectangular in shape, and it is to be fenced off by a fence of total length 50 m.

What plot dimensions will provide the goats with the largest grazing area?

What range of values can the rectangle width ($x$ metres) take in order for the enclosed area to be at least 250 m²?

$50 - 2x$

$x$

If the total fence length is 50 m, the dimensions of the rectangle are $x$ by $(50 - 2x)$.
Let the area enclosed be $A$ square metres. Then
$$A = x(50 - 2x)$$
$$\Rightarrow \quad A = 50x - 2x^2$$

Construct a table, and plot a graph from it.

| $x$ | 0 | 5 | 10 | 15 | 20 | 25 |
|---|---|---|---|---|---|---|
| $50x$ | 0 | 250 | 500 | 750 | 1000 | 1250 |
| $-2x^2$ | 0 | −50 | −200 | −450 | −800 | −1250 |
| $A$ | 0 | 200 | 300 | 300 | 200 | 0 |

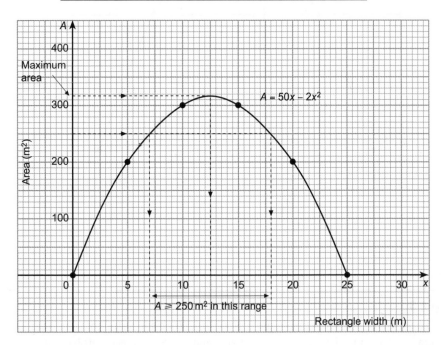

The solutions can be read from the graph.

The maximum enclosed area is when $x = 12.5$ m, giving dimensions of 12.5 m by 25 m, and an area of 313 m$^2$ (to 3 s.f.).

If $A \geqslant 250$ m$^2$, $x$ must be in the range $7 \leqslant x \leqslant 18$ approximately.

# Exercise 80

For Questions 1–4, draw a graph for each equation after compiling a suitable table between the stated $x$ values.

**1** $y = x^2 + 2$    $-3 \leqslant x \leqslant 3$          **2** $y = x^2 - 2$    $-3 \leqslant x \leqslant 3$

**3** $y = x^2 + 2x$    $-3 \leqslant x \leqslant 3$        **4** $y = x^2 - 2x$    $-3 \leqslant x \leqslant 3$

For Questions 5–8, draw a graph for each equation between the stated $x$ values after copying and completing these tables.

**5** $y = x^2 + x + 2$    $-3 \leqslant x \leqslant 2$

| $x$ | $-3$ | $-2$ | $-1$ | $0$ | $1$ | $2$ |
|---|---|---|---|---|---|---|
| $x^2$ | 9 | | | 0 | | 4 |
| $x$ | $-3$ | | | 0 | | 2 |
| $2$ | 2 | | | 2 | | 2 |
| $y$ | 8 | | | 2 | | 8 |

**6** $y = x^2 + x - 2$    $-3 \leqslant x \leqslant 2$

| $x$ | $-3$ | $-2$ | $-1$ | $0$ | $1$ | $2$ |
|---|---|---|---|---|---|---|
| $x^2$ | 9 | | | 0 | | 4 |
| $x$ | $-3$ | | | 0 | | 2 |
| $-2$ | $-2$ | | | $-2$ | | $-2$ |
| $y$ | 4 | | | $-2$ | | 4 |

**7** $y = 2x^2 + 3x + 2$    $-3 \leqslant x \leqslant 2$

| $x$ | $-3$ | $-2$ | $-1$ | $0$ | $1$ | $2$ |
|---|---|---|---|---|---|---|
| $y$ | 11 | | | 2 | | 16 |

**8** $y = 2x^2 - 3x - 2$    $-2 \leqslant x \leqslant 3$

| $x$ | $-2$ | $-1$ | $0$ | $1$ | $2$ | $3$ |
|---|---|---|---|---|---|---|
| $y$ | 12 | | $-2$ | | | 7 |

**9** A water tank has a square base of side length $x$ metres, and a height of 2 m.
   **a** Show that the volume $V$, in cubic metres, of water in a full tank is given by the formula $V = 2x^2$.
   **b** Copy and complete this table, and use it to draw the graph of $V$ against $x$.

| $x$ (m) | 0 | 0.4 | 0.8 | 1.2 | 1.6 | 2.0 |
|---|---|---|---|---|---|---|
| $V$ (m³) | 0 | | | | | 8 |

   **c** Use your graph to estimate the dimensions of the base that give a volume of 4 m³.
   **d** What volume of water could be held by the tank if its base area is 0.36 m²?
   **e** A hotel needs a water tank to hold at least 3 m³. If the tank is to fit into the loft, its side length cannot be more than 1.8 m. What range of $x$ values enables the tank to fit into the roof space?

**10** On a Big-Dipper ride at a funfair, the height $y$ metres of a carriage above the ground $t$ seconds after the start is given by the formula $y = 0.5t^2 - 3t + 5$ for $0 \leqslant t \leqslant 6$.

Start

**a** Copy and complete this table, and use it to draw the graph of $y$ against $t$.

**b** Use your graph to find in this time period the height above the ground at the start of the ride.

**c** What is the minimum height above the ground and at what time does this occur?

**d** What is the height above the starting point after 6 s?

**e** Between what times is the carriage at least 3 m above the ground?

| $t$ (s) | 0 | 1 | 2 | 3 | 4 | 5 | 6 |
|---------|---|-----|---|-----|---|-----|---|
| $y$ (m) |   | 2.5 |   | 0.5 |   | 2.5 |   |

# Exercise 80★

For Questions 1–4, draw a graph for each equation after compiling a suitable table between the stated $x$ values.

**1** $y = -x^2 + 2$     $-3 \leqslant x \leqslant 3$

**2** $y = -x^2 - 1$     $-3 \leqslant x \leqslant 3$

**3** $y = -x^2 + 4x$     $-1 \leqslant x \leqslant 5$

**4** $y = -x^2 - 3x$     $-4 \leqslant x \leqslant 1$

For Questions 5 and 6, draw the graph for the equation between the stated $x$ values after completing the table.

**5** $y = -2x^2 + 2x + 5$     $-2 \leqslant x \leqslant 3$

| $x$ | $-2$ | $-1$ | 0 | 1 | 2 | 3 |
|-----|------|------|---|---|---|-----|
| $y$ | $-7$ |      |   |   |   | $-7$ |

**6** $y = -3x^2 - 4x + 3$     $-3 \leqslant x \leqslant 2$

| $x$ | $-3$ | $-2$ | $-1$ | 0 | 1 | 2 |
|-----|-------|------|------|---|---|------|
| $y$ | $-12$ |      |      |   |   | $-17$ |

**7** The population $P$ (in millions) of bacteria on a piece of cheese after $t$ days is given by the equation $P = kt^2 + t + 1$, where $k$ is a constant that is valid for $2 \leqslant t \leqslant 12$.

**a** Study this table carefully to find the value of $k$, and then copy and complete the table.

| $t$ (days) | 2 | 4 | 6 | 8 | 10 | 12 |
|------------|-----|---|---|---|----|-----|
| $P$ (millions) | 10 |   |   |   |    | 265 |

**b** Draw a graph of $P$ against $t$.

**c** Use your graph to estimate the bacteria population after 5 days.

**d** How many days does it take for the bacteria population to exceed $10^8$?

**8** The depth of water, $y$ m, at the entrance of a tidal harbour $t$ hours after midday is given by the formula $y = 4 + 3t - t^2$ where $0 \leqslant t \leqslant 4$.

**a** Copy and complete this table, and use it to draw a graph of $y$ against $t$.

| $t$ (hours after 12:00) | 0 | 1 | 1.5 | 2 | 3 | 4 |
|---|---|---|---|---|---|---|
| $y$ (m) | | 6 | | | 4 | |

**b** Use your graph to find the depth of water at the harbour entrance at midday.

**c** At what time is the harbour entrance dry?

**d** What is the maximum depth of water at the entrance and at what time does this occur?

**e** A large ferry requires at least 5 m of water if it is to be able to enter a harbour. Between what times of the day can it safely enter the harbour? Give your answers to the nearest minute.

**9** An open box is made from a thin square metal sheet measuring 10 cm by 10 cm. Four squares of side $x$ centimetres are cut away, and the remaining sides are folded upwards to make a box of depth $x$ centimetres.

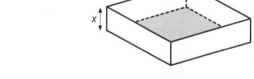

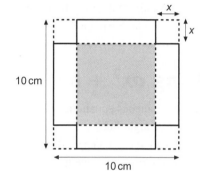

**a** Show that the external surface area $A$ cm$^2$ is given by the formula $A = 100 - 4x^2$ where $0 \leqslant x \leqslant 5$.

**b** Draw the graph of $A$ against $x$ by first constructing a table of values.

**c** Use your graph to find values of $x$ which will produce a box with an external surface area of between 50 cm$^2$ and 75 cm$^2$, inclusive.

**10** The total stopping distance $y$ metres of a car in dry weather travelling at a speed of $x$ m.p.h. is given by the formula $y = 0.015x^2 + 0.3x$ where $20 \leqslant x \leqslant 80$.

**a** Copy and complete this table and use it to draw a graph of $y$ against $x$.
(Fill a sheet of A4 graph paper.)

| $x$ (miles/hour) | 20 | 30 | 40 | 50 | 60 | 70 | 80 |
|---|---|---|---|---|---|---|---|
| $y$ (m) | | 22.5 | | 52.5 | | 94.5 | |

**b** Use your graph to find the stopping distance for a car travelling at 55 m.p.h.

**c** At what speed does a car have a stopping distance of 50 m?

**d** The stopping distance is measured from when an obstacle is observed to when the car is stationary. Hence, a driver's reaction time before applying the brakes is an important factor.

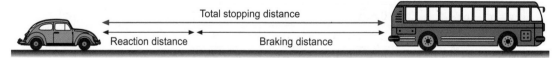

Total stopping distance

Reaction distance    Braking distance

Sam is driving at 75 m.p.h. when she sees a stationary school bus ahead of her. Given that she just manages to stop before hitting the bus, and her braking distance is 83.5 m, calculate her reaction time. (1 mile $\simeq$ 1600 m)

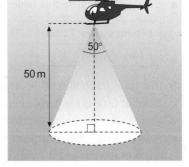

## Investigate

A rescue helicopter has a searchlight.
- ◆ Find the radius of the illuminated circle when the light is 50 m above the ground.
- ◆ What is the radius when the light is 100 m above the ground?
- ◆ Calculate the illuminated area for both 50 m and 100 m.
- ◆ Show that the relationship of the illuminated area $A$ and the vertical height of the beam $H$ is given by $A = \pi(H \tan 25°)^2$.
- ◆ Investigate their relationship by drawing a suitable graph.
- ◆ Consider beam angles other than 50°.

# Solving $ax^2 + bx + c = 0$ using graphs

Solving simultaneous equations by graphs is a useful technique. The solutions are the intersection point of two straight lines.

The same method can be applied for the intersection points of a line (the $x$-axis where $y = 0$) and a curve $y = ax^2 + bx + c$ to give the solutions (roots) to $ax^2 + bx + c = 0$.

### Example 4

Use the graph of $y = x^2 - 5x + 4$ to solve the equation $0 = x^2 - 5x + 4$.

The curve $y = x^2 - 5x + 4$ cuts the $x$-axis ($y = 0$) at $x = 1$ and $x = 4$. Thus at these points, $0 = x^2 - 5x + 4$, and the solutions are $x = 1$ or $x = 4$.

Check: If $x = 1$, $y = 1^2 - 5 \times 1 + 4$
$$= 0$$
If $x = 4$, $y = 4^2 - 5 \times 4 + 4$
$$= 0$$

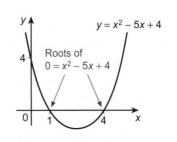

# Exercise 81

Draw the graphs between the suggested $x$-values, and use them to solve the equations.

| | Graph | $x$-values | Equation |
|---|---|---|---|
| 1 | $y = x^2 - 5x + 6$ | $0 \leqslant x \leqslant 5$ | $0 = x^2 - 5x + 6$ |
| 2 | $y = x^2 - 6x + 5$ | $0 \leqslant x \leqslant 6$ | $0 = x^2 - 6x + 5$ |
| 3 | $y = x^2 - 2x - 3$ | $-2 \leqslant x \leqslant 4$ | $0 = x^2 - 2x - 3$ |
| 4 | $y = x^2 - 7x + 10$ | $0 \leqslant x \leqslant 6$ | $0 = x^2 - 7x + 10$ |
| 5 | $y = x^2 - 3x$ | $-2 \leqslant x \leqslant 4$ | $0 = x^2 - 3x$ |
| 6 | $y = x^2 - 5x$ | $-1 \leqslant x \leqslant 6$ | $0 = x^2 - 5x$ |

# Exercise 81★

Draw the graphs between the suggested $x$-values, and use them to solve the equations. Check your solutions.

| | Graph | $x$-values | Equation |
|---|---|---|---|
| **1** | $y = 2x^2 - 5x + 2$ | $0 \leqslant x \leqslant 4$ | $0 = 2x^2 - 5x + 2$ |
| **2** | $y = 2x^2 - 3x - 5$ | $-2 \leqslant x \leqslant 3$ | $0 = 2x^2 - 3x - 5$ |
| **3** | $y = 3x^2 - 8x + 4$ | $0 \leqslant x \leqslant 4$ | $0 = 3x^2 - 8x + 4$ |
| **4** | $y = 4x^2 - 1$ | $-2 \leqslant x \leqslant 2$ | $0 = 4x^2 - 1$ |

Find the equations of these curves in the form $y = ax^2 + bx + c$.

**5**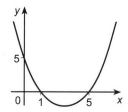

**6**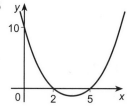

**7**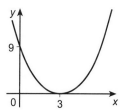

**8**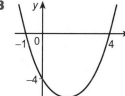

**9** Sketch the graphs of $y = ax^2 + bx + c$ (if $a > 0$) when the equation $0 = ax^2 + bx + c$ has two solutions, one solution, and no solutions.

## Activity 23

A forest contains $F$ foxes and $R$ rabbits. Their numbers change throughout the course of a given year as shown in the graph of $F$ against $R$. $t$ is the number of months after 1 January.

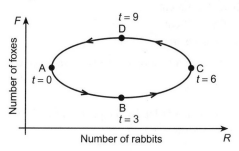

♦ Copy and complete this table.

| Year interval | Fox numbers | Rabbit numbers | Reason |
|---|---|---|---|
| Jan–Mar (A–B) | Decreasing | Increasing | Fewer foxes to eat rabbits |
| Apr–Jun (B–C) | | | |
| Jul–Sep (C–D) | | | |
| Oct–Dec (D–A) | | | |

♦ Sketch two graphs of $F$ against $t$ and $R$ against $t$ for the interval $0 \leqslant t \leqslant 12$, placing the horizontal axes as shown for comparison.

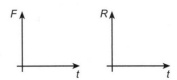

## Exercise 82 (Revision)

Draw the graphs of these equations between the stated $x$ values by first compiling suitable tables of values.

1   $y = x^2 + x - 2$     $-3 \leqslant x \leqslant 3$          2   $y = x^2 + 2x - 3$     $-4 \leqslant x \leqslant 2$

3   Draw the graph of $y = x^2 - 4x + 2$ for $-1 \leqslant x \leqslant 5$ and use this graph to find approximate solutions to the equation $x^2 - 4x + 2 = 0$. Check your answers.

4   Draw the graph of $y = x^2 - 2x - 5$ for $-2 \leqslant x \leqslant 4$ and use this graph to find approximate solutions to the equation $x^2 - 2x - 5 = 0$. Check your answers.

5   The area $A$ cm$^2$ of a semicircle formed from a circle of diameter $d$ cm is given by the approximate formula $A \approx 0.4d^2$.

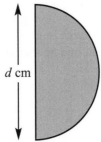

$d$ cm

Draw a graph of $A$ against $d$ for $0 \leq d \leq 6$ and use this graph to find
   a   the area of a semicircle of diameter 3.5 cm
   b   the diameter of a semicircle of area 8 cm$^2$
   c   the perimeter of the semicircle whose area is 10 cm$^2$

6   The distance $s$ m fallen by a pebble from a clifftop after $t$ seconds is given by the equation $s = 4.9t^2$, for $0 \leqslant t \leqslant 4$.
   a   Draw a graph of $s$ against $t$.
   b   Use your graph to estimate the distance fallen by the pebble after 2.5 s.
   c   At what time has the pebble fallen 50 m?

# Exercise 82★ (Revision)

Draw the graphs of these equations between the stated $x$ values by first compiling tables of values.

**1** $y = 2x^2 + 3x - 4$    $-3 \leqslant x \leqslant 3$            **2** $y = -3x^2 + 2x + 4$    $-3 \leqslant x \leqslant 3$

**3** Draw the graph of $y = 2x^2 + 3x - 6$ for $-3 \leqslant x \leqslant 3$ and use this graph to find approximate solutions to the equation $2x^2 + 3x - 6 = 0$. Check your answers.

**4** Draw the graph of $y = -2x^2 - 3x + 6$ for $-3 \leqslant x \leqslant 3$ and use this graph to find approximate solutions to the equation $-2x^2 - 3x + 6 = 0$. Check your answers.

**5** Lee is designing a small bridge to cross a jungle river. She decides that it will be supported by a parabolic arch. The equation $y = -\frac{1}{2}x^2$ is used as the mathematical model to design the arch, where the axes are as shown in the diagram.

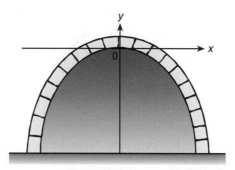

   **a** Draw the graph of the arch for $-4 \leqslant x \leqslant 4$.

   **b** Draw the line $y = -6$ to represent the water level of the river, and use your graph to estimate the width of the arch 10 m above the river, if the scale of the graph is 1 unit to 5 m.

**6** The equation for the flight path of a golfer's shot is $y = 0.2x - 0.001x^2$, for $0 \leqslant x \leqslant 200$, where $y$ m is the ball's height, and $x$ m is the horizontal distance moved by the ball.

   **a** Draw a graph of $y$ against $x$ by first compiling a suitable table of values between the stated $x$ values.

   **b** Use your graph to estimate the maximum height of the ball.

   **c** Between what distances is the ball at least 5 m above the ground?

## Circles

Angles can be calculated in shapes involving parallel lines, triangles, quadrilaterals and other straight-sided shapes called polygons. This section shows you how to calculate angles in circles.

> **Remember**
>
> ♦ A straight line can intersect a circle in three ways. It can be a **diameter**, a **chord** or a **tangent**.
>
>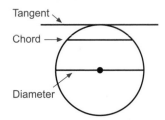
> Tangent
> Chord
> Diameter
>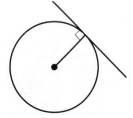
>
>
> ♦ A **tangent** 'touches' the circle. It is perpendicular to the radius at the point of contact.
>
> ♦ Always look for **isosceles triangles** in circle problems.

## Angles in a semicircle and tangents

> **Remember**
>
> The angle in a semicircle is always a right angle.
>
>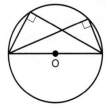

## Activity 24

Proving the result

♦ Calculate the sizes of the angles in each circle.

$C_1$

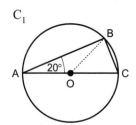

$C_2$

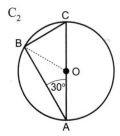

$C_3$

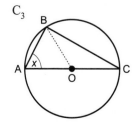

♦ Copy this table to record your answers and complete the table one row at a time.

|  | ∠BAO | ∠ABO | ∠AQB | ∠BOC | ∠OCB + ∠OBC | ∠OBC | ∠ABC |
|---|---|---|---|---|---|---|---|
| $C_1$ | 20° | | | | | | |
| $C_2$ | 30° | | | | | | |
| $C_3$ | $x$ | | | | | | |

Using $x$ in the final case generalises the process used for $C_1$ and $C_2$. If you add reasons for each step, you have a formal proof to show $\angle ABC = 90°$ when AC is a diameter.

Let             $\angle BAC = x$

Then            $\angle ABO = x$                                 ($\triangle ABO$ is isosceles, and $\therefore$ $\angle BAO = \angle ABO$)

                $\angle BOC = 2x$                                (Exterior angle of $\triangle ABO$)

    $\angle OCB + \angle OBC = 180° - 2x$                        (Angle sum of $\triangle BCO$)

                $\angle OBC = 90° - x$                           ($\triangle BCO$ is isosceles and $\therefore$ $\angle OBC = \angle OCB$)

                $\angle ABC = (90° - x) + x = 90°$.

---

**Remember**

To enable other people to read your work, you need to be consistent with your mathematical language. On diagrams, always label points with capital letters, lengths with lower-case letters, and angles with lower-case letters.

---

In Exercises 83 and 83*, you are asked to give reasons. This is an important part of any explanation or proof.

## Key Points

Examples of reasons in explanations or proofs:

♦ Angle sum of $\triangle$ ...

♦ Angles on straight line at ...

♦ Angles at the point ...

♦ Alternate angles, ... ∥ ...

♦ Radius ⊥ tangent at ...

♦ $\triangle$ ... is isosceles. $\therefore$ ... = ...

♦ Vert. opp. angles at ...

♦ Exterior angle of $\triangle$ ...

♦ Angle in a semicircle

The reason *follows* the statement or equation, and abbreviations can be used.

# Exercise 83

For Questions 1–24, find the size of each lettered angle.

**1**

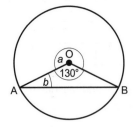

**2**

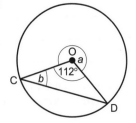

**3**

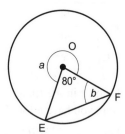

**4**

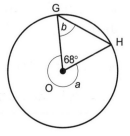

**5**

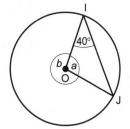

**6**

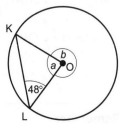

**7**

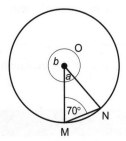

**8**

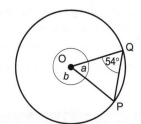

**9**

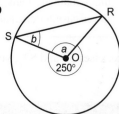

**10**

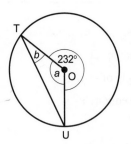

**11**

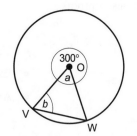

**12**

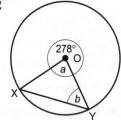

**13**

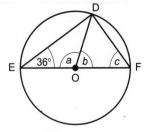

**14**

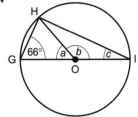

**15**

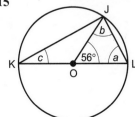

**16**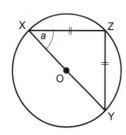

**17**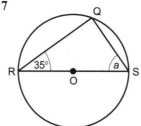

**18**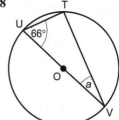

**19**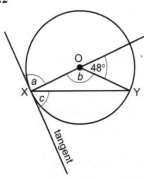

**20**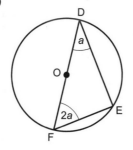

**21**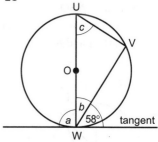

**22**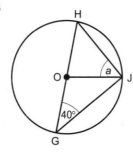

**23**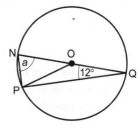

**24**

**25** Giving a reason for each step of your working, find these angles.

   **a** ∠DEO

   **b** ∠EDI

   **c** ∠EOG

   **d** ∠GOI

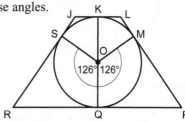

**26** JLPR is a quadrilateral, and KQ is a straight line. Find these angles.

   **a** ∠OML

   **b** ∠MPQ

   **c** ∠KLM

   What type of quadrilateral is JLPR?

   Comment on the 'kites' within the diagram.

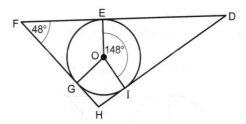

# Exercise 83★

**1**

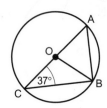

a Find ∠OBC.
b Find ∠ABO.
c Add a reason to your answers to
parts **a** and **b**.

**2**

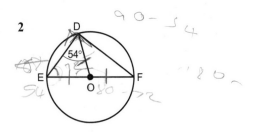

a Find ∠DEO.
b Find ∠DOF.
c Find ∠DFO.
d Add a reason to your answers in
parts **a**, **b** and **c**.

**3** ∠OFG = x.

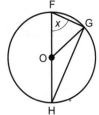

a Find ∠OGF in terms of x.
b Find ∠GOH in terms of x.
c Add a reason to your answers
in parts **a** and **b**.

**4** ∠MPO = y.

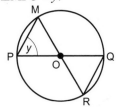

a Find these angles in terms of y:
∠PMO, ∠POM, ∠ROQ and ∠ORQ.
b Add a reason to each of your
answers in part. Hence show that
PM is parallel to RQ.

**5** AB is a tangent to the circle, AB = AC,
and ∠AOB = 67.5°.
  a Find ∠ABO.
  b From triangle CBA, find ∠ABC.
Give a reason with each answer.
Hence show that OB bisects ∠ABC.

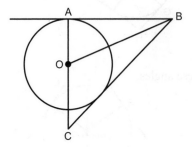

**6** BT is a tangent to the circle. Let
∠ATB = x. Giving a reason with each
answer, find ∠APB, ∠TPB and ∠TBP.
Hence show that ∠ATB = ∠PBA.

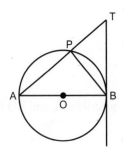

**7 a** Find, in terms of $x$, the values of ∠OUT, ∠TOU and ∠TUV.

  **b** Add a reason to your answers in part **a**.

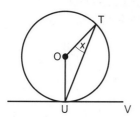

**8 a** Find, in terms of $y$, the values of ∠XZW, ∠XZY and ∠WXY.

  **b** Add a reason to your answers in part **a**.

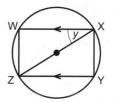

**9** Find, in terms of $x$, the values of ∠OBA, ∠AOB and ∠COB. Give a reason with each step of your working.

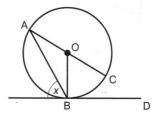

**10** Find, in terms of $x$, the values of ∠OFG, ∠EFO, ∠EOF and ∠GOF. Give a reason for each step of your working.

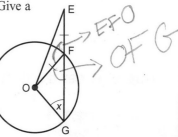

**11** Let OM = $x$ centimetres, and let the radius of the circle be $r$ centimetres.

  **a** From triangle OAM, find AM in terms of $x$ and $r$.

  **b** From triangle OBM, find BM in terms of $x$ and $r$.

  **c** Hence show that AM = BM.

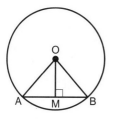

**12** LM = 6 cm and MN = 8 cm.

  **a** Calculate the radius of the circle.

  **b** Calculate the area of triangle LMN.

  **c** Calculate the length of the perpendicular from M to LN.

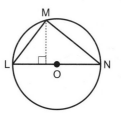

**13** OR = 7.5 cm and SR = 9 cm.

   **a** Calculate ST.

   **b** Calculate the area of triangle RTS.

   **c** Calculate the length of the perpendicular
      from S to RT.

**14** AB = BC. AC = 6 cm. The radius of
    the circle = 5 cm. Let M be the mid-
    point of AC.

   **a** Calculate OM.

   **b** Calculate the area of triangle ABC.

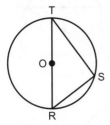

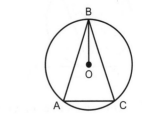

**15 a** Calculate the radius of the circle.

   **b** Calculate the area of the quadrilateral DEFO.

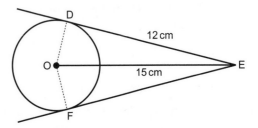

## Angle at centre is twice angle at circumference

**Activity 25**

Proving the result

♦ Copy and complete the table one row at a time by
  calculating the sizes of the angles on each diagram.

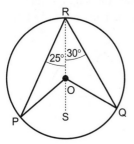

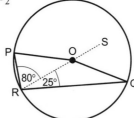

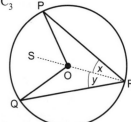

| | $\angle$ORP | $\angle$ORQ | $\angle$RPO | $\angle$POS | $\angle$RQO | $\angle$QOS | $\angle$PRQ | $\angle$POQ |
|---|---|---|---|---|---|---|---|---|
| $C_1$ | 25° | 30° | | | | | | |
| $C_2$ | 80° | 25° | | | | | | |
| $C_3$ | $x$ | $y$ | | | | | | |

♦ What is the relationship between ∠PRQ and ∠POQ?

Using $x$, you have generalised. Adding reasons gives the formal proof of another 'theorem'.

| | | |
|---|---|---|
| Let | ∠ORP = $x$ and ∠ORQ = $y$ | |
| Then | ∠RPO = $x$ | (△RPO is isosceles, and ∴ ∠RPO = ∠PRQ) |
| So | ∠POS = $2x$ | (Exterior angle of △PRO) |
| | ∠RQO = $y$ | (△RQO is isosceles, and ∴ ∠RQS = ∠PRQ) |
| So | ∠QOS = $2y$ | (Exterior angle of △QRO) |
| Thus | ∠PRQ = $x + y$ and ∠POQ = $2x + 2y$ | |
| that is, | ∠POQ = $2 \times$ ∠PRQ | |

---

### Example 1

Show that ∠APB = ∠AQB.

Draw lines AO and BO to make an angle at the centre.
The chord AB divides the circle into two segments.
P and Q are in the same segment.

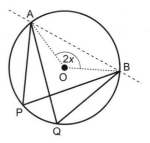

| | |
|---|---|
| Let | ∠AOB = $2x$ |
| Then | ∠APB = $x$ |
| and | ∠AQB = $x$ |
| So | ∠APB = ∠AQB |

---

### Example 2

Show that ∠ABC + ∠CDA = 180°.

Draw the radii AO and CO.

| | |
|---|---|
| Let | ∠ABC = $x$ |
| Then | ∠AOC = $2x$ |
| Let | ∠CDA = $y$ |
| Then | ∠AOC reflex = $2y$ |
| | $2x + 2y = 360°$ |
| | $x + y = 180°$ |

As the four angles of a quadrilateral sum to 360°, the other two angles must sum to 180° as well.

## Remember

◆ The angle subtended at the centre of a circle is twice the angle at the circumference.

◆ Angles in the same segment are equal.

◆ Opposite angles of a cyclic quadrilateral sum to 180°.

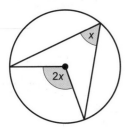

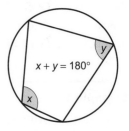

# Exercise 84

Find the size of each lettered angle.

**1**

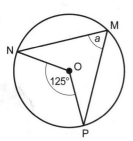

**2**

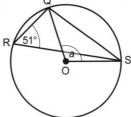

**3**

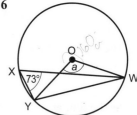

**4**

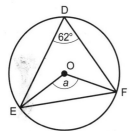

**5**

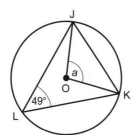

**6**

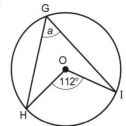

**7**

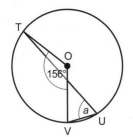

**8**

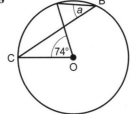

**9**

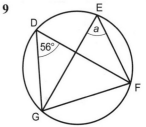

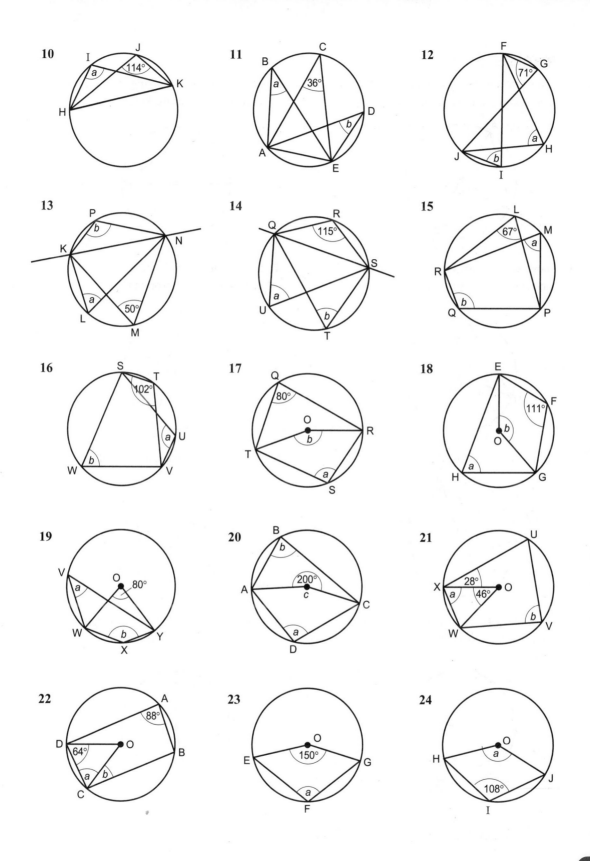

# Exercise 84★

1 Show that $x = 96°$, giving a reason for each step of your working.

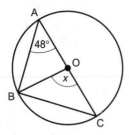

2 Show that $x = 35°$, giving a reason for each step of your working.

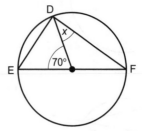

3 Show that $x = 50°$, giving a reason for each step of your working.

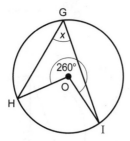

4 Show that $x = 296°$, giving a reason for each step of your working.

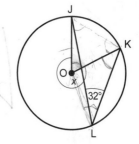

5 Find, giving reasons, the size of angle $x$.

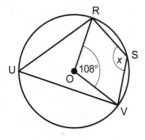

6 Find, giving reasons, the size of angle $x$.

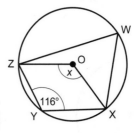

7 Show that $x = 160°$, giving a reason for each step of your working.

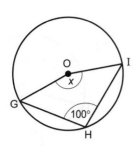

8 Show that $x = 97°$, giving a reason for each step of your working.

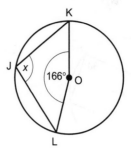

**9** Find, giving reasons, the sizes of angles *x* and *y*.

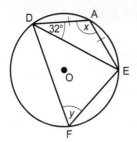

**10** Find, giving reasons, the sizes of angles *x* and *y*.

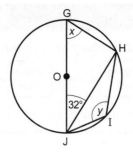

**11** Find, giving reasons, the size of angle *x*.

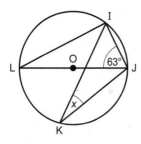

**12** Find, giving reasons, the size of angle *x*.

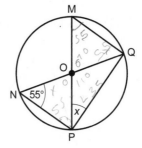

**13** Prove that OBC is an equilateral triangle.

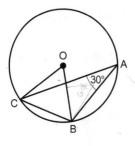

**14** Prove that DE is parallel to FO.

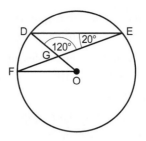

**15** Prove that triangle UVW is isosceles.

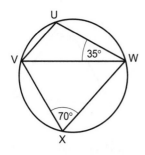

**16** Given that OP is parallel to RQ, prove that *x* = 126°.

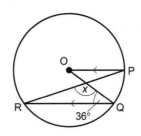

# Similar triangles

## Key Points

♦ **Similar** shapes have these properties.

Corresponding angles equal ◄───► Similar shape ◄───► Ratios of the corresponding sides are equal

♦ If any one of these facts is true, then the other two must also be true.

## Activity 26

♦ Measure each of the angles in these three triangles.

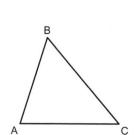

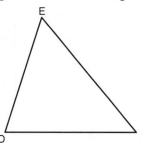

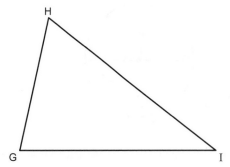

You should find that the **corresponding** angles in triangles ABC and DEF are equal.
This is because these two triangles are **similar** in shape.

♦ Now measure each of the nine sides, and use your measurements to calculate these ratios.

$$\frac{AC}{DF}, \frac{AB}{DE}, \frac{BC}{EF}, \frac{AB}{GH}, \frac{AC}{GI}, \frac{EF}{HI}$$

You should find that only the first three ratios give the same result. This is because only triangles ABC and DEF are similar in shape.

---

**Example 3**

Which of these triangles are similar to each other?

$T_1$

$T_2$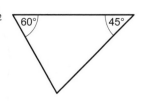

$T_3$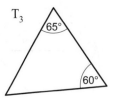

The angle sum of a triangle is 180°.

Therefore in $T_1$ the angles are 55°, 60° and 65°, in $T_2$ the angles are 45°, 60° and 75°, and in $T_3$ the angles are 55°, 60° and 65°.

Thus the triangles in $T_1$ and $T_3$ are similar in shape.

---

## Example 4

The ancient Egyptians used similar triangles to work out the heights of their pyramids. The unit they used was the cubit, a measure based on the length from a man's elbow to his fingertips.

The shadow of a pyramid reached C, which was 500 cubits from B. The Egyptian surveyor found that a pole of length 4 cubits had to be placed at Y, 20 cubits from C, for its shadow to reach C as well. What was the height of the pyramid AB?

As AB and XY are both vertical, the triangles CAB and CXY are similar in shape.

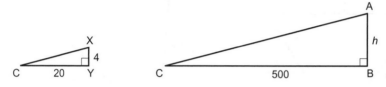

So the ratios of their corresponding sides are equal.

$$\frac{AB}{XY} = \frac{CB}{CY} \qquad \frac{AB}{4} = \frac{500}{20} = 25 \qquad AB = 4 \times 25 = 100$$

So the height of the pyramid is 100 cubits.

Activity 27 shows another method of calculating a height, again using similar triangles.

## Activity 27

To find the height of a wall AB, place a mirror on the ground at any point R. An observer stands in line with the wall at point Y so that the top of the wall can be seen in the mirror.

♦ Explain why the triangles ABR and YXR are similar in shape.

♦ Show that the ratio of the corresponding sides is given by $\frac{AB}{XY} = \frac{AR}{RY}$.

♦ Use this method to find out the height of your classroom or some other tall object.

♦ Repeat the experiment by placing the mirror in various positions. Describe your method, and comment on the accuracy of your results.

♦ Use the method of Example 4 to check some of your measurements. Compare the two methods.

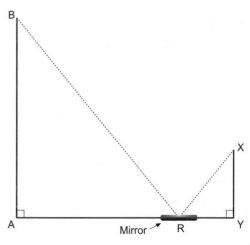

# Key Points

When using similar triangles, follow these steps.

♦ Show that the two triangles are similar in shape.

♦ If necessary, redraw the triangles in a corresponding position. Identify the corresponding vertices.

♦ Write down the ratios of the corresponding sides.

## Example 5

Use the method of similar triangles to work out $x$ and $y$.

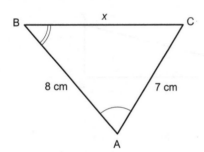

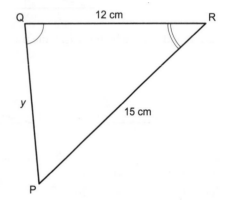

$\hat{A} = \hat{Q}$, $\hat{B} = \hat{R}$ (and $\therefore \hat{C} = \hat{P}$). $\quad \therefore \Delta s \; {}^{ABC}_{QRP}$ are similar.

$$\therefore \frac{AB}{QR} = \frac{AC}{QP} = \frac{BC}{RP} \qquad \text{So,} \frac{8}{12} = \frac{7}{y} = \frac{x}{15}$$

$$\frac{8}{12} = \frac{x}{15} \implies \frac{8 \times 15}{12} = x \implies x = 10 \text{ cm}$$

$$\frac{8}{12} = \frac{7}{y} \implies \frac{12}{8} = \frac{y}{7} \implies \frac{12 \times 7}{8} = y \implies y = 10.5 \text{ cm}$$

## Investigate

♦ Investigate these two results.

Tangents from an external point are equal in length.

The perpendicular bisector of any chord passes through the centre of the circle.

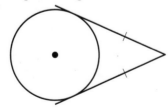

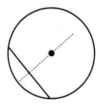

♦ What principles could be applied to prove them?

*UNIT 4 ♦ Shape and space*

# Exercise 85

In Questions 1–8, find which two of the three triangles are similar in shape.

**1**

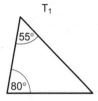 T₁

 T₂

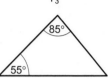 T₃

**2**

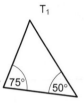 T₁

 T₂

 T₃

**3**

T₁

T₂

T₃

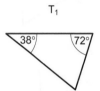  

**4**

T₁ T₂ T₃

**5**

T₁

  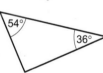

T₂ T₃

**6**

T₁ T₂ T₃

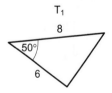 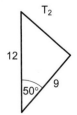 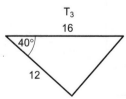

**7**

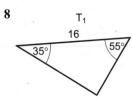
T₁
4
3
5

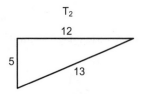
T₂
12
5
13

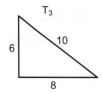
T₃
6
10
8

**8**

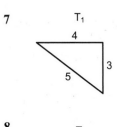
T₁
16
35°
55°

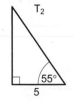
T₂
55°
5

T₃
8
6
10

**9** These two triangles are similar.
Find $x$.

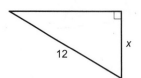
8
4
12
$x$

**10** These two triangles are similar.
Find $y$.

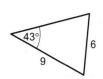
43°
6
9

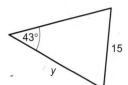
43°
15
$y$

**11** Triangles ABC and DEF are similar. Find $x$ and $y$.

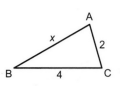
A
$x$
2
B   4   C

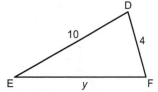
D
10
4
E   $y$   F

**12** Triangles MNO and PQR are similar. Find $v$ and $w$.

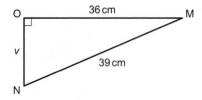
O   36 cm   M
$v$
39 cm
N

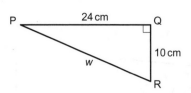
P   24 cm   Q
10 cm
$w$
R

**13 a** Find BE when AB = 7 cm,
AC = 10.5 cm and DC = 4.5 cm.

 **b** Find BE when AB = 7 cm,
BC = 5 cm and DC = 24 cm.

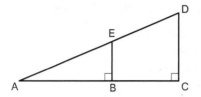

**14 a** Find RS when PQ = 6 cm, PR = 5 cm,
QR = 4 cm and RT = 6 cm.

 **b** Find PS when QP = 4 cm, PR = 3 cm,
QR = 2 cm and ST = 10 cm.

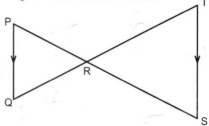

**15** At 4.30pm, a tree casts a shadow of 10 m. Abeni is 1.50 m tall, and his shadow is 1.20 m long.
How high is the tree?

**16** Badia and Caimile are sitting on a see-saw. Caimile is sitting 1.5 m
from the pivot, and his end is on the ground. The pivot is 90 cm high.
If Badia is sitting 5 m from the pivot, how high is she in the air?

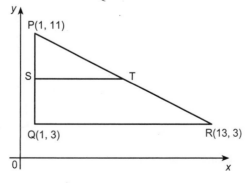

# Exercise 85★

**1** Triangles ABC and AEF are similar, and $\dfrac{AE}{AB} = \dfrac{1}{2}$. Calculate the coordinates of E and F.

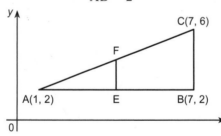

**2** Triangles PQR and PST are similar and $\dfrac{PS}{PQ} = \dfrac{1}{2}$. Calculate the coordinates of S and T.

**3** A = (−3, 2) and B = (9, 8).
Find the coordinates of
 **a** the mid-point of AB
 **b** the point P such that AP = $\frac{1}{3}$AB
 **c** the point Q such that AP = $\frac{2}{3}$AB.

**4** A = (1, 7) and B = (16, −3).
Find the coordinates of
 **a** the mid-point of AB
 **b** the point P such that AP = $\frac{1}{5}$AB
 **c** the point Q such that AP = $\frac{3}{5}$AB.

**5** **a** Prove that △GHI and △FHJ are similar.
 **b** Calculate FJ.
 **c** Calculate GF.

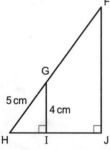

HG : HF = 1 : 4

**6** **a** Prove that △ADE and △ABC are similar.
 **b** Calculate DE.
 **c** Calculate BD.

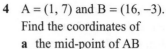

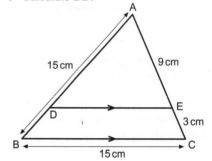

**7** These two triangles are similar.
 **a** Find the length of AC.
 **b** Explain how it is possible for there to be two correct answers to part **a**, and find the second answer.

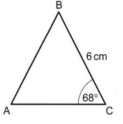

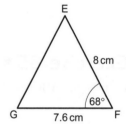

**8** These two triangles are similar.
 **a** Find the length of XZ.
 **b** Explain how it is possible for there to be two correct answers to part **a**, and find the second answer.

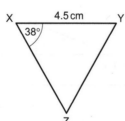

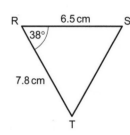

**9** In this diagram, which three triangles are similar?

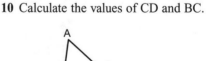

 **a** Calculate the values of PS and RS.
 **b** Hence calculate the area of the triangle PQS.

**10** Calculate the values of CD and BC.

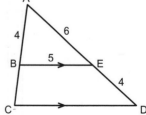

**11** A crude method of estimating large heights is to take a 'line of sight' over two poles.

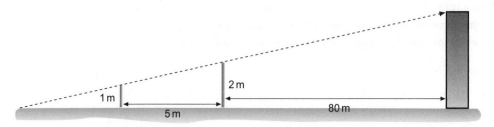

Poles of length 1 m and 2 m are placed 5 m apart, with the longer pole 80 m from the base of a clock tower, and a line of sight is taken. Calculate the height of the tower.

**12** You can estimate the width of a river without getting your feet wet.
You need just one landmark on the far bank.
Mark L, M, N and P by line of sight, and then use similar triangles.
If LP = 10 m, calculate the width of the river.

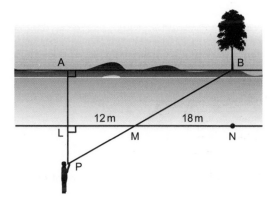

**13** Using the ancient Egyptian method of Example 4 (page 227), a pole of length 5 cubits was placed 100 cubits from the base of a tree.
The tree cast its shadow to a point 10 cubits from the pole. Calculate the height of the tree.

**14** Using the ancient Egyptian method of Example 4 (page 227), a pole of length 5 cubits was used to measure the height of a statue.
The length of the shadow was 105 cubits.
If the height of the statue was 40 cubits, how far from the base of the statue was the pole placed?

**15** Joe, whose eyeline is 1.50 m from the ground, uses the mirror method of Activity 27 to measure the height of the Millennium Wheel in London.
The Millennium Wheel is 120 m high.
He places his mirror 200 m from the base of the wheel.
How far from the mirror does he have to stand?

**16** The tallest structure in the world currently is a television transmitting tower near Fargo in North Dakota in the USA. It is 629 m high.
Bessie is using the mirror method of Activity 27 to sight the tower.
Her eyeline is 1.60 m from the ground, and she must stand no more than 2 m from the mirror to be able to see clearly.
What is the furthest distance from the foot of the tower that she can place the mirror?
Give your answer to 2 significant figures.

**17** The volume of a cone is given by the formula $V = \frac{1}{3}\pi r^2 h$,
where $r$ is the radius of the cone and $h$ is its height.
Use this formula, and similar triangles, to calculate the
volume of this truncated cone, correct to 3 significant figures.

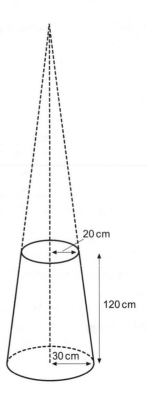

20 cm

120 cm

30 cm

**18** ABCD is a rectangle. Calculate the sum of the angles ∠AED and ∠BEF.

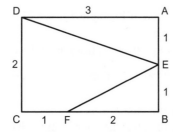

# Pythagoras' theorem

The Greek philosopher and mathematician Pythagoras found a connection between the lengths of the
sides of right-angled triangles. It is probably the most famous mathematical theorem in the world.

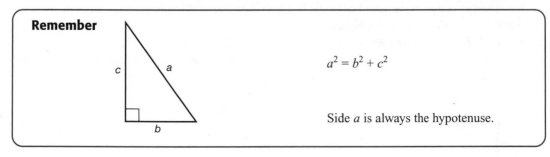

**Remember**

$a^2 = b^2 + c^2$

Side $a$ is always the hypotenuse.

**Example 6**

Calculate side $a$.

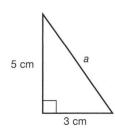

5 cm
$a$
3 cm

From Pythagoras' theorem,
$$a^2 = b^2 + c^2$$
$$a^2 = 3^2 + 5^2$$
$$= 34$$
$$a = \sqrt{34}$$
$$\therefore a = 5.83 \text{ cm (3 s.f.)}$$

**Example 7**

Calculate side $b$.

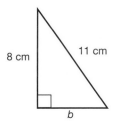

8 cm
11 cm
$b$

From Pythagoras' theorem,
$$a^2 = b^2 + c^2$$
$$11^2 = b^2 + 8^2$$
$$b^2 = 11^2 - 8^2$$
$$= 57$$
$$b = \sqrt{57}$$
$$b = 7.55 \text{ cm (3 s.f.)}$$

## Activity 28

**Proof of Pythagoras' theorem**

There are many elegant proofs of Pythagoras' theorem.

One of the easiest to understand involves a square of side $(a + b)$.

Inside the large square is a smaller one of side $c$.

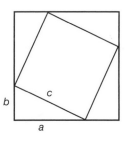

Given that the area of the large square is $(a + b)^2$ and is clearly equal to the area of the four identical triangles plus the area $c^2$ of the smaller square, form an equation.

Now simplify it to show that $c^2 = a^2 + b^2$, and hence prove that Pythagoras was correct!

# Exercise 86

Find length $a$ in these right-angled triangles to 3 s.f.

**1**

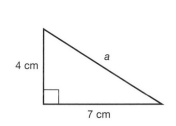

9 cm
$a$
5 cm

**2**

4 cm
$a$
7 cm

**3**

$a$
12 cm
8 cm

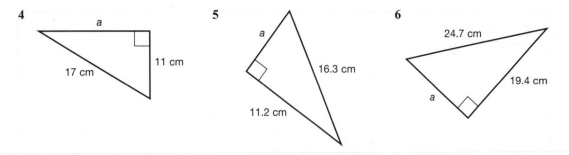

**7** Find the diagonal length of a square field of side 50 m.

**8** A fishing boat sails from Bastia in Corsica. It travels 15 km due East, then 25 km due South. How far is it from Bastia at this position?

**9** A 3.5 m ladder rests against a vertical wall such that its foot is 1.5 m away from the wall. How far up the wall is the top of the ladder?

**10** A large rectangular Persian rug has a diagonal length of 12 m and a width of 6 m. Find the length of the rug.

**11** Let OQ = $y$ centimetres, and let the radius of the circle be $r$.
Use Pythagoras's theorem to answer parts **a**, **b** and **c**.
  **a** From triangle OPQ, find PQ in terms of $y$ and $r$.
  **b** From triangle ORQ, find RQ in terms of $y$ and $r$.
  **c** Hence show that PQ = RQ.

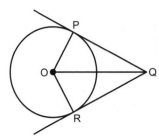

## Exercise 86*

Find length $a$ in these right-angled triangles to 3 s.f.

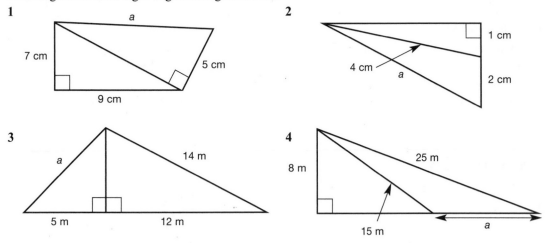

**5** Calculate the distance between the points (7, 4) and (−5, −3).

**6** Calculate the area of a square whose diagonals are 20 cm long.

**7** Thuso sails his boat from Mogadishu directly North-East for 50 km, then directly South-East for 100 km. He then sails directly back to Mogadishu at 13:00 hrs at a speed of 25 km/h. What time does he arrive?

**8** Find the length AB in this rectangular block.

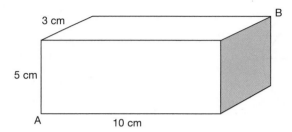

3 cm

5 cm

A

10 cm

B

**9** A fierce guard dog is tethered by a 15 m chain to a post that is 6 m from a straight path. For what distance along the path is a trespasser in danger from the dog?

**10** A ladder is resting against a vertical wall such that its foot is 1.5 m away from the base of the wall. When George steps on the ladder, it slips down the wall 0.25 m and its foot is now 2 m away from the wall.

Find the length of the ladder.

**11** OA = 50 m. OB = 80 m. The 50 m start of a ski jump at A has a vertical height of 30 m.
  **a** Use similar triangles to calculate the vertical height of the 80 m start at B.
  **b** Use Pythagoras' theorem to calculate the horizontal distance between the two starts.

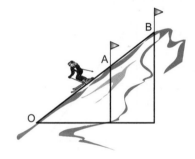

# Exercise 87 (Revision)

**1** Find angles *a*, *b* and *c*.

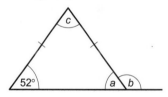

52°

*a* *b*

*c*

**2** Find angles *a*, *b*, *c* and *d*.

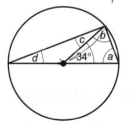

*c* *b*

34° *a*

*d*

**3** Find angles *a*, *b* and *c*.

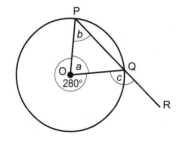

P

*b*

O *a*

280°

Q

*c*

R

**4** Find angles *a*, *b*, *c* and *d*.

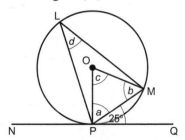

L

*d*

O

*c*

*b* M

*a* 25°

N P Q

**5** Calculate *a*.

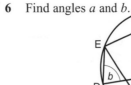

**6** Find angles *a* and *b*.

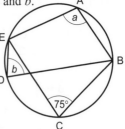

**7** Calculate the lengths of PQ and AC.

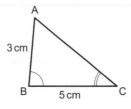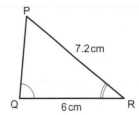

**8** Find length *x* in each of these triangles.

**a**

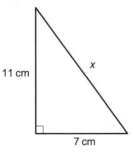

**b**

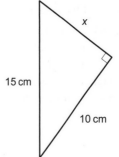

**9** Find the diagonal length of a square of area 1000 cm².

**10** The diagram shows the side of a 'lean-to' shed. Find
  **a** the height of the door AB
  **b** the area of the side of the shed.

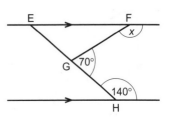

## Exercise 87★ (Revision)

**1** Find angles *a*, *b*, *c* and *d*.

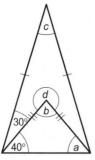

**2** Find angle *x*, giving a reason with each step of your working.

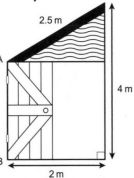

**UNIT 4 ◆ Shape and space**

**3** The diagonals of a rhombus have lengths 10 cm and 24 cm.
   **a** Calculate the length of the side of the rhombus.
   **b** Calculate the area of the rhombus.

**4** What do you call a regular polygon if
   **a** its interior angle is twice the exterior angle?
   **b** its exterior angle is twice the interior angle?

**5** Find, giving reasons with each step, angles $a$, $b$, $c$ and $d$.

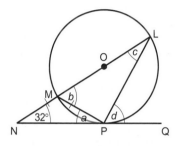

**6** **a** Prove that triangles ABC and CDE are similar.
   **b** If DC : DB = 5 : 13, AB = 9.6 cm and CE = 4 cm, calculate DE and AE.

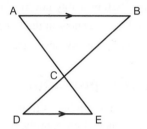

**7** A hockey goal is 4 yards wide. A shot is taken from 16 yards away. The goalkeeper's pads are 2 feet wide. The goalkeeper advances 3 yards from his line to narrow the angle (1 yard = 3 feet). Find the length $x$, in feet.

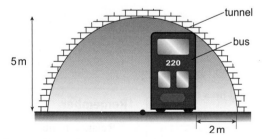

Not drawn to scale

Goalkeeper's pads 2 feet wide

3 yds

$x$   $x$

Goal 4 yds

**8** A tunnel has a semicircular cross-section and a diameter of 10 m. If the roof of a bus just touches the roof of the tunnel when its wheels are 2 m from one side, how high is the bus?

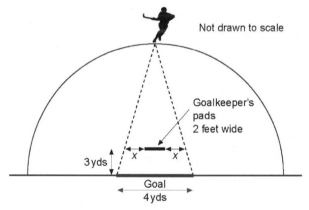

tunnel

bus

5 m

220

2 m

**9** Calculate the height of an equilateral triangle which has the same area as a circle of circumference 10 cm.

## Probability

Probability theory enables us to analyse random events to assess the *likelihood* that an event will occur. One of the earliest known works on probability was written in the 16th century by Italian mathematician (and gambler) Cardano, '*On Casting the Die*'.

Consider these statements. They all involve a degree of uncertainty, which could be estimated through experiment or using previous knowledge.

- ♦ I doubt if I will ever win the lottery.
- ♦ My dog will probably not live beyond 15 years.
- ♦ It is unlikely to snow in the Sahara Desert.
- ♦ Roses will probably never grow at the South Pole.

## Experimental probability

If in an experiment a number of trials are carried out to see how often event $A$ happens, it is possible to find the experimental probability, p($A$), of $A$ occurring.

**Remember**

- ♦ p($A$) means the probability of event $A$ happening.
- ♦ p($\overline{A}$) means the probability of event $A$ *not* happening.
- ♦ $\text{p}(A) = \dfrac{\text{number of times } A \text{ occurs}}{\text{total number of trials}}$.

**Example 1**

Event $A$ is that a particular warbler lands on Mrs Leung's bird table before 9am each day. It does this on 40 days over a period of 1 year (365 days).

**a** Estimate the probability that tomorrow event $A$ happens.

$$\text{p}(A) = \frac{40}{365} = \frac{8}{73}$$

**b** Estimate the probability that tomorrow event $A$ does *not* happen.

$$\text{p}(\overline{A}) = \frac{325}{365} = \frac{65}{73}$$

Notice that $\text{p}(A) + \text{p}(\overline{A}) = 1$.

## Relative frequency

Relative frequency can help build up a picture of a probability as an experiment takes place.

It *usually* leads to a more accurate conclusion as the number of trials increases.

**Remember**

$$\text{Relative frequency} = \frac{\text{number of successes}}{\text{total number of trials}}.$$

## Example 2

Bill is curious about the chances of a piece of toast landing buttered-side up.

He suspects that this event, $A$, is unlikely to happen.

He then conducts eight trials, with the results shown in this table.

| Trial number | 1 | 2 | 3 | 4 | 5 | 6 | 7 | 8 |
|---|---|---|---|---|---|---|---|---|
| Butter lands upwards | ✗ | ✔ | ✗ | ✗ | ✔ | ✔ | ✗ | ✗ |
| Relative frequency | $\frac{0}{1}=0$ | $\frac{1}{2}$ | $\frac{1}{3}$ | $\frac{1}{4}$ | $\frac{2}{5}$ | $\frac{3}{6}$ | $\frac{3}{7}$ | $\frac{3}{8}$ |

He plots the results on a **relative frequency diagram**.

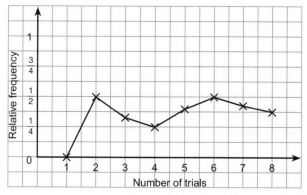

From these eight trials, Bill estimates that the probability of his toast landing buttered-side up is $\frac{3}{8}$. How could he improve his estimation of p($A$)? Comment on Bill's initial assumption.

## Exercise 88

1   Heidi is suspicious that a particular dice is biased towards the odd numbers so she carries out a number of trials. The results are given in this table.

| Trial number | 1 | 2 | 3 | 4 | 5 | 6 | 7 | 8 | 9 | 10 |
|---|---|---|---|---|---|---|---|---|---|---|
| Odd number | ✗ | ✔ | ✔ | ✗ | ✔ | ✗ | ✔ | ✗ | ✔ | ✔ |

a   Draw a relative frequency diagram to investigate Heidi's suspicion.
b   To what conclusion do these results lead? How could the experiment be improved?

2   April in the UK has a reputation for being a particularly wet month. The data in this table was collected by a weather station in Edinburgh for the first 20 days in one April.

| Day number | 1 | 2 | 3 | 4 | 5 | 6 | 7 | 8 | 9 | 10 |
|---|---|---|---|---|---|---|---|---|---|---|
| Rain | ✔ | ✗ | ✗ | ✔ | ✔ | ✔ | ✗ | ✗ | ✔ | ✔ |
| Day number | 11 | 12 | 13 | 14 | 15 | 16 | 17 | 18 | 19 | 20 |
| Rain | ✗ | ✔ | ✔ | ✔ | ✔ | ✗ | ✗ | ✔ | ✔ | ✗ |

a   Draw a relative frequency diagram to investigate the experimental probability of rain in Edinburgh in the first 20 days of April.
b   What conclusion can you draw from this data?

3  A spinner for a word game is an irregular pentagon that has sections divided into five parts denoted by letters A, B, C, D and E. Sanjeev is a keen player, and he experiments to see if he can calculate an estimate of the probability of a vowel being spun. The results are shown as ticks (a vowel) and crosses (no vowel).

✔ ✔ ✘ ✘ ✘
✘ ✘ ✘ ✔ ✔
✔ ✔ ✘ ✘ ✔
✘ ✘ ✔ ✘ ✔

Draw a relative frequency diagram to investigate the experimental probability of the spinner landing on a vowel.

# Exercise 88★

1  Pierre is a basketball shooter, and he practises hard to improve this particular skill.
He makes 12 attempts from the left-hand side of the court, and 12 from the right.
His results are shown in these tables.

**Left-hand side**

| ✘ | ✘ | ✔ | ✘ | ✔ | ✔ |
|---|---|---|---|---|---|
| ✔ | ✔ | ✘ | ✔ | ✔ | ✔ |

**Right-hand side**

| ✘ | ✘ | ✘ | ✔ | ✘ | ✘ |
|---|---|---|---|---|---|
| ✘ | ✘ | ✔ | ✔ | ✘ | ✔ |

Events $L$ and $R$ are defined as successful shots from the left-hand side and right-hand side of the court, respectively.
a  Draw a relative frequency diagram to estimate p($L$) and p($R$).
b  What advice would you now give to Pierre on this evidence?

2  A famous experiment in probability is Buffon's needle, in which a sewing needle of length $x$ centimetres is dropped onto a sheet of paper with parallel lines drawn on it that are $x$ centimetres apart. Event $A$ is defined as the needle landing across a line.
a  Draw a relative frequency diagram to find an estimate of p($A$) using at least 50 trials.
b  Complicated probability theory predicts that $p(A) = \dfrac{2}{\pi}$. Compare your result with this one, and record more trials to see if this takes your result any closer to the expected probability of event $A$.

3  A bag contains 100 marbles of similar size and texture. The marbles are either white or purple, and the number of each is not known. A marble is randomly taken from the bag and replaced before another is withdrawn. 20 marbles are sampled in this way, with the results as shown in this table.

| W | W | P | P | W | W | W | P | W | W |
|---|---|---|---|---|---|---|---|---|---|
| P | P | W | W | W | W | W | P | P | P |

Events $W$ and $P$ are defined as the withdrawal of white and purple marbles, respectively.
a  Use a relative frequency diagram to estimate the values of p($W$) and p($P$). Comment.
b  Estimate how many marbles of each colour are in the bag.

# Theoretical probability

If all possible outcomes are equally likely, it is possible to find out how many of these *ought* to be event $A$, that is, to calculate the theoretical probability, $p(A)$.

---

**Remember**

$$p(A) = \frac{\text{number of desired outcomes}}{\text{total number of possible outcomes}}.$$

---

**Example 3**

A fair dice is rolled. Calculate the probability of a prime number being thrown.

The relevant prime numbers are 2, 3 and 5. Event $A$ is the event of a prime being observed.

$p(A) = \dfrac{3}{6}$      (3 is the number of desired outcomes, and 6 is the total number of possible outcomes)

$\phantom{p(A)} = \dfrac{1}{2}$

**Example 4**

A double-headed coin is tossed. If event $A$ is that of a head being thrown, calculate $p(A)$ and $p(\overline{A})$.

$p(A) = \dfrac{2}{2} = 1$      (A certainty)               $p(\overline{A}) = \dfrac{0}{2} = 0$      (An impossibility)

---

These results imply two important results in probability.
The first is that all probabilities can be measured on a scale from 0 to 1 inclusively.

**Key Point**

If $A$ is an event, $0 \leqslant p(A) \leqslant 1$.

## Investigate

Copy this scale across your page.

Label the scale, marking approximately where you think the probability of these five events $A$–$E$ should be placed.

- ◆ A hockey captain wins the toss ($A$).
- ◆ A heart is drawn from a pack of cards ($B$).
- ◆ A heart is not drawn from a pack of cards ($C$).
- ◆ You will be abducted by aliens on your way home from school today ($D$).
- ◆ Your teacher will be wearing shoes for your next geography lesson ($E$).

The second result is that, if $A$ is an event, it either occurs ($A$) or it does not ($\overline{A}$). It is certain that nothing else can happen.

**Example 5**

A card is randomly selected from a pack of 52 playing cards.
Calculate the probability that a queen is *not* chosen.

Let event $Q$ be that a queen is chosen.

$$p(\overline{Q}) = 1 - p(Q)$$
$$= 1 - \frac{4}{52}$$
$$= \frac{48}{52} = \frac{12}{13}$$

## Exercise 89

1  Nelson is a keen collector of tropical fish. In his tank, there are four guppies, three angel fish, two cat fish, and one Siamese fighting fish. The tank has to be cleaned, so he randomly scoops one of these fish up in his net. Calculate the probability that it is
   **a**  a guppy                          **b**  an angel fish
   **c**  a tiger fish                      **d**  not a Siamese fighting fish

2  A letter is chosen randomly from a collection of Scrabble tiles that spell the word PERIODONTOLOGY. Calculate the probability that it is
   **a**  an O                              **b**  a T
   **c**  a vowel                           **d**  an N or another non-vowel

3  A card is randomly selected from a pack of 52 playing cards. Calculate the probability that it is
   **a**  a red card                        **b**  a king
   **c**  a number card that is a multiple of 3   **d**  an ace, jack, queen or king

4  The bar chart shows the sock colours worn by pupils in class 5C. If a pupil is chosen at random from 5C, calculate the probability that he or she will be wearing
   **a**  grey (G) socks
   **b**  white (W) socks
   **c**  red (R) or black (B) socks
   **d**  not red socks

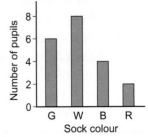

**5** A fair ten-sided dice with numbers from 1 to 10 on it is thrown. Calculate the probability of obtaining

    **a** a 1                **b** an even number

    **c** a number which has an integer square root     **d** a number of at most 7 and at least 4

**6** David and Melissa play battleships. On their $10 \times 10$ grid of squares, both have these in their fleet.

Battleship

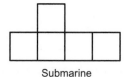
Submarine

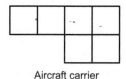
Aircraft carrier

David shoots first, choosing a square randomly. Calculate the probability that he

    **a** hits Melissa's aircraft carrier           **b** hits nothing

    **c** hits her battleship or submarine        **d** does not hit her submarine

# Exercise 89★

**1** A black dice and a white dice are thrown together, and their scores are added. Copy and complete the 'probability space' table showing all 36 possible outcomes.

    **a** Use your table to calculate the probability of obtaining

      (i) a total of 6

      (ii) a total of more than 10

      (iii) a total less than 4

      (iv) a prime number

    **b** What is the most likely total?

|   | White |   |   |   |   |   |
|---|---|---|---|---|---|---|
|   | 1 | 2 | 3 | 4 | 5 | 6 |
| 1 | 2 | 3 | 4 |   |   |   |
| 2 | 3 | 4 |   |   |   |   |
| 3 | 4 |   |   |   |   |   |
| 4 |   |   |   |   |   |   |
| 5 |   |   |   |   |   |   |
| 6 |   |   |   |   |   |   |

(Black labels the rows.)

**2** Four marbles are in a red bag. They are numbered 2, 3, 5 and 7. A green bag contains four more marbles numbered 11, 13, 17 and 19. Two marbles, one from each bag, are randomly selected and the *difference* in the two scores is noted.

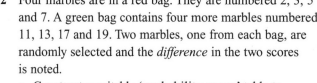

    **a** Construct a suitable 'probability space' table to calculate the probability of obtaining

      (i) a score of 6         (ii) a score of at most 8

      (iii) a score of at least 12     (iv) a square number

    **b** What are the least likely scores?

**3** A regular five-sided spinner is spun twice, and the scores are multiplied. 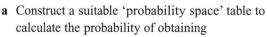 Copy and complete the 'probability space' table. Use the table to calculate the probability of scoring

    **a** an odd number

    **b** a number less than 9

    **c** a number of at least 15

    **d** a triangular number

|   | First spin |   |   |   |   |
|---|---|---|---|---|---|
|   | 1 | 2 | 3 | 4 | 5 |
| 1 | 1 | 2 | 3 |   |   |
| 2 | 2 | 4 |   |   |   |
| 3 | 3 |   |   |   |   |
| 4 |   |   |   |   |   |
| 5 |   |   |   |   |   |

(Second spin labels the rows.)

**4** Three vets record the number of allergic reactions experienced by puppies given the same vaccination.

| Vet | No. of puppies vaccinated | No. of allergic reactions |
|-----|---------------------------|---------------------------|
| X | 50 | 3 |
| Y | 60 | 7 |
| Z | 70 | 10 |

a Calculate the probability that a puppy injected by vet X or Y will experience an allergic reaction.
b Calculate the probability that a puppy injected by vet Y or Z will *not* experience any reaction.
c If 7650 puppies are given this injection in a particular year, estimate how many of them will show signs of an allergy.

**5** A regular three-sided spinner numbered 2, 4 and 6 is spun, and a six-sided dice is cast. The highest number obtained is noted, and if the two numbers are equal, that number is taken. Using a probability space or other method, calculate the probability of obtaining
a a multiple of 3
b a number less than 4
c a non-prime number
d two consecutive numbers

**6** Five beads numbered 1, 2, 3, 4 and 5 are placed in bag X. Three beads numbered 1, 2 and 3 are placed in bag Y. One bead is withdrawn from X and one from Y. These represent the coordinates (for example (1, 3)) of a point on the positive $x$-axis and $y$-axis, respectively. Calculate the probability that after one selection from each bag, the selected point
a lies on the line $y = x$
b lies on the line $x = 2$
c lies on the line $y = 2x - 5$
d lies on the curve $y = x^2 - 6$

**7** A pond contains 20 tadpoles, of which $f$ are frog tadpoles and the others are toad tadpoles. If 10 more frog tadpoles are added to the pond, the probability of catching a frog tadpole is doubled. Find $f$.

**8** A dartboard is in the shape of an equilateral triangle inside which is inscribed a circle. A dart is randomly thrown at the board (assume that it hits the board).

a Given that $\tan 60° = \sqrt{3}$ and $\sin 60° = \dfrac{\sqrt{3}}{2}$, show that the probability of the dart hitting the board inside the circle is $\dfrac{\pi}{3\sqrt{3}}$.

b If 100 darts are thrown at the board, and they all hit the board, how many would you expect to land outside the circle?

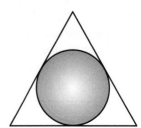

The information in the table was compiled by the League of Dangerous Sports.
Rank the sports in terms of their safety.
Comment.

| Activity | Deaths over 5-year period | Participation of adults (millions) |
|---|---|---|
| Air sports | 51 | 1 |
| Badminton | 3 | 59 |
| Boating/sailing | 69 | 23 |
| Cricket | 2 | 20 |
| Fishing | 50 | 37 |
| Football | 14 | 128 |
| Golf | 1 | 110 |
| Gymnastics | 1 | 14 |
| Hockey | 2 | 9 |
| Horse riding | 62 | 39 |
| Motor sports | 65 | 11 |
| Mountaineering | 51 | 6 |
| Running | 9 | 200 |
| Rugby | 2 | 12 |
| Swimming/diving | 191 | 370 |
| Tennis | 1 | 45 |

# Exercise 90 (Revision)

1 Stan wants to estimate his chances of scoring a goal from a penalty.
  He does this by taking 12 penalties in succession, with these results.

| ✔ | ✘ | ✔ | ✔ | ✘ | ✔ | ✔ | ✔ | ✘ | ✘ | ✔ | ✔ |
|---|---|---|---|---|---|---|---|---|---|---|---|

  Use a relative frequency diagram to estimate his chances of scoring. Comment.

2 The probability that a new truck gets a puncture in a tyre during its first 30 000 km is $\frac{2}{15}$.
  What is the probability of a puncture-free first 30 000 km for this vehicle?

3 One letter is randomly chosen from this sentence: 'All the world's a stage and all the men and
  women merely players.' What is the probability of the letter being
  **a** an 'a'?      **b** a 't'?      **c** a vowel?      **d** an 'x'?

**4** A $1 coin and a $2 coin are tossed.

Write down all the possible outcomes, and calculate the probability of obtaining

**a** two tails                    **b** a head and a tail

**5** Umberto has $1, $10, $20 and $50 notes in his wallet. He has one of each type. He randomly removes two notes together. Find the probability that these two notes total

**a**    $11

**b**    $70

**c**    $80

**d**    at least $11

**6** Frances buys ten raffle tickets from 500 sold. If she does not win anything with any of the first six tickets drawn, what is the probability that she will win with the seventh?

## Exercise 90★ (Revision)

**1** Germaine is a keen bird-watcher and spots an Australian magpie at the same place in a rain forest from 1–10 January in three successive years. She keeps a record, shown below.

| Jan | 1 | 2 | 3 | 4 | 5 | 6 | 7 | 8 | 9 | 10 |
|---|---|---|---|---|---|---|---|---|---|---|
| Year | | | | | | | | | | |
| 2002 | 1 | 0 | 1 | 0 | 1 | 1 | 1 | 0 | 1 | 1 |
| 2003 | 0 | 0 | 1 | 1 | 0 | 1 | 1 | 1 | 1 | 0 |
| 2004 | 0 | 0 | 1 | 0 | 0 | 1 | 0 | 0 | 1 | 1 |

**a** Use a relative-frequency graph to estimate the probability of seeing an Australian magpie in this place for each year.

**b** Comment.

**2** A black dice and a red dice are thrown at the same time. Their scores are multiplied together. Use a probability space diagram to calculate the probability of obtaining

**a** a 4                    **b** an even number                    **c** at least 16

**3** A region of Eastern China is called 'GUANGXI ZHUANGZU ZIZHIQU'. One letter is randomly chosen from this name.

**a** Find the probability of the letter being

(i)    an A

(ii)    a Z

(iii)    a B

**b** What is the most likely letter to be picked?

**4** The sets A and B consist of the following numbers:

A = {1, 3, 5, 7, 9, 11}            B = {1, 5, 9, 13, 17, 21}

A whole number from 1 to 25 inclusive is randomly chosen.

Find the probability that this number is in the set

**a** A

**b** B′

**c** A∩B

**d** A∪B

**5** A spinner is spun and a dice is thrown.
£5 is won when the dice score is at least the score on the
spinner. How much would be won if this game were
played 12 times in succession?
(Assume that there is no charge to play the game.)

**6** Three coins are tossed simultaneously. List all the possible outcomes in a 'probability space' table
and use it to calculate the probability of obtaining

   **a** three heads           **b** two heads and a tail         **c** at least two heads

# Summary 4

**Inverse percentages**

♦ A laptop computer is sold by Suzi for $850, giving her a profit of 25%.

Find the original price she paid.

Let $x$ = original price.

$$x \times 1.25 = 850$$

$$\Rightarrow x = \frac{850}{1.25}$$

$$\Rightarrow x = \$680$$

♦ A motorbike is sold by Max for $3250, giving him a loss of 35%.

Find the original price he paid.

Let $y$ = original price.

$$y \times 0.65 = 3250$$

$$\Rightarrow y = \frac{3250}{0.65}$$

$$\Rightarrow y = \$5000$$

**Estimation**

Estimate the following: — Rounding to 1 s.f.

♦ $\dfrac{9.3 \times 101.7}{1.9} \approx \dfrac{9 \times 100}{2} \approx 450$

Rounding to 2 s.f.

♦ $\dfrac{25.7 \times 223.1}{20.3} \approx \dfrac{26 \times 220}{20} \approx 286$

Standard form

♦ $793\,000\,000 \times 4\,200\,000 \approx 8 \times 10^8 \times 4 \times 10^6 = 8 \times 4 \times 10^{(8+6)} = 32 \times 10^{14} = 3.2 \times 10^{15}$

♦ $793\,000\,000 \div 4\,200\,000 \approx 8 \times 10^8 \div 4 \times 10^6 = 8 \div 4 \times 10^{(8-6)} = 2 \times 10^2$

Adding and subtracting: make sure terms have matching indices.

♦ $(7.93 \times 10^8) + (4.2 \times 10^6) = 7.93 \times 10^8 + 0.042 \times 10^8 = 7.972 \times 10^8 \approx 8 \times 10^8$

♦ $(7.93 \times 10^8) - (4.2 \times 10^6) = 7.93 \times 10^8 - 0.042 \times 10^8 = 7.888 \times 10^8 \approx 8 \times 10^8$

If a question is given in standard form, leave the final answer in standard form too.

♦ $(2 \times 10^5) \div (5 \times 10^{-4}) = (2 \div 5) \times (10^{(5-(-4))}) = 0.4 \times 10^9 = 4 \times 10^8$

♦ $\sqrt{(4.9 \times 10^{15})} = \sqrt{(49 \times 10^{14})} = \sqrt{49} \times \sqrt{10^{14}} = 7 \times 10^7$

## Limits of accuracy

This carpet's dimensions are $2\,\text{m} \times 5\,\text{m}$ (to the nearest 0.1 m).
What are the limits of accuracy for the carpet's perimeter and its area?

The largest possible dimensions are 2.05 m by 5.05 m.
The smallest possible dimensions are 1.95 m by 4.95 m.

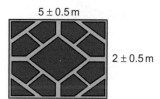

Largest perimeter $= 2 \times 2.05\,\text{m} + 2 \times 5.05\,\text{m} = 14.2\,\text{m}$
Smallest perimeter $= 2 \times 1.95\,\text{m} + 2 \times 4.95\,\text{m} = 13.8\,\text{m}$
So, the *exact* perimeter is between 13.8 m and 14.2 m.
Largest area $= 2.05\,\text{m} \times 5.05\,\text{m} = 10.35\,\text{m}^2$
Smallest area $= 1.95\,\text{m} \times 4.95\,\text{m} = 9.65\,\text{m}^2$
So, the *exact* area is between $9.65\,\text{m}^2$ and $10.35\,\text{m}^2$.

# Algebra

## Rearranging formulae

Identical process to solving equations: **perform the same operation on both sides** of the formula.

♦ Make $u$ the subject of $v^2 = u^2 + 2as$.

$v^2 = u^2 + 2as$     (Subtract $2as$ from both sides)
$u^2 = v^2 - 2as$     (Square root of both sides)
$u = \sqrt{v^2 - 2as}$

♦ Make $e$ the subject of $v = \dfrac{1+e}{1-e}$.

$v = \dfrac{1+e}{1-e}$.     (Multiply both sides by $1-e$)
$v(1-e) = 1+e$     (Multiply out bracket)
$v - ev = 1 + e$     (Collect all terms containing $e$ on one side)
$v - 1 = ev + e$     (Factorise)
$v - 1 = e(v + 1)$     (Divide by $v + 1$)
$e = \dfrac{v-1}{v+1}$

### Using formulae

Rearrange the formula to make the unknown the subject.
Substitute in the values given to find the unknown value.

# Graphs

These graphs are often used to model real-life situations.

## Quadratic graphs $y = ax^2 + bx + c$

### Parabolas

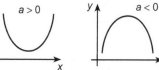

## Solution of $0 = ax^2 + bx + c$

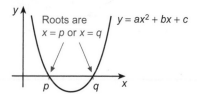

# Shape and space

## Circle theorems

Triangle OAB is isosceles.

Angles off the diameter in a semicircle = 90°.

Angles in same segment off a chord are equal.

Angle at centre is twice angle formed at circumference in same segment off a chord.

$a + c = 180°$
$b + d = 180°$

Opposite angles in a cyclic quadrilateral sum to 180°.

Angles off a tangent to the radius = 90°.

## Similar triangles

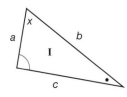

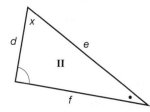

Triangle I is similar to triangle II, as corresponding angles are equal.

$$\Rightarrow = \frac{a}{d} = \frac{b}{e} = \frac{c}{f}$$

## Pythagoras' theorem

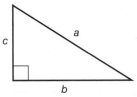

$$a^2 = b^2 + c^2$$

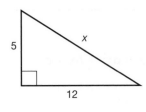

$x^2 = 12^2 + 5^2$
$\quad = 169$
$\therefore x = \sqrt{169}$
$\Rightarrow x = 13$

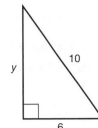

$10^2 = 6^2 + y^2$
$\therefore y^2 = 10^2 - 6^2$
$\quad = 64$
$\therefore y = \sqrt{64}$
$\Rightarrow y = 8$

## Probability

The probability of an event $E$ happening: $p(E) = \dfrac{\text{number of desired outcomes}}{\text{total number of possible outcomes}}$

Impossible $0 \leqslant p(E) \leqslant 1$ certain

The probability of an event not happening is $p(\overline{E})$.

$p(E) + p(\overline{E}) = 1$

Experimental probability is measured by **relative frequency** $= \dfrac{\text{number of successes}}{\text{total number of trials}}$

♦ A fair dice is thrown.

Event $A$ is a 5 being thrown.

$p(A) = \dfrac{1}{6}$

$p(\overline{A}) = \dfrac{5}{6}$

# Examination practice 4

UNIT 4 ♦ Examination practice

**1 a** Estimate the value to 1 significant figure of

(i) $\dfrac{1.98 \times \sqrt{79}}{59.8 - 30.6}$

(ii) $\dfrac{1.21 \times 10^4 - 2.2 \times 10^3}{9.8 \times 10^{-2}}$

**b** If $\dfrac{34.7 \times 7.3}{(5.22 + 9.02)} = 17.7886$

find the value of

(i) $\dfrac{3.47 \times 0.73}{(5.22 + 9.02)}$

(ii) $\dfrac{347 \times 73}{(0.522 + 0.902)}$

**2 a** A chair is bought at an auction for £50, and a week later is sold for £71. Find the percentage profit.

**b** A table is bought at an auction for £275. What must the selling price be, for a profit of 15% to be made?

**c** A dealer bids for a picture at an auction. She thinks she will be able to sell it later for £350. If she plans to make a profit of 25% when she sells it, what price should she bid for it at auction?

**3** A building plot is stated to be 10 m wide and 30 m long, correct to the nearest metre.

**a** Write down the minimum width and the minimum length.

**b** Work out the minimum area.

**4** Sarah uses the formula

$$t = \frac{2s}{u + v}$$

She has to calculate the value of $t$ when $s = 623.25$, $u = 11.37$ and $v = 87.22$. Sarah estimates the value of $t$ **without using her calculator**.

**a** (i) Write down the numbers Sarah could use in the formula to estimate the value of $t$.

(ii) Work out the estimate for the value of $t$ that these numbers would give.

A calculator is to be used to work out the actual value of $t$.

**b** To what degree of accuracy would you give your calculator answer? Give a reason for your answer.

*LONDON*

**5**

The cost $C$, in pounds, of a fence with $n$ panels is given by the formula

$$C = 11n + 4(n + 1)$$

**a** Expand the brackets and express the formula as simply as possible.

**b** Make $n$ the subject of the formula.

*LONDON*

**6**

The air temperature, $T°C$, outside an aircraft flying at a height of $h$ feet is given by the formula

$$T = 26 - \frac{h}{500}$$

The air temperature outside an aircraft is $-52°C$.

Calculate the height of the aircraft.

*LONDON*

**7** Dave launches a model aeroplane powered by a rubber band. The height, $h$ m, of the plane $t$ seconds after launch is given by

$$h = \frac{t^2}{4} - t + 1.5$$

valid for $0 \leqslant t \leqslant 5$.

**a** Make a table of values giving $h$ for $0 \leqslant t \leqslant 5$ and draw a graph of $h$ against $t$.

**b** Use your graph to find the height above the ground at the launch.

**c** What is the minimum height above the ground and the time at which this occurs?

**d** Between what times is the plane less than one metre above the ground?

**8** The formula $V = 150N + 2500$ gives the water-tank volume $V$ in litres that is necessary to supply water to $N$ apartments in a new block of flats.

**a** Find $V$ if $N = 20$.

**b** If $V = 4000$, find the value of $N$.

**c** Rearrange the formula to make $N$ the subject.

**d** A water tank has a volume of 5500 litres. What is the greatest number of apartments that the tank could serve?

**9** The cost $C$ in pounds of each layer of a circular wedding cake is related to its diameter $d$ in centimetres by the formula $C = 0.1 \times d^2$.

**a** Calculate the cost of a wedding cake consisting of three layers of diameters 20 cm, 15 cm and 10 cm.

**b** A single-layered cake costs £90. Calculate the cake's diameter.

**10** $ABCD$ is a circle, centre $O$.
$XAB$ is a straight line.
Angle $BCD = 96°$.

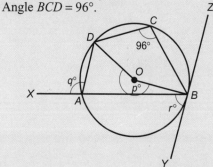

**a** Find the values of $p$ and $q$.
$YBZ$ is the tangent to the circle at $B$.
Angle $AOB = 144°$.

**b** Find the value of $r$.

*OCR*

**11**

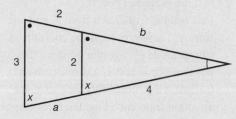

Find lengths $a$ and $b$.

**12**

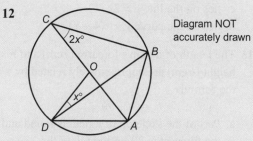

Diagram NOT accurately drawn

$A$, $B$, $C$ and $D$ are points on the circumference of a circle centre $O$.
$AC$ is a diameter of the circle.
Angle $BDO = x°$.
Angle $BCA = 2x°$.
Express, in terms of $x$, the size of
(i) angle $BDA$,
(ii) angle $AOD$,
(iii) angle $ABD$.

*EDEXCEL*

**13** A roulette wheel includes 36 alternate-coloured segments (red and black), numbered from 1 to 36. If the black is number 1, and a single ball is rolled, state the probability of scoring

**a** a prime black number

**b** a square red number

**14** Five beads consecutively numbered from 1 to 5 are placed in bag X.

Three beads consecutively numbered from 1 to 3 are placed in bag Y.

One bead is withdrawn from X and one from Y. These represent the coordinates (for example (1, 3)) of a point on the positive $x$-axis and $y$-axis respectively. Calculate the probability that after one selection from each bag, the selected point

**a** lies on the line $y = 2x$

**b** lies on the line $y = 3$

**c** lies on the line $y = 2x - 3$

**d** lies on the curve $y = x(4 - x)$

**15** The length of a man's forearm ($f$ cm) and his height ($h$ cm) are approximately related by the formula

$$h = 3f + 90$$

**a** Part of the skeleton of a man is found and the forearm is 20 cm long. Use the formula to estimate the man's height.

**b** A man's height is 162 cm. Use the formula to estimate the length of his forearm.

**c** George is 1 year old and he is 70 cm tall. Find the value the formula gives for the length of his forearm and state why this value is impossible.

*LONDON*

**16** Use Pythagoras' theorem to find the area of an equilateral triangle of side 20 cm.

# Number 5

## Proportion

### Comparative costs

To help shoppers compare the prices of packaged foods, shopkeepers show the **cost per 100 grams** or **per litre** on the packs.

---

**Example 1**

A 100 g jar of Brazilian coffee costs €4.50, and a 250 g jar of the same coffee costs €12. Which is the best buy?

100 g jar:

$$\text{cost/g} = \frac{450}{100} \text{ cents/g}$$

$$= 4.50 \text{ c/g}$$

250 g jar:

$$\text{cost/g} = \frac{1200}{250} \text{ cents/g}$$

$$= 4.80 \text{ c/g}$$

The 100 g jar of coffee is better value.

**Example 2**

**a** Change 1 hour to seconds.

1 hour = 3600 s.

**b** Using the fact that 1 mile = 1.609 km, change 30 miles to metres.

30 miles = 30 × 1.609 km = 30 × 1.609 × 1000 m = 48 270 m.

**c** Use your answers to parts **a** and **b** to convert 30 miles/hour to metres per second.

$$30 \text{ miles/hour} = \frac{48\,270 \text{ m}}{3600 \text{ s}} = 13.4 \text{ m/s (to 3 s.f.)}$$

$$\left( \text{As 'per' means divide, metres per second} = \frac{\text{metres}}{\text{seconds}}. \right)$$

---

# Activity 29

Use the information given to find the 'best buy' by calculating cents/g and comparing the four types of chicken portion. Comment on your results. The number in brackets is the percentage of the chicken that is **not** edible.

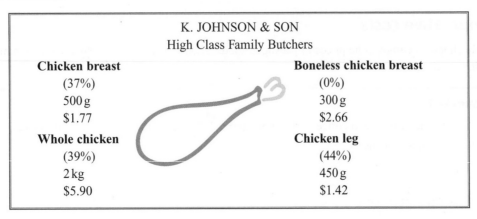

K. JOHNSON & SON
High Class Family Butchers

**Chicken breast**
(37%)
500 g
$1.77

**Boneless chicken breast**
(0%)
300 g
$2.66

**Whole chicken**
(39%)
2 kg
$5.90

**Chicken leg**
(44%)
450 g
$1.42

# Exercise 91

Where appropriate, give your answers to 3 significant figures.

1 Which tin of paint is the better value:
  tin X, 1 litre for $4.50, or
  tin Y, 2.5 litres for $10.50?

2 Which bag of rice is the better value:
  bag P, 1 kg for $3.25, or
  bag Q, 5 kg for $16.50?

3 Which roll of the same type of curtain material is the better value:
  roll A, 10 m for $65, or
  roll B, 16 m for $100?

4 Which bottle of cranberry juice is the better value:
  bottle I, 330 ml for $1.18, or
  bottle II, 990 ml for $3.20?

5 Convert 110 km to metres, and 1 hour to seconds. Then change 110 km/h, the maximum speed on a motorway, to metres per second.

6 Convert 6 km to metres and 1 hour to seconds. Then change 6 km/h (walking speed) to metres per second.

7 On average, a human consumes 1 m$^3$ (1000 litres) of fluid a year.
  Calculate a person's average daily fluid consumption in litres per day.

8 The Earth is hit by a meteor storm travelling at 70 km per second.
  Change 70 km to miles, given that 1 km = 0.6215 miles, and then calculate 70 km/s in miles per hour.

9 In 18 days, a queen termite lays 1 million eggs. How many eggs per minute is this?

10 By the age of 18, the average American child has seen 350 000 commercials on television.
  How many is this per day?

**11** A house, 14 m long, is drawn to a scale of $1:50$. Find its length on the plan, in centimetres.

**12** A 20 m-wide building plot is drawn to a scale of $1:200$. How wide is it on the plan, in centimetres?

**13** To a scale of $1:100$, a door is shown as 8 mm wide. Find its actual width, in centimetres.

**14** To a scale of $1:500$, a garden is shown as 6 cm long. Find its actual length, in metres.

**15** Write down the scale of this drawing in the form $1:n$.

The reticulated python, at 10 m long, is the longest snake in the world.

## Activity 30

How far, in metres, do these travel in 0.2 s?

A man walks
4 km in 1 hour.

A greyhound runs
400 m in 19.5 s.

A rifle bullet takes
1.79 s to travel 1 mile.

The Earth turns at
just over 1000 miles/h.

A thunderclap travels
at $3.3 \times 10^4$ cm/s.

Lightning travels
at $3 \times 10^{10}$ cm/s.

## Exercise 91★

Where appropriate give your answers to 3 significant figures.

**1** Mrs Becker wants to tile a kitchen floor. She considers three types of square tile:
marble, 20 cm × 20 cm for $3.00 each, or
slate, 25 cm × 25 cm for $4.50 each, or
limestone, 0.3 m × 0.3 m for $6.50 each.
For each type of tile, calculate the cost in dollars per square metre and hence find the cheapest choice for Mrs Becker.

**2** A garden centre sells various sizes of monkey-puzzle trees.

A 1.5 m tree costs $45, while a 2 m tree is $55 and a 3.5 m tree is sold at $110.

By calculating the cost for each tree in dollars per metre, list the trees in order of value for money.

For Questions 3–6, use this currency conversion table.

| Country | Currency | Rate per US$ |
|---------|----------|--------------|
| Brazil | real | 2.75 |
| Eurozone | euro | 0.81 |
| India | rupee | 40.18 |
| Kenya | shilling | 72.22 |
| Malaysia | ringgit | 3.73 |
| Sweden | krona | 7.52 |
| Thailand | baht | 36.80 |
| UK | pound | 0.56 |

**3** Convert 1000 Swedish krona into Kenyan shillings.

**4** Convert 1000 Thai baht into euros.

**5** The same model of a DVD player costs the following in various parts of the world:

| Eurozone | 486 euros |
|----------|-----------|
| Malaysia | 2200 ringgits |
| UK | 325 pounds |

Convert each price into US$ and state which country sells the item most cheaply.

**6** Stefan travels from the USA to three countries: Brazil, UK and India. He has $1200 to convert and wishes to do so in the ratios of 1:3:2 for each country respectively.

Find out how much of each currency he will have for his journey.

**7** In a record rainfall, 100 cm fell in 24 hours. How many millimetres fell per minute?

**8** In 1999, the shadow of the total eclipse travelled at 1700 miles/hour.
Given that 1 km = 0.6215 miles, change this to metres per second.

**9** A Boeing 737 uses 15 400 litres of fuel in 5.5 hours.
Given that 1000 litres = 1 m$^3$, find the consumption in cubic metres per minute.

**10** While hovering, a Harrier jump jet uses 91 kg of fuel per minute.
Find the fuel consumption in grams per second.

**11** In this photograph, the diameter 7 cm represents $1.4 \times 10^6$ km, the actual diameter of the Sun. Calculate the scale of the photograph in the form $1:n$. Estimate the actual length of the tongue.

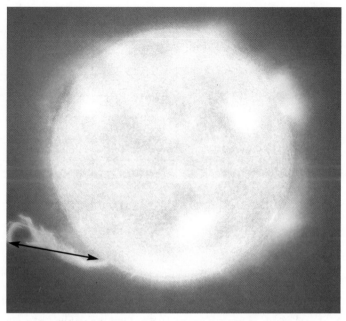

A spectacular incandescent tongue of gas shoots out from the Sun.

**12** Four million tonnes of hydrogen are consumed by our Sun every second. How long would it take to consume $6 \times 10^{21}$ tonnes of hydrogen, the mass of the Earth?

## Investigate

On 1 January 1999 the European common currency, the euro, was launched. Investigate the percentage change in the euro over its first 12 months.

| 1 euro on 1 Jan 1999 | £1 on 1 Jan 2000 |
|---|---|
| UK £ 0.7040 | European Euro 1.572 |
| German DM 1.956 | German DM 3.075 |
| French Fr 6.560 | French Fr 10.31 |
| Italian Lira 1936 | Italian Lira 3043 |
| Spanish Ptas 166.4 | Spanish Ptas 261.6 |
| US$ 1.169 | US$ 1.617 |

## Science problems

### Activity 31

Work out the density of each element in this table in g/cm³ correct to 2 significant figures.

| Metal | Mass (g) | Volume (cm³) |
|---|---|---|
| Aluminium | 40.2 | 14.9 |
| Brass | 122 | 14.3 |
| Gold | 281 | 14.8 |
| Platinum | 317 | 15.1 |

Copy and complete this table.

Write your entry in the final row as an equation.

| Substance | Mass (g) | Volume (cm³) | Density (g/cm³) |
|---|---|---|---|
|  | 135 | 50 |  |
|  | 459 | 54 |  |
|  |  | 10 | 21 |
| Oak | 46.2 | 55 |  |
| Petrol | 46.8 |  | 0.72 |
| Air | 1200 |  | 0.0012 |
| Water |  |  |  |
| – | $M$ | $V$ |  |

## Exercise 92

1   $1\,cm^3$ of iron has a mass of $7.8\,g$. Find the mass of an iron hammer head of volume $20\,cm^3$.

2   $1\,cm^3$ of silver has a mass of $10.5\,g$. Find the volume of a silver cup of mass $420\,g$.

3   An aluminium kettle of volume $80\,cm^3$ has a density of $2.6\,g/cm^3$. Find its mass.

4   A $198\,g$ paperweight is made from impure lead of density $11\,g/cm^3$. Find its volume.

5   The amount of heat needed to raise the temperature of $1\,g$ of water by $1\,°C$ is $4.2\,J$ (joules). How much heat is needed to raise the temperature of $20\,g$ of water by $1\,°C$?

6   How much heat is needed to raise the temperature of $1\,g$ of water by $30\,°C$?

7   What length of wire has a resistance of $5\,\Omega$ (ohms) if $9\,m$ of it has a resistance of $30\,\Omega$?

8   A piece of wire, $2.8\,m$ long, has a resistance of $7\,\Omega$. Find the resistance of a $3.6\,m$ length.

## Exercise 92★

1   The hub of a car wheel has a density of $2.5\,g/cm^3$. Find its volume if it weighs $4750\,g$.

2   A piece of mahogany has a density of $0.75\,g/cm^3$.
Find the mass, in kilograms, of a mahogany plank measuring $2\,cm$ by $9\,cm$ by $2.5\,m$.

3   What length of wire has a resistance of $8\,\Omega$ if $2.8\,m$ of it has a resistance of $7\,\Omega$?

4   An $11.9\,g$ brooch has a volume of $3.4\,cm^3$. Find its density.

5   A pound coin has a mass of $9.75\,g$ and a volume of $0.75\,cm^3$. Find its density.

6   The density of air is $0.001\,g/cm^3$. Find the mass of air in a classroom of dimensions $4\,m$ by $5\,m$ by $6\,m$. Give your answer in kilograms.

7   The amount of heat needed to raise the temperature of $1\,g$ of water by $1\,°C$ is $4.2\,J$ (joules). How much heat is needed to raise the temperature of $20\,g$ of water by $30\,°C$?

**8** The Earth was formed 4500 million years ago. If this time is represented by a straight line 5.40 m long, where should the following events be placed?

| Life form | Approximate number of years ago when first appeared |
|---|---|
| First living cells | $3200 \times 10^6$ |
| First land animals | $400 \times 10^6$ |
| First mammals | $225 \times 10^6$ |
| Giant dinosaurs | $135 \times 10^6$ |
| Man | $2 \times 10^6$ |

# Positive integer powers of numbers

Powers are used to write certain numbers in a convenient way, and rules for combining them or simplifying expressions involving powers are shown in Example 3.

---

**Example 3**

To multiply powers, add the indices:

$$3^2 \times 3^3 = 3^{2+3} = 3^5 = 243$$

To divide powers, subtract the indices:

$$\frac{8^6}{8^4} = 8^6 \div 8^4 = 8^{6-4} = 8^2 = 64$$

For a power of a power, multiply the indices:

$$(2^3)^2 = 2^6 = 64$$

For large numbers, write in standard form:

$$30^5 = (3 \times 10)^5 = 3^5 \times 10^5$$
$$= 243 \times 100\,000 = 24\,300\,000$$

Or, write as factors:

$$4^4 = (2^2)^4 = 2^8 = 256$$

---

**Remember**

- $a^m \times a^n = a^{m+n}$
- $a^m \div a^n = a^{m-n}$
- $(a^m)^n = a^{m \times n}$

# Exercise 93

Write these as a single power and then calculate the answer.

**1** $2^2 \times 2^2$      **2** $3^2 \times 3^3$      **3** $2 \times 2 \times 2 \times 2 \times 2$      **4** $5 \times 5 \times 5 \times 5$

**5** $2^4 \div 2^2$      **6** $4^4 \div 4^2$      **7** $5^5 \div 5^2$      **8** $8^6 \div 8^3$

**9** $\dfrac{3^8}{3^2}$      **10** $\dfrac{6^5}{6^2}$      **11** $(2^2)^5$      **12** $(2^4)^2$

**13** $0.1 \times (0.1)^2$      **14** $0.2 \times (0.2)^2$      **15** $2.1^{10} \div 2.1^8$      **16** $1.3^5 \div 1.3^3$

**17** $\dfrac{4^2 \times 4^5}{4^3}$      **18** $\dfrac{7^4 \times 7^2}{7^3}$      **19** $20^2 \times 20^2$      **20** $30^2 \times 30^2$

# Exercise 93★

For Questions 1–16, write as a single power to find the answer.

**1** $8^4 \times 8^5 \div 8^6$      **2** $9^4 \times 9^5 \div 9^6$      **3** $(7^2)^3 \div 7^3$      **4** $(4^2)^4 \div 4^4$

**5** $5^5 \div 25$      **6** $4^6 \div 64$      **7** $216 \div 6^2$      **8** $2^{10} \div 512$

**9** $125^2 \div 5^3$      **10** $100^3 \div 10^5$      **11** $(10^3)^3 \div 1000$      **12** $(2^5)^3 \div 4^3$

**13** $8^4 \div 4^6$      **14** $27^3 \div 9^4$      **15** $\dfrac{125^3}{25^3}$      **16** $\dfrac{64^2}{2 \times (2^5)^2}$

**17** How many zeros follow the digit 1 in $10^{(3^2)}$?

**18** How many zeros follow the digit 1 in $10^{(10^2)}$?

**19** How many times does $(1.2)^4$ divide into $12^4$?

**20** Given that $2^{20} = 1\,048\,576$, calculate $2^{21}$ and $2^{19}$.

**21** Which is larger, $8^{10}$ or $4^{14}$?

**22** Which is larger, $3^{26}$ or $2^{39}$?

# Simple recurring decimals

All fractions can be written as decimals.

These either **terminate**, e.g. $\dfrac{3}{8} = 0.375$, or produce a set of **recurring** digits, e.g. $\dfrac{1}{3} = 0.333333\ldots$.

Some fairly simple-looking fractions produce a large number of digits which recur, e.g. $\dfrac{1}{97} = 0.010$
309 278 350 515 463 917 525 773 195 876 288 659 793 814 432 989 690 721 649 484 536 082 474 226 804
123 711 340 206 185 567…, and all of these 95 digits are repeated again and again and again!

> **Remember**
>
> **Recurring decimals** are written using the 'dot notation' to show the repeating digits.
>
> $\dfrac{2}{9} = 0.222\,222\ldots\ = 0.\dot{2}$                $\dfrac{5}{6} = 0.833\,3333\ldots = 0.8\dot{3}$
>
> $\dfrac{34}{99} = 0.343\,4343\ldots = 0.\dot{3}\dot{4}$          $\dfrac{25}{999} = 0.025\,0250\ldots = 0.\dot{0}2\dot{5}$

## Exercise 94

Express these recurring decimals as fractions.

**1** $0.\dot{3}$     **2** $0.\dot{7}$     **3** $0.\dot{1}\dot{5}$     **4** $0.8\dot{1}$     **5** $0.62\dot{1}$     **6** $0.40\dot{3}$

## Exercise 94★

Express these recurring decimals as fractions.

**1** $0.0\dot{1}$     **2** $0.05\dot{3}$     **3** $0.\dot{4}32\dot{1}$     **4** $0.987\dot{6}$     **5** $0.7\dot{3}$     **6** $0.5\dot{8}$

## Exercise 95 (Revision)

Where appropriate, give your answers to 3 significant figures.

For Questions 1–4, first write as a single power and then calculate the answer.

**1** $2^5 \times 2^3$     **2** $7^4 \div 7^2$     **3** $(3^3)^2$     **4** $\dfrac{5^5}{25}$

**5** Which is the better buy in tinned pineapples:
tin A, 450 g at $2.70, or
tin B, 240 g at $1.36?

**6** Which DVD rental shop gives better value:
shop X, 1 DVD at $7 for a week, or
shop Y, 2 DVDs at $12 for five days?

**7** Express as an exact fraction, the recurring decimal $0.\dot{5}\dot{6}$.

**8** Express as an exact fraction the recurring decimal $0.\dot{3}0\dot{1}$.

**9** In 2001, a revolutionary catamaran, shown in the photo, was designed to set a new world record by circumnavigating the world in 65 days.

Assuming that the distance to be travelled was 40 000 km, find its speed in kilometres per hour.

**10** The scale of the catamaran photograph is 1 : 600. Find the height of the mast above the water line.

**11** Change 26 km/h to metres per second.

**12** The average human in the developed world consumes about 900 kg of food per year. What is this in grams per minute?

**13** A metal spanner has a volume of 8.4 cm$^3$ and a density of 9.5 g/cm$^3$. Find its mass.

**14** A piece of wire is 6 m long, and it has a resistance of 33 Ω (ohms). Find the resistance of an 8 m length.

## Exercise 95★ (Revision)

Where appropriate, give your answers to 3 significant figures.

For Questions 1–3, first write as a single power and then calculate the answer.

**1** $\dfrac{20^4}{20^6} \times 20^5$　　　　　　**2** $(1.2^2)^4 \div 1.2^6$　　　　　　**3** $2^{14} \div 16$

For Questions 4–7, use this currency conversion table.

| Country | Currency | Rate per US$ |
|---|---|---|
| Australia | dollar | 1.29 |
| Mexico | peso | 10.45 |
| Norway | krone | 6.84 |
| Switzerland | franc | 1.27 |

**4** Convert 1000 Australian dollars into US dollars.

**5** Convert 500 Mexican pesos into Swiss francs.

**6** Franz goes on a skiing holiday to Norway. He has 675 Swiss francs and wishes to change 70% of this sum into krone. How much Norwegian money will he have?

**7** Chantal has 750 Mexican pesos and wishes to change this into Australian and Swiss currency in the ratio of 1 : 2 respectively.
How much of each currency will she have?

**8** Express the recurring decimal $0.7\dot{8}$ as an exact fraction.

**9** Express the recurring decimal $0.005\dot{7}$ as an exact fraction.

**10** The density of water is $1\,\text{g/cm}^3$. Find the mass of 1 litre and the mass of $1\,\text{m}^3$.

**11** Each year, about 4.8 billion aluminium cans are dumped in landfill sites in the UK.
Given that one aluminium can weighs $20\,\text{g}$, find, correct to 1 significant figure, an estimate of the total mass in tonnes.
The density of aluminium is $2.7\,\text{g/cm}^3$. What is the volume, in cubic metres?

**12** Every 25 minutes $4.1 \times 10^9$ litres of water cascades over the Niagara Falls.
Find the rate at which water cascades in cubic metres per second.

# Algebra 5

## Multiplying brackets

Finding the area of a rectangle entails multiplying two numbers together.

Multiplying $(x + 2)$ by $(x + 4)$ can also be done by finding the area of a rectangle.

This rectangular poster has sides $(x + 2)$ and $(x + 4)$.
Notice that the diagram shows the area of each part.

The total area is

$(x + 2)(x + 4) = x^2 + 4x + 2x + 8 = x^2 + 6x + 8.$

♦ Draw similar diagrams to calculate these:

$\qquad (x + 5)(x + 2) \qquad (10 + x)(x + 7) \qquad (x + 1)(x + 1)$

♦ A very common mistake is to say that $(x + 2)^2 = x^2 + 2^2$.
Show that $(x + 2)^2 \neq x^2 + 2^2$ by substituting various numbers for $x$.
Are there any values of $x$ for which $(x + 2)^2 = x^2 + 2^2$?
What does $(x + 2)^2$ equal?
Remember that $(x + 2)^2 = (x + 2)(x + 2)$.

♦ With imagination, this method can be extended to deal with negative numbers.

|  | $x$ | $-5$ |
|---|---|---|
| $x$ | $x^2$ | $-5x$ |
| $2$ | $2x$ | $-10$ |

$\qquad \therefore (x + 2)(x - 5) = x^2 - 5x + 2x - 10 = x^2 - 3x - 10$

♦ Use diagrams to calculate these:

$\qquad (x + 4)(x - 3) \qquad (x - 1)(x - 6) \qquad (x - 3)^2 \qquad (x + 2)(x - 2)$

### First – Outside – Inside – Last

Brackets can be multiplied without drawing diagrams:

$\qquad (x + 2) \times a = xa + 2a$

$\therefore \qquad (x + 2) \times (x + 4) = x(x + 4) + 2(x + 4)$

giving $\qquad (x + 2)(x + 4) = x^2 + 4x + 2x + 8 = x^2 + 6x + 8$

# Key Points

The mnemonic **FOIL** will remind you of what to do when multiplying out brackets.

**FOIL** stands for First, Outside, Inside, Last.

From each bracket,

- multiply the **F**irst terms
- multiply the **O**utside terms
- multiply the **I**nside terms
- multiply the **L**ast terms.

Then add the four terms and simplify.

## Example 1

Multiply out and simplify $(x + 1)(x + 2)$.

| Multiply the **F**irst terms | $x^2$ |
|---|---|
| Multiply the **O**utside terms | $2x$ |
| Multiply the **I**nside terms | $x$ |
| Multiply the **L**ast terms | $2$ |
| Add these terms to give | $(x + 1)(x + 2) = x^2 + 2x + x + 2 = x^2 + 3x + 2$ |

## Example 2

Multiply out and simplify $(x + 3)(x - 2)$.

| Multiply the **F**irst terms | $x^2$ | |
|---|---|---|
| Multiply the **O**utside terms | $-2x$ | (Note how the negative signs are dealt with) |
| Multiply the **I**nside terms | $3x$ | |
| Multiply the **L**ast terms | $-6$ | |
| Add these terms to give | $(x + 3)(x - 2) = x^2 + (-2x) + 3x + (-6)$ | |
| | $\qquad\qquad = x^2 + x - 6$ | |

## Example 3

Multiply out and simplify $(2x - 3)(3x - 5)$.

| Multiply the **F**irst terms | $6x^2$ |
|---|---|
| Multiply the **O**utside terms | $-10x$ |
| Multiply the **I**nside terms | $-9x$ |
| Multiply the **L**ast terms | $+15$ |
| Add these terms to give | $(2x - 3)(3x - 5) = 6x^2 + (-10x) + (-9x) + 15$ |
| | $\qquad\qquad = 6x^2 - 19x + 15$ |

# Exercise 96

Multiply out and simplify these expressions.

**1** $(x + 4)(x + 1)$       **2** $(x + 2)(x + 6)$       **3** $(x - 7)(x + 3)$

**4** $(x - 5)(x + 4)$       **5** $(x + 2)(x - 6)$       **6** $(x + 7)(x - 1)$

**7** $(x - 3)(x - 5)$       **8** $(x - 2)(x - 8)$       **9** $(x + 3)^2$

**10** $(x + 5)^2$       **11** $(x - 4)^2$       **12** $(x - 7)^2$

**13** $(x + 5)(x - 5)$       **14** $(x + 4)(x - 4)$       **15** $(x + 2)(x - 8)$

**16** $(3x + 2)(x - 4)$       **17** $(3x - 2)(5x + 1)$       **18** $(4x - 1)(3x + 5)$

**19** $(x^2 - 5)(x + 2)$       **20** $(x^2 + 3)(x - 4)$

# Exercise 96★

For Questions 1–22, multiply out and simplify the expression.

**1** $(x + 7)(x - 3)$       **2** $(x - 2)(x + 9)$       **3** $(x - 3)(x + 3)$

**4** $(x + 6)(x - 6)$       **5** $(x + 12)^2$       **6** $(x - 9)^2$

**7** $(3x - 4)(4x - 3)$       **8** $(7x + 3)(2x + 5)$       **9** $(x - a)(x + b)$

**10** $(x - a)(x + a)$       **11** $(4x - 5)^2$       **12** $(7x + 4)^2$

**13** $(3x^2 + 1)(5x + 7)$       **14** $(5x^2 + 2)(4x - 3)$       **15** $(x + 3)^2 - (x - 1)^2$

**16** $(4x + 1)(x + 3) - (2x + 5)^2$       **17** $\left(\dfrac{a}{2} - \dfrac{b}{5}\right)^2$       **18** $\left(\dfrac{x}{4} + \dfrac{y}{3}\right)^2$

**19** $x(5x^3 + 3x^2)(2x + 1)$       **20** $\left(x + \dfrac{1}{x}\right)\left(x - \dfrac{1}{x}\right)$       **21** $\left(\dfrac{a}{b} + \dfrac{b}{a}\right)^2 - \left(\dfrac{a}{b} - \dfrac{b}{a}\right)^2$

**22** $(5x + 2)(4x + 7) - (5x + 2)(4x + 6)$

**23** Solve $2x^2 + (x + 4)^2 = (3x + 2)(x - 2)$.

**24** If $(x + a)(x - 3) = x^2 + 2x - 15$, what is the value of $a$?

**25** If $(x + a)^2 + b = x^2 + 6x + 10$, find the values of $a$ and $b$.

## Activity 33

♦ Draw diagrams to show how to multiply out these expressions.

$(x^2 + 2x + 3)(x + 1)$       $(x + y + 3)(x - 2y)$       $(x^2 + 2x - 3)(x^2 - 2x + 3)$

Work out how to do the multiplication *without* using diagrams.

## Exercise 97

**1** A rectangle with length $(x + 2)$ cm and width $(x + 1)$ cm has a square of length $x$ cm cut out of it.

   **a** Find and simplify an expression for the area of the original rectangle.

   **b** Hence find an expression for the shaded area.

   **c** The shaded area is 11 cm². Find the value of $x$.

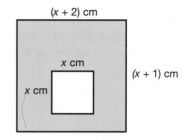

**2** A circle of radius $(x + 4)$ m has a circle of radius $x$ m cut out of it.

   **a** Find and simplify an expression for the area of the original circle.

   **b** Hence find an expression for the shaded area.

   **c** The shaded area is $32\pi$ m². Find the value of $x$.

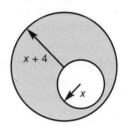

**3** A concrete block is in the shape of a cuboid with length $(x + 3)$ cm, width $(x + 2)$ cm and height 5 cm.

   **a** Find and simplify an expression for the volume of the block.

   **b** Find and simplify an expression for the surface area of the block.

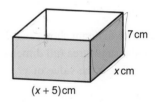

**4** An open box has a rectangular base with dimensions $x$ cm and $(x + 5)$ cm.

The height of the box is 7 cm.

   **a** Find and simplify an expression for the volume of the box.

   **b** Find and simplify an expression for the total surface area (inside and outside) of the box.

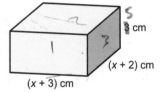

**5** These two pictures have the same area. Find $x$.

**6** A right-angled triangle has lengths as shown.

   **a** Use Pythagoras' theorem to form an equation in $x$.

   **b** Solve your equation to find $x$.

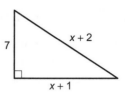

# Exercise 97★

**1** A circle of radius $(x + 6)$ cm has a circle of radius $(x)$ cm cut out of it.

   **a** Find and simplify an expression for the area of the original circle.

   **b** The area that remains is $45\pi$ cm². Find $x$.

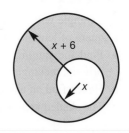

**2** A rectangle with length $(4x + 5)$ cm and width $(x + 8)$ cm has four squares of side $x$ cm cut out of its corners.

   **a** Find and simplify an expression for the area of the original rectangle.

   **b** The area that remains is 95.5 cm². Find the value of $x$.

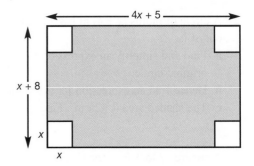

**3** A right-angled triangle has lengths as shown. Find $x$.

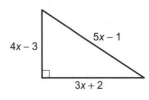

**4** The diagram shows two flower beds which have the same area (all dimensions are in metres). What is the value of $x$?

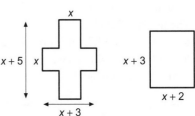

**5** **a** If $n$ is an integer, explain why $2n + 1$ must be an odd number.

   **b** Show that when two odd numbers are multiplied together, the answer is always odd.

   **c** Show that if you add 1 to the product of two consecutive odd numbers, the answer is always a perfect square.

**6** A metal sheet of width 20 cm is to be bent into a chute with width 10 cm and height 6 cm with symmetrical cross-section as shown.

   **a** Find an expression in terms of $x$ for the length CB.

   **b** Find an expression in terms of $x$ for the length AB.

   **c** Use Pythagoras' theorem to find the value of $x$.

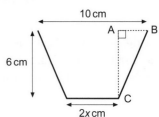

# Factorising quadratic expressions with two terms

Factorising expressions such as $x^2 + 2x$ to give $x(x + 2)$ is easy, because $x$ is a common factor.

**Example 4**

Factorise $x^2 - 12x$.

$x$ is a common factor, and so

$x^2 - 12x = x(x - 12)$

**Example 5**

Expand $(x - 3)(x + 3)$ using **FOIL**.

$x^2 + 3x - 3x - 9 = x^2 - 9$

$\therefore$ factorising $x^2 - 9$ gives $(x - 3)(x + 3)$

## Investigate

- ♦ Does it matter what order the brackets are in when you are factorising $x^2 - 9$?
- ♦ What is the connection between the numbers in the brackets ($-3$ and $+3$) and the number 9? Is this always the case?
- ♦ Can you factorise $x^2 + 9$ in the same way?

## Exercise 98

Factorise these expressions.

**1** $x^2 - 3x$      **2** $x^2 - x$      **3** $x^2 + 2x$      **4** $x^2 + 4x$

**5** $x^2 - 31x$      **6** $x^2 - 17x$      **7** $x^2 + 42x$      **8** $x^2 + 38x$

**9** $x^2 - 16$      **10** $x^2 - 25$      **11** $x^2 - 49$      **12** $x^2 - 81$

## Exercise 98★

Factorise these expressions.

**1** $x^2 - 312x$      **2** $x^2 - 273x$      **3** $x^2 + 51x$      **4** $x^2 + 74x$

**5** $x^2 - 64$      **6** $x^2 - 100$      **7** $x^2 - 121$      **8** $x^2 - 196$

# Factorising quadratic expressions with three terms

Expanding $(x + 2)(x - 5)$ using **FOIL** gives $x^2 - 3x - 10$.

Factorising is the reverse process. $x^2 - 3x - 10$ factorises to $(x + 2)(x - 5)$.

**Example 6**

Find $a$ if $x^2 + 5x + 6 = (x + 3)(x + a)$.

Using FOIL, the last terms in each bracket are multiplied to give 6.

$\therefore 3 \times a = 6$, and $a = 2$.

Check: $(x + 3)(x + 2) = x^2 + 2x + 3x + 6$
$= x^2 + 5x + 6$

**Example 7**

Find $a$ if $x^2 + x - 12 = (x + 4)(x + a)$.

Using FOIL, the last terms in each bracket are multiplied to give $-12$.

$\therefore 4 \times a = -12$ and $a = -3$.

Check: $(x + 4)(x - 3) = x^2 + 4x - 3x - 12$
$= x^2 + x - 12$

## Exercise 99

Find $a$.

**1** $x^2 + 3x + 2 = (x + 2)(x + a)$

**2** $x^2 + 6x + 8 = (x + 4)(x + a)$

**3** $x^2 + 7x + 12 = (x + 3)(x + a)$

**4** $x^2 - x - 6 = (x - 3)(x + a)$

**5** $x^2 + 3x - 4 = (x + 4)(x + a)$

**6** $x^2 - 2x - 3 = (x + 1)(x + a)$

**7** $x^2 - 7x + 10 = (x - 5)(x + a)$

**8** $x^2 - 7x + 12 = (x - 4)(x + a)$

**9** $x^2 + 4x + 4 = (x + 2)(x + a)$

**10** $x^2 + 2x + 1 = (x + 1)(x + a)$

**11** $x^2 - 1 = (x + 1)(x + a)$

**12** $x^2 - 4 = (x - 2)(x + a)$

## Exercise 99★

Find $a$.

**1** $x^2 + 4x + 3 = (x + 1)(x + a)$

**2** $x^2 + x - 6 = (x - 2)(x + a)$

**3** $x^2 + x - 12 = (x - 3)(x + a)$

**4** $x^2 - 5x + 6 = (x - 2)(x + a)$

**5** $x^2 - 12x + 35 = (x - 5)(x + a)$

**6** $x^2 + 4x - 12 = (x + 6)(x + a)$

**7** $x^2 + 2x - 15 = (x + 5)(x + a)$

**8** $x^2 + 12x + 36 = (x + 6)(x + a)$

**9** $x^2 - 64 = (x + 8)(x + a)$

**10** $x^2 - 81 = (x - 9)(x + a)$

**11** $x^2 + 4\frac{1}{2}x + 2 = (x + 4)(x + a)$

**12** $x^2 - 6\frac{1}{3}x + 2 = (x - 6)(x + a)$

When the number in the first bracket is not given, then try all the factors of the last number in the expression. The two factors chosen must add to the number in front of the $x$ term.

### Example 8

Factorise $x^2 + 5x + 6$.

The last sign is +, and so both brackets will have the same sign as $+5x$, giving $(x + \quad)(x + \quad)$.

The missing numbers are both positive, multiply to give +6, and add to +5.

The two numbers are +3 and +2.

Thus $x^2 + 5x + 6 = (x + 3)(x + 2)$.

### Example 9

Factorise $x^2 - 7x + 6$.

The last sign is +, and so both brackets will have the same sign as $-7x$, giving $(x - \quad)(x - \quad)$.

The missing numbers are both negative, multiply to give +6, and add to $-7$.

The two numbers are $-1$ and $-6$.

Thus $x^2 - 7x + 6 = (x - 1)(x - 6)$.

## Exercise 100

Factorise these. (Notice that the last sign is always +.)

**1** $x^2 - 3x + 2$      **2** $x^2 + 4x + 4$      **3** $x^2 - 4x + 3$

**4** $x^2 - 5x + 4$      **5** $x^2 - 7x + 12$      **6** $x^2 + 7x + 10$

**7** $x^2 + 8x + 16$      **8** $x^2 + 6x + 9$      **9** $x^2 - 9x + 8$

**10** $x^2 - 7x + 10$      **11** $x^2 - 2x + 1$      **12** $x^2 - 4x + 4$

## Exercise 100★

Factorise these. (Notice that the last sign is always +.)

**1** $x^2 + 10x + 21$      **2** $x^2 + 9x + 18$      **3** $x^2 - 8x + 12$

**4** $x^2 - 11x + 30$      **5** $x^2 - 16x + 64$      **6** $x^2 - 14x + 49$

**7** $x^2 - 18x + 72$      **8** $x^2 - 19x + 88$      **9** $x^2 + 14x + 45$

**10** $x^2 + 20x + 96$      **11** $x^2 + 24x + 144$      **12** $x^2 + 30x + 225$

## Exercise 101

Factorise these. (Notice that the last sign is always $-$.)

| | | |
|---|---|---|
| **1** $x^2 + x - 6$ | **2** $x^2 + 2x - 8$ | **3** $x^2 - 3x - 10$ |
| **4** $x^2 - 4x - 5$ | **5** $x^2 - 4x - 12$ | **6** $x^2 - 2x - 8$ |
| **7** $x^2 - 9x - 10$ | **8** $x^2 - 7x - 18$ | **9** $x^2 + 5x - 14$ |
| **10** $x^2 + 2x - 15$ | **11** $x^2 + 7x - 8$ | **12** $x^2 + 3x - 10$ |

## Exercise 101★

Factorise these. (Notice that the last sign is always $-$.)

| | | |
|---|---|---|
| **1** $x^2 + x - 30$ | **2** $x^2 + 6x - 27$ | **3** $x^2 - 2x - 24$ |
| **4** $x^2 - 3x - 28$ | **5** $x^2 + 7x - 60$ | **6** $x^2 + x - 90$ |
| **7** $x^2 - 9x - 70$ | **8** $x^2 - 5x - 50$ | **9** $x^2 - 7x - 120$ |
| **10** $x^2 - 12x - 64$ | **11** $x^2 + 10x - 75$ | **12** $x^2 + 6x - 72$ |

## Exercise 102

Factorise these. (Notice that the signs are mixed.)

| | | |
|---|---|---|
| **1** $x^2 - 3x + 2$ | **2** $x^2 - x - 6$ | **3** $x^2 + 2x - 3$ |
| **4** $x^2 + x - 12$ | **5** $x^2 + 13x + 12$ | **6** $x^2 + 3x + 2$ |
| **7** $x^2 - 8x + 12$ | **8** $x^2 - 6x + 5$ | **9** $x^2 - 8x + 16$ |
| **10** $x^2 - 6x + 9$ | | |

## Exercise 102★

Factorise these. (Notice that the signs are mixed.)

| | | |
|---|---|---|
| **1** $x^2 + 8x - 20$ | **2** $x^2 + 6x - 16$ | **3** $x^2 - 7x - 18$ |
| **4** $x^2 - 2x - 35$ | **5** $x^2 + 13x + 36$ | **6** $x^2 + 13x + 40$ |
| **7** $x^2 - 12x + 32$ | **8** $x^2 - 2x - 48$ | **9** $x^2 + 8x - 48$ |
| **10** $x^2 - 26x + 48$ | **11** $3 + 2x - x^2$ | |

# Solving quadratic equations by factorisation

If $a \times b = 0$, what can be said about either $a$ or $b$?

A little thought should convince you that either $a = 0$ or $b = 0$ (or both are zero).

## Investigate

> If $a \times b = 12$, what can be said about either $a$ or $b$?

---

**Example 11**

Solve $(x + 2)(x - 3) = 0$.

Either $(x + 2) = 0$ or $(x - 3) = 0$.
If $(x + 2) = 0$, then $x = -2$.
If $(x - 3) = 0$, then $x = 3$.
There are *two* solutions:
$x = -2$ or $x = 3$

**Example 12**

Solve $(x - 5)^2 = 0$.

$(x - 5)^2 = 0$ is the same as $(x - 5)(x - 5) = 0$.
If the first bracket $(x - 5) = 0$, then $x = 5$.
If the second bracket $(x - 5) = 0$, then $x = 5$.
There is *one* solution: $x = 5$.

---

## Exercise 103

Solve these equations.

**1** $(x + 1)(x + 2) = 0$

**2** $(x + 2)(x + 3) = 0$

**3** $(x + 4)(x - 1) = 0$

**4** $(x + 5)(x - 4) = 0$

**5** $0 = (x - 7)(x - 2)$

**6** $0 = (x - 5)(x - 3)$

**7** $(x + 8)^2 = 0$

**8** $(x - 9)^2 = 0$

**9** $x(x - 10) = 0$

**10** $x(x + 7) = 0$

## Exercise 103★

Solve these equations.

**1** $(x + 8)(x - 4) = 0$

**2** $(x + 12)(x - 3) = 0$

**3** $0 = (x + 21)(x - 5)$

**4** $0 = (x - 9)(x + 6)$

**5** $x(x - 8) = 0$

**6** $x(x + 10) = 0$

**7** $(2x + 3)(4x - 3) = 0$

**8** $(3x - 2)(5x + 1) = 0$

**9** $(x + 1)(x - 1)(2x + 5) = 0$

**10** $(x - a)(x - b)(x - c) = 0$

---

**Example 13**

Solve $x^2 + 5x + 6 = 0$.

$x^2 + 5x + 6 = 0$ factorises to
$(x + 3)(x + 2) = 0$.  (See Example 8)
$\therefore x = -3$ or $x = -2$.

**Example 14**

Solve $x^2 - 7x + 6 = 0$.

$x^2 - 7x + 6 = 0$ factorises to
$(x - 1)(x - 6) = 0$.  (See Example 9)
$\therefore x = 1$ or $x = 6$.

### Example 15

Solve $x^2 - 5x = 6$.

$x^2 - 5x = 6$ must first be rearranged to $x^2 - 5x - 6 = 0$.

Then $x^2 - 5x - 6 = 0$ factorises to $(x + 1)(x - 6) = 0$. (See Example 10) $\therefore x = -1$ or $x = 6$.

## Exercise 104

Factorise and solve these for $x$.

**1** $x^2 - 3x + 2 = 0$      **2** $x^2 - 5x + 6 = 0$      **3** $x^2 + x - 2 = 0$

**4** $x^2 - 2x - 3 = 0$      **5** $x^2 + 6x + 8 = 0$      **6** $x^2 + 3x + 2 = 0$

**7** $x^2 - x - 12 = 0$      **8** $x^2 + x - 12 = 0$      **9** $x^2 - 8x + 15 = 0$

**10** $x^2 - 10x + 24 = 0$      **11** $x^2 + 8x + 16 = 0$      **12** $x^2 - 10x + 25 = 0$

## Exercise 104★

Factorise and solve these for $x$.

**1** $x^2 - 9x + 20 = 0$      **2** $x^2 - 12x + 35 = 0$      **3** $x^2 - 5x - 24 = 0$

**4** $x^2 - 3x - 28 = 0$      **5** $x^2 + 21x + 108 = 0$      **6** $x^2 + 21x + 110 = 0$

**7** $x^2 - 18x + 56 = 0$      **8** $x^2 - 18x + 65 = 0$      **9** $x^2 + 22x + 96 = 0$

**10** $x^2 + 23x + 90 = 0$      **11** $24x^2 - 48x - 72 = 0$      **12** $x^2 + 5x - 6 = 8$

**13** $x^2 + 7x - 78 = 42$

### Remember

To solve a quadratic equation, rearrange it so that the right-hand side is zero.

Then factorise the left-hand side to solve the equation.

If the quadratic expression has only two terms, the working is easier.

### Example 16

Solve $x^2 - 12x = 0$.

$x^2 - 12x$ factorises to $x(x - 12)$. (See Example 4)

So $x^2 - 12x = 0 \Rightarrow x(x - 12) = 0$

Now, either $x = 0$ or $x - 12 = 0$, giving the two solutions $x = 0$ or $x = 12$.

### Example 17

Solve $x^2 - 9 = 0$.

$x^2 - 9$ could be factorised and the working continued, but the following is easier.

$x^2 - 9 = 0 \Rightarrow x^2 = 9$

Square-rooting both sides gives $x = \pm\sqrt{9}$, so $x = 3$ or $-3$. (Don't forget the negative square root!)

# Exercise 105

Solve for $x$.

**1** $x^2 - 2x = 0$          **2** $x^2 - 5x = 0$          **3** $x^2 + 7x = 0$

**4** $x^2 + 3x = 0$          **5** $x^2 - 25x = 0$        **6** $x^2 - 17x = 0$

**7** $x^2 + 23x = 0$       **8** $x^2 + 31x = 0$        **9** $x^2 - 4 = 0$

**10** $x^2 - 36 = 0$        **11** $x^2 - 25 = 0$        **12** $x^2 - 16 = 0$

# Exercise 105★

Solve for $x$.

**1** $x^2 - 125x = 0$      **2** $x^2 - 117x = 0$      **3** $x^2 + 231x = 0$

**4** $x^2 + 321x = 0$      **5** $x^2 - 64 = 0$        **6** $x^2 - 81 = 0$

**7** $x^2 - 169 = 0$       **8** $x^2 - 144 = 0$       **9** $x^2 - 7 = 0$

**10** $x^2 - a^2 = 0$        **11** $x^2 + 9 = 0$

# Problems leading to quadratic equations

### Example 18

The product of two consecutive even numbers is 120. What are the numbers?

Let the first number be $x$. The second even number is two more than $x$ and can be written as $x + 2$.

Then
$$x \times (x + 2) = 120 \qquad \text{(Multiply out the bracket)}$$
$$x^2 + 2x = 120 \qquad \text{(Rearrange to equal zero)}$$
$$x^2 + 2x - 120 = 0 \qquad \text{(Factorise)}$$
$$(x - 10)(x + 12) = 0$$
$$x = 10 \text{ or } -12$$

So, the numbers are 10 and 12 or −12 and −10.

There are two possible answers. Answering the question by 'trial and improvement' would find the positive answer, but probably not the negative answer.

### Example 19

The length of a rectangular patio is 3 m more than the width.
If the area is $28 \, \text{m}^2$, find the length and width of the patio.

Let $x$ be the width in metres. Then the length is $(x + 3)$ metres.
The area is
$$x \times (x + 3) = 28 \qquad \text{(Multiply out the bracket)}$$
$$x^2 + 3x = 28 \qquad \text{(Rearrange to equal zero)}$$
$$x^2 + 3x - 28 = 0 \qquad \text{(Factorise)}$$
$$(x + 7)(x - 4) = 0$$
$$x = -7 \text{ or } x = 4$$

As the answer cannot be negative, the width is 4 m and the length is 7 m.

# Exercise 106

**1** When $x$ is added to its square, $x^2$, the answer is 12. Find the values of $x$.

**2** When $x$ is added to its square, $x^2$, the answer is 30. Find the values of $x$.

**3** When $x$ is subtracted from its square, the answer is 20. Find the values of $x$.

**4** When $x$ is subtracted from its square, the answer is 42. Find the values of $x$.

**5** I think of a number. I then square it and add twice the original number. The answer is 35. What was the original number?

**6** I think of a number. I then square it and subtract twice the original number. The answer is 24. What was the original number?

**7** A rectangle has a length of $(x + 3)$ cm and a width of $x$ cm.
  **a** Write down an expression for the area of the rectangle.
  **b** If the area is 18 cm$^2$, find the value of $x$.

**8** A triangle has a base of $2x$ cm and a height of $(x + 1)$ cm.
  **a** Write down an expression for the area of the triangle.
  **b** If the area is 42 cm$^2$, find the value of $x$.

**9** The length of a mobile phone is 6 cm more than its width. The area of the face is 40 cm$^2$. Find the length and width.

**10** The length of a credit card is 3 cm more than its width. The area is 40 cm$^2$. Find the length and width.

**11** The rectangles shown have the same area. Find $x$.

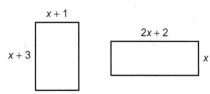

**12** The right-angled triangles shown have the same area. Find $x$.

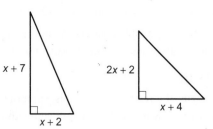

# Exercise 106★

1 The product of two consecutive odd numbers is 143. Find the two numbers.

2 The product of two numbers is 96. One number is 4 more than the other number. Find the two numbers.

3 The length of a picture is 10 cm more than the width. The area is 1200 cm$^2$. Find the dimensions of the picture.

4 The length of a swimming pool is 80 m more than its width. The area is 2000 m$^2$. Find the dimensions of the pool.

5 A ball is thrown vertically upwards so that its height above the ground after $t$ seconds is $(15t - 5t^2)$ m. At what times is it 10 m above the ground?

6 On a roller coaster, a carriage rolls down a sloping track and travels $(5t^2 + 5t)$ m in $t$ seconds. How long does it take to travel 30 m?

7 The sum of the squares of two consecutive integers is 145. Find the two integers.

8 The sum of the squares of two consecutive odd integers is 130. Find the two integers.

9 The sum of the first $n$ integers 1, 2, 3, 4, …, $n$ is given by the formula $\frac{1}{2}n(n + 1)$. How many integers must be taken to add up to 210?

10 An $n$-sided convex polygon has $\frac{1}{2}n(n - 3)$ diagonals. How many sides has a polygon with 135 diagonals?

11 The sides of two cubes differ by 2 cm and their volumes differ by 152 cm$^3$. Find the length of the side of the smaller cube.

12 Sammy spends €10 on some cans of drink. In another shop, she sees that the same cans are each 10c cheaper, and she calculates that she could have bought five more cans for the same money. How many cans did she buy?

# Exercise 107 (Revision)

For Questions 1–3, multiply out and simplify.

1 $(x - 7)(x - 3)$        2 $(x + 2)^2$        3 $(2x + 3)(x - 5)$

4 A rectangle with length $(x + 3)$ cm and width $(x + 2)$ cm has a square of length $x$ cm cut out of it.
 a Find and simplify an expression for the area of the original rectangle.
 b Hence find an expression for the shaded area.
 c The shaded area is 26 cm$^2$. Find the value of $x$.

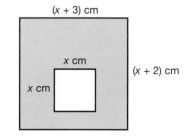

Factorise.

5 $x^2 - 36$        6 $x^2 + 4x + 3$        7 $x^2 + 2x - 8$

Solve for $x$.

8 $x^2 - 4x - 12 = 0$    9 $x^2 - 5x = 0$    10 $x^2 - 36 = 0$    11 $x^2 - x = 20$

12 The width of a television screen is 10 cm more than the height. The area of the screen is 600 cm$^2$. Find the dimensions of the screen.

# Exercise 107★ (Revision)

For Questions 1–3, multiply out and simplify.

**1** $(x + 9)(x - 12)$       **2** $(2x - 3)^2$       **3** $(3x - 1)(2x + 3)$

**4** The foot of a ladder is 2 m away from a vertical wall.
The height of the top of the ladder is $\frac{1}{2}$ m less than the length of the ladder. Find the length of the ladder.

2 m

**5** An abstract painting is 96 cm square, and shows three circles that touch each other and the sides of the square.
The top two circles have the same radius.
What is the radius of the third circle?

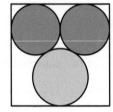

Solve for $x$.

**6** $x^2 - 121 = 0$      **7** $x^2 - 7x = 0$      **8** $x^2 - x = 56$      **9** $x^2 - 15x + 54 = 0$

**10** A square picture has a border 5 cm wide. The picture area is $\frac{4}{9}$ of the total area.
Find the area of the picture.

# Sequences 5

## Continuing sequences

A set of numbers that follows a definite pattern is called a **sequence**. Many problems in mathematics can be solved by using sequences.

Activity 34

**Triangular patterns**: Seema is decorating the walls of a hall with balloons in preparation for a disco. She wants to place the balloons in a triangular pattern.

To make a 'triangle' with one row, she needs **one** balloon.

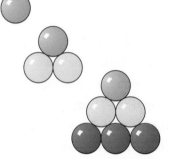

To make a triangle with two rows, she needs **three** balloons.

To make a triangle with three rows, she needs **six** balloons.

The numbers of balloons needed form the sequence 1, 3, 6, .... Describe in words how to continue the sequence, and find the next three terms.

Seema thinks she can work out how many balloons are needed for $n$ rows by using the formula $\frac{1}{2}n(n+1)$. Her friend Julia thinks the formula is $\frac{1}{2}n(n-1)$. Which formula is correct? If Seema has 100 balloons, find, using 'trial and improvement', how many rows she can make and how many balloons will be left over.

**Pyramidal patterns**: In the centre of the hall, Seema wants a triangular pyramid of balloons.

To make a 'pyramid' with one layer, she needs **one** balloon.

To make a pyramid with two layers, she needs **four** balloons.

To make a pyramid with three layers, she needs **ten** balloons.

The numbers of balloons needed form the sequence 1, 4, 10, .... Describe in words how to continue the sequence, and find the next three terms.

The formula for the number of balloons needed for $n$ layers is $\frac{1}{6}n(n+1)(n+2)$. Find, using 'trial and improvement', the number of layers that Seema can make if she uses 200 balloons.

283

## Remember

These are some important sequences.

- ◆ **Natural numbers**     1, 2, 3, 4, ...
- ◆ **Even numbers**     2, 4, 6, 8, ...
- ◆ **Odd numbers**     1, 3, 5, 7, ...
- ◆ **Triangle numbers**     1, 3, 6, 10, ...        (Seema's first sequence)
- ◆ **Square numbers**     1, 4, 9, 16, ...        (The squares of the natural numbers)
- ◆ **Powers of 2**     1, 2, 4, 8, ...        (Numbers of the form $2^n$)
- ◆ **Powers of 10**     1, 10, 100, ...        (Numbers of the form $10^n$)
- ◆ **Prime numbers**     2, 3, 5, 7, ...        (Notice that 1 is not a prime number)

Other sequences may be based on these important sequences.

# Exercise 108

For Questions 1–10, write down the first four terms of the sequence.

**1** Starting with 2, keep adding 2.

**2** Starting with 1, keep adding 2.

**3** Starting with –9, keep adding 3.

**4** Starting with –8, keep adding 4.

**5** Starting with 15, keep subtracting 5.

**6** Starting with 10, keep subtracting 3.

**7** Starting with 2, keep multiplying by 2.

**8** Starting with 1, keep multiplying by 3.

**9** Starting with 12, keep dividing by 2.

**10** Starting with 32, keep dividing by 2.

For Questions 11–20, describe the rule for going from one term to the next, and write down the next three numbers in the sequence.

**11** 3, 7, 11, 15, ..., ..., ...

**12** 5, 9, 13, 17, ..., ..., ...

**13** 13, 8, 3, –2, ..., ..., ...

**14** 5, 2, –1, –4, ..., ..., ...

**15** 3, 6, 12, 24, ..., ..., ...

**16** 2, 6, 18, 54, ..., ..., ...

**17** 64, 32, 16, 8, ..., ..., ...

**18** 288, 144, 72, 36, ..., ..., ...

**19** 0.2, 0.5, 0.8, 1.1, ..., ..., ...

**20** 1, 0.1, 0.01, 0.001, ..., ..., ...

# Exercise 108★

For Questions 1–8, write down the first four terms of the sequence.

**1** Starting with –1, keep adding 1.5.

**2** Starting with –0.8, keep adding 0.2.

**3** Starting with 3, keep subtracting 1.25.

**4** Starting with 5, keep subtracting 2.5.

**5** Starting with 1, keep multiplying by 2.5.

**6** Starting with 1, keep multiplying by 1.5.

**7** Starting with 3, keep dividing by –3.

**8** Starting with 1, keep dividing by –2.

**9** The first two terms of a sequence are 1, 1. The next term is found by adding together the last two terms. Find the first six terms of the sequence.

**10** The first two terms of a sequence are 1, 2. The next term is found by multiplying together the last two terms. Find the first six terms of the sequence.

For Questions 11–20, describe the rule for going from one term to the next, and write down the next three numbers in the sequence.

**11** $3, 5\frac{1}{2}, 8, 10\frac{1}{2}, \ldots, \ldots, \ldots$

**12** $1, 0.8, 0.6, 0.4, \ldots, \ldots, \ldots$

**13** $243, 81, 27, 9, \ldots, \ldots, \ldots$

**14** $1, \frac{1}{2}, \frac{1}{4}, \frac{1}{8}, \ldots, \ldots, \ldots$

**15** $2, 4, 16, 256, \ldots, \ldots, \ldots$

**16** $40, 8, 1.6, 0.32, \ldots, \ldots, \ldots$

**17** $1, -\frac{1}{2}, \frac{1}{4}, -\frac{1}{8}, \ldots, \ldots, \ldots$

**18** $-\frac{1}{9}, \frac{1}{3}, -1, 3, \ldots, \ldots, \ldots$

**19** $1, 3, 7, 15, 31, \ldots, \ldots, \ldots$

**20** $1, 4, 10, 22, 46, \ldots, \ldots, \ldots$

# Formula for sequences

Sometimes the sequence is given by a formula. This means that any term can be found without working out all the previous terms

---

**Example 1**

Find the sequence given by $n$th term $= 2n - 1$. Also find the 100th term.

Substituting $n = 1$ into the formula gives the first term as $\qquad$ $2 \times 1 - 1 = 1$
Substituting $n = 2$ into the formula gives the second term as $\qquad$ $2 \times 2 - 1 = 3$
Substituting $n = 3$ into the formula gives the third term as $\qquad$ $2 \times 3 - 1 = 5$
Substituting $n = 4$ into the formula gives the fourth term as $\qquad$ $2 \times 4 - 1 = 7$
So the sequence is 1, 3, 5, 7… or the odd numbers.

Substituting $n = 100$ into the formula gives the 100th term as $2 \times 100 - 1 = 199$

**Example 2**

A sequence is given by $n$th term $= 4n + 2$. Find the value of $n$ for which the $n$th term equals 50.

$4n + 2 = 50 \Rightarrow 4n = 48 \Rightarrow n = 12$
So the 12th term equals 50.

---

# Exercise 109

In Questions 1–12, find the first four terms of the sequence.

**1** $n$th term $= 2n + 1$

**2** $n$th term $= 3n + 2$

**3** $n$th term $= 5n - 1$

**4** $n$th term $= 4n - 3$

**5** $n$th term $= 33 - 3n$

**6** $n$th term $= 28 - 2n$

**7** $n$th term $= n^2 + 1$

**8** $n$th term $= n^2 - 1$

**9** $n$th term $= 3n$

**10** $n$th term $= 2n$

**11** $n$th term $= \frac{n + 1}{n}$

**12** $n$th term $= \frac{n - 1}{n + 1}$

In Questions 13–18, find the value of $n$ for which the $n$th term has the value given in brackets.

13 $n$th term = $4n + 4$   (36)

14 $n$th term = $3n + 10$   (46)

15 $n$th term = $6n - 12$   (30)

16 $n$th term = $5n - 13$   (32)

17 $n$th term = $22 - 2n$   (8)

18 $n$th term = $30 - 3n$   (15)

19 If the $n$th term = $\dfrac{1}{n-1}$, which is the first term less than $\dfrac{1}{20}$?

20 If the $n$th term = $\dfrac{1}{n+1}$, which is the first term less than $\dfrac{1}{30}$?

# Exercise 109★

In Questions 1–12, find the first four terms of the sequence.

1 $n$th term = $5n - 6$

2 $n$th term = $6n - 8$

3 $n$th term = $100 - 3n$

4 $n$th term = $84 - 4n$

5 $n$th term = $\frac{1}{2}(n + 1)$

6 $n$th term = $\frac{1}{3}(n - 2)$

7 $n$th term = $n^2 + n + 1$

8 $n$th term = $\frac{1}{2}(n^2 - n)$

9 $n$th term = $2n + n$

10 $n$th term = $n(2n + 1)$

11 $n$th term = $\dfrac{2n+1}{2n-1}$

12 $n$th term = $\dfrac{3-2n}{2+n}$

In Questions 13–18, find the value of $n$ for which the $n$th term has the value given in brackets.

13 $n$th term = $7n + 9$   (65)

14 $n$th term = $8n + 6$   (62)

15 $n$th term = $3n - 119$   (–83)

16 $n$th term = $4n - 97$   (–41)

17 $n$th term = $12 - 5n$   (–38)

18 $n$th term = $18 - 6n$   (–48)

19 If the $n$th term = $\dfrac{1}{2n-1}$, which is the first term less than 0.01?

20 If the $n$th term = $\dfrac{1}{3n+1}$, which is the first term less than 0.02?

## Investigate

A sequence is given by the formula    $n$th term = $an + b$.

Investigate the connection between $a$ and $b$ and the numbers in the sequence.

# The difference method

When it is difficult to spot a pattern in a sequence, the difference method can often help.
Underneath the sequence, write down the differences between each pair of terms.
If the differences show a pattern, then the sequence can be extended.

**Example 3**

Find the next three terms in the sequence 2, 5, 10, 17, 26, ..., ..., ...

| Sequence | | 2 | | 5 | | 10 | | 17 | | 26 |
|---|---|---|---|---|---|---|---|---|---|---|
| Differences | | | 3 | | 5 | | 7 | | 9 | |

$= 5 - 2$    $= 10 - 5$

The differences increase by 2 each time, so the table can now be extended.

$= 26 + 11$    $= 37 + 13$

| Sequence | 2 | | 5 | | 10 | | 17 | | 26 | | **37** | | **50** | | **65** |
|---|---|---|---|---|---|---|---|---|---|---|---|---|---|---|---|---|
| Differences | | 3 | | 5 | | 7 | | 9 | | **11** | | **13** | | **15** | | |

If the pattern in the differences is still not clear, add a third row giving the differences between the terms in the second row. More rows can then be added until a pattern is found, though not all sequences will result in a pattern.

**Example 4**

Find the next three terms in the sequence 3, 13, 29, 51, 79, ...

| Sequence | 3 | | 13 | | 29 | | 51 | | 79 |
|---|---|---|---|---|---|---|---|---|---|
| Differences | | 10 | | 16 | | 22 | | 28 | |
| | | | 6 | | 6 | | 6 | | |

Now the table can be extended.

| Sequence | 3 | | 13 | | 29 | | 51 | | 79 | | **113** | | **153** | | **199** |
|---|---|---|---|---|---|---|---|---|---|---|---|---|---|---|---|---|
| Differences | | 10 | | 16 | | 22 | | 28 | | **34** | | **40** | | **46** | | |
| | | | 6 | | 6 | | 6 | | **6** | | **6** | | **6** | | | |

# Exercise 110

Find the next three terms of the following sequences, using the difference method.

**1** 2, 5, 8, 11, 14

**2** 4, 9, 14, 19, 24

**3** 8, 5, 2, −1, −4

**4** 11, 7, 3, −1, −5

**5** 1, 6, 14, 25, 39

**6** 2, 8, 16, 26, 38

**7** 5, −2, −6, −7, −5

**8** 4, 0, −2, −2, 0

**9** 1, 4, 5, 4, 1

**10** 3, 5, 6, 6, 5

## Exercise 110★

Find the next three terms of the following sequences, using the difference method.

**1** 1, 3, 8, 16, 27, 41

**2** 2, 4, 10, 20, 34, 52

**3** 1, 6, 9, 10, 9, 6

**4** 5, 8, 10, 11, 11, 10

**5** 1, −4, −7, −8, −7, −4

**6** 3, −1, −4, −6, −7, −7

**7** 1, 3, 6, 11, 19, 31, 48

**8** 3, 5, 10, 20, 37, 63, 100

**9** 1, 2, 4, 6, 7, 6, 2

**10** 2, 1, 1, 0, −4, −13, −29

### Investigate

Investigate how to use a spreadsheet to work out sequences using the difference method.

# Finding a formula for a sequence

### Activity 35

Seema decides to decorate the walls of the hall with different patterns of sausage-shaped balloons. She starts with a triangular pattern.

Seema wants to work out how many balloons she will need to make 100 triangles.

♦ If *t* is the number of triangles and *b* is the number of balloons, copy and complete this table.

| *t* | 1 | 2 | 3 | 4 | 5 | 6 |
|-----|---|---|---|---|---|---|
| *b* | | | | | | |

$t = 1$  $t = 2$

Notice that the sequence in the row labelled *b* goes up by 2 each time. Add another row to the table labelled 2*t*.

| **2t** | 2 | 4 | 6 | 8 | 10 | 12 |
|--------|---|---|---|---|----|----|

♦ Write down the formula that connects *b* and 2*t*.
How many balloons does Seema need to make 100 triangles?

**UNIT 5 ◆ Sequences**

◆ Draw some other patterns that Seema might use, and find a formula for the number of balloons needed. Here are some possible patterns.

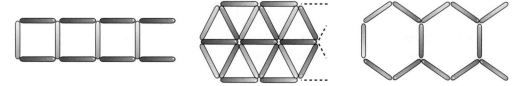

## Investigate

If Seema decided to use a regular pentagon as her basic shape, the pattern would not follow a straight line but would curve round until it joined up with itself.
How many balloons would be needed for this pattern?

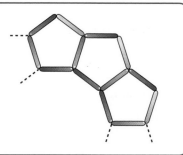

## Activity 36

12 is a square perimeter number, because 12 pebbles can be arranged as the perimeter of a square.

The first square perimeter number is 4.

◆ Copy and complete this table, where $s$ is the square perimeter number and $p$ is the number of pebbles needed.

| $s$ | 1 | 2 | 3 | 4 | 5 | 6 |
|---|---|---|---|---|---|---|
| $p$ | 4 | | | | | |

◆ Use the method of Activity 35 to find a formula for the number of pebbles in the $n$th perimeter number.

## Investigate

Investigate triangular, pentagonal and hexagonal perimeter numbers, finding formulae for the $n$th perimeter number in each case.

Sometimes it is easy to see the formula.

**Example 5**

Find a formula for the $n$th term of the sequence $\frac{1}{3}, \frac{2}{4}, \frac{3}{5}, \frac{4}{6}, \frac{5}{7}$

The top row of numbers is given by $n$.

The bottom row is always 2 more than the top row, so is given by $n + 2$.

So $n$th term $= \frac{n}{n+2}$

**Remember**

If the first row of differences is constant and equal to $a$, then the formula for the $n$th term will be $an + b$, where $b$ is another constant.

**Example 6**

Find a formula for the $n$th term of the sequence 40, 38, 36, 34, ...

The first row of differences is constant and equal to $-2$, so the formula is $-2n + b$.

When $n = 1$, the formula must give the first term as 40. So $-2 \times 1 + b = 40$, and $b = 42$.

The formula for the $n$th term is $42 - 2n$.

# Exercise 111

For Questions 1–8, find a formula for the $n$th term of the sequence.

**1** $1, \frac{1}{2}, \frac{1}{3}, \frac{1}{4}, \frac{1}{5}, \dots$    **2** $\frac{1}{3}, \frac{1}{4}, \frac{1}{5}, \frac{1}{6}, \frac{1}{7}, \dots$    **3** $\frac{1}{2}, \frac{1}{4}, \frac{1}{6}, \frac{1}{8}, \frac{1}{10}, \dots$    **4** $1, \frac{1}{3}, \frac{1}{5}, \frac{1}{7}, \frac{1}{9}, \dots$

**5** 4, 7, 10, 13, ...    **6** 5, 7, 9, 11, ...    **7** 30, 26, 22, 18, ...    **8** 26, 23, 20, 17, ...

**9** Anna has designed a range of Christmas candle decorations using a triangle of wood and some candles. She makes them in various sizes. The one shown here is the three-layer size, because it has three layers of candles.

  **a** Copy and complete this table, where $l$ is the number of layers and $c$ is the number of candles.

| $l$ | 1 | 2 | 3 | 4 | 5 | 6 |
|-----|---|---|---|---|---|---|
| $c$ | 1 |   |   |   |   |   |

  **b** Find a formula connecting $l$ and $c$.

  **c** Mr Rich wants a Christmas candle decoration with exactly 100 candles.
Explain why this is impossible.
What is the largest number of layers than can be made if 100 candles are available?

**10** Julia is investigating rectangle perimeter numbers with one pair of constant sides of three pebbles.

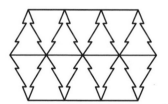

**a** Copy and complete this table, where $n$ is the number in the sequence and $p$ is the number of pebbles

| $n$ | 1 | 2 | 3 | 4 | 5 | 6 |
|-----|---|---|---|---|---|---|
| $p$ | 8 |   |   |   |   |   |

**b** Find a formula connecting $n$ and $p$.

**c** Given only 100 pebbles, what is the largest rectangle in her sequence that Julia can construct?

# Exercise 111★

For Questions 1–8, find a formula for the $n$th term of the sequence.

**1** $\dfrac{2}{1}, \dfrac{3}{2}, \dfrac{4}{3}, \dfrac{5}{4}, \dfrac{6}{5}, \dots$

**2** $0, \dfrac{1}{3}, \dfrac{2}{4}, \dfrac{3}{5}, \dfrac{4}{6}, \dots$

**3** $\dfrac{2}{1}, \dfrac{4}{3}, \dfrac{6}{5}, \dfrac{8}{7}, \dfrac{10}{9}, \dots$

**4** $\dfrac{1}{3}, \dfrac{3}{5}, \dfrac{5}{7}, \dfrac{7}{9}, \dfrac{9}{11}, \dots$

**5** $3, 7, 11, 15, \dots$

**6** $1, 4, 7, 10, \dots$

**7** $6, 3, 0, -3, \dots$

**8** $9, 5, 1, -3, \dots$

**9** Pippa has designed a tessellation based on this shape.

Here are the first three members of the tessellation sequence.

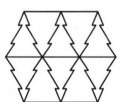

**a** Copy and complete this table, where $n$ is the number in the sequence and $s$ is the number of shapes used.

| $n$ | 1 | 2 | 3 | 4 | 5 | 6 |
|-----|---|---|---|---|---|---|
| $s$ | 6 |   |   |   |   |   |

**b** Find a formula giving $s$ in terms of $n$.

**c** How many shapes will be needed to make the 50th member of the sequence?

**10** Marc is using patterns of pebbles to investigate rectangle perimeter numbers where the inner rectangle is twice as long as it is wide.

**a** Copy and complete this table, where $n$ is the number in the sequence and $p$ is the number of pebbles.

| $n$ | 1 | 2 | 3 | 4 | 5 | 6 |
|-----|----|---|---|---|---|---|
| $p$ | 10 |   |   |   |   |   |

**b** Find a formula connecting $n$ and $p$.

**c** What is the largest member of the sequence that Marc can build with only 200 pebbles?

## Exercise 112 (Revision)

**1** Find the first, tenth and hundredth terms of the sequence given by $n$th term $= 5n - 7$.

**2** Find the next three members of the sequence $-7, -4, -1, 2, \ldots$

**3** Find a formula for the $n$th term of the sequence $\dfrac{1}{2}, \dfrac{2}{3}, \dfrac{3}{4}, \dfrac{4}{5}, \ldots$

**4** Find a formula for the $n$th term of the sequence $10, 16, 22, 28, \ldots$

**5** Phil is training for a long-distance run.
On the first day of his training he runs 1000 m. Each day after that he runs an extra 200 m.
**a** How far has he run on the fifth day?
**b** How far has he run on the $n$th day?
**c** One day he runs 8 km. How many days has he been training?

**6** The first four terms of two sequences are given in these two tables.

Sequence 1

| $n$ | 1 | 2 | 3 | 4 |
|-----|---|----|----|----|
| $s$ | 3 | 12 | 27 | 48 |

Sequence 2

| $n$ | 1 | 2 | 3 | 4 |
|-----|---|---|----|----|
| $s$ | 3 | 7 | 11 | 15 |

**a** A formula for one of these sequences is $3n^2$.
Which sequence is this? What is the tenth term of this sequence?
**b** Find a formula for the $n$th term of the other sequence.

**7** **a** What name is given to the sequence $1, 3, 5, 7, 9, \ldots$?
**b** Copy and complete:
$1 + 3 = \ldots$
$1 + 3 + 5 = \ldots$
$1 + 3 + 5 + 7 = \ldots$
$1 + 3 + 5 + 7 + 9 = \ldots$
**c** Find the sum $1 + 3 + 5 + \cdots + 19 + 21$
**d** Find a formula for the sum of the first $n$ members of the sequence.
**e** The sum of $m$ members of this sequence is 841. What is $m$?

**8 a** Find the $n$th term of the sequence $(1 + 2)$, $(2 + 3)$, $(3 + 4)$, …

**b** Find the $n$th term of the sequence $(2 + 1)$, $(4 + 1)$, $(6 + 1)$, …

**c** Explain why the two sequences are the same.

**d** Explain why every term is odd.

**9** Sequence A:  1, 2, 3, 4, 5, …

Sequence B:  1, 4, 9, 16, 25, …

Sequence C:  2, 6, 12, 20, 30, …

**a** Find the $n$th term of sequence A.

**b** Find the $n$th term of sequence B.

**c** Sequence C is obtained from sequences A and B. Find the $n$th term of sequence C.

**d** Find the 100th term of sequence C.

**10** A modular storage system is built up from cubes as shown.

Each cube is made by screwing together identical lengths of wood.

**a** Copy and complete this table, where $c$ is the number of cubes and $w$ is the number of lengths of wood used.

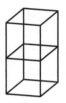

| $c$ | 1 | 2 | 3 | 4 | 5 | 6 |
|---|---|---|---|---|---|---|
| $w$ | 12 | | | | | |

**b** Find a formula giving $w$ in terms of $c$.

**c** A kit contains 80 lengths of wood. Explain why there must be some lengths left over when the storage system is built.

## Exercise 112★ (Revision)

**1** Find the first, tenth and hundredth terms of the sequence given by $n$th term $= 12 - 7n$.

**2** Find the next three members of the sequence 10, 7, 4, 1, …

**3** Find a formula for the $n$th term of the sequence $1, \frac{2}{3}, \frac{3}{5}, \frac{4}{7}, \frac{5}{9},$ …

**4** Find a formula for the $n$th term of the sequence 13, 8, 3, –2, …

**5** On Jamie's fifth birthday, she was given pocket money of 50p/month, to increase by 20p each month.

**a** How much pocket money did Jamie receive on her sixth birthday?

**b** Find a formula that gives Jamie's pocket money on the $n$th month after her fifth birthday.

**c** How old was Jamie when her pocket money became £17.30 per month?

**6 a** Find the next four terms in the sequence 1, 5, 9, 13, …

**b** The first and third terms of this sequence are square numbers.

Find the positions of the next two members of the sequence that are square numbers.

**c** Form a new sequence from the numbers giving the positions of the square numbers (that is, starting 1, 3, …). Use this sequence to find the position of the fifth square number in the original sequence.

**7** **a** Find the $n$th term of the sequence $(1 \times 2)$, $(2 \times 3)$, $(3 \times 4)$, …

**b** Find the $n$th term of the sequence $(1 + 1)$, $(4 + 2)$, $(9 + 3)$, …

**c** Explain why the two sequences are the same.

**d** Explain why every term is even.

**8** Sequence A:   2, 3, 4, 5, 6, …

Sequence B:   0, 1, 2, 3, 4, …

Sequence C:   0, 3, 8, 15, 24, …

**a** Find the $n$th term of sequence A.

**b** Find the $n$th term of sequence B.

**c** Sequence C is obtained from sequences A and B. Find the $n$th term of sequence C.

**d** Show how sequence C can be obtained from the sequence whose $n$th term $= n^2$.

**9** An engineer is designing a bridging kit using beams that are all the same length.
The bridge can be built to various lengths, as shown.

 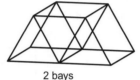 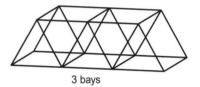

1 bay                  2 bays                  3 bays

**a** Copy and complete this table, where $b$ is the number of bays of the bridge and $n$ is the number of beams used.

| $b$ | 1 | 2 | 3 | 4 | 5 | 6 |
|---|---|---|---|---|---|---|
| $n$ | 9 | | | | | |

**b** Find a formula for $n$ in terms of $b$.

**c** How many beams are needed to make a bridge with 10 bays?

**10** **a** For this sequence of pebbles, copy and complete the table, where $m$ is the number in the sequence and $p$ is the number of pebbles.

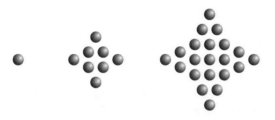

| $m$ | 1 | 2 | 3 | 4 | 5 |
|---|---|---|---|---|---|
| $p$ | 1 | 8 | | | |

**b** By adding a row for $3m^2$, find a formula connecting $m$ and $p$.

**c** A pattern uses 645 pebbles. Which member of the sequence is this?

## Basic transformations

Reflection, rotation and translation are three basic transformations. They change the position of an object, but not its size or shape.

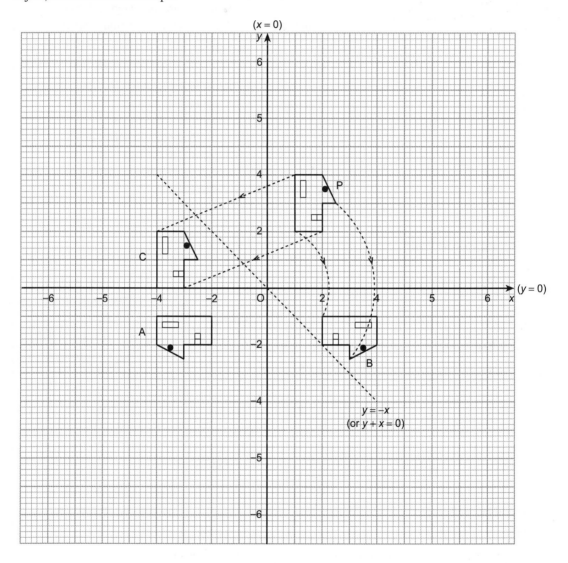

## Remember

| Transformation | Image of P | Definition | Notes |
|---|---|---|---|
| Reflection | A | • In the line $y = -x$ | Line of reflection is also called the 'mirror line'. |
| Rotation | B | • 90° clockwise<br>• About (0, 0) | Rotations are measured anticlockwise. A 90° clockwise rotation is actually a rotation of −90°. |
| Translation | C | • 5 left, 2 down<br>• Or $\begin{pmatrix} -5 \\ -2 \end{pmatrix}$ | The minus sign indicates a shift in the negative direction, which is:<br>• for the top number: to the left parallel to the *x*-axis<br>• for the bottom number: down parallel to the *y*-axis. |

## Defining translations

*By length and a direction*

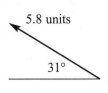

5.8 units

31°

A ruler and protractor would be needed to perform this translation.

−ve = left     +ve = right

$\begin{pmatrix} x \\ y \end{pmatrix}$

−ve = down     +ve = up

The *order* of the two numbers in a vector is very important. $\begin{pmatrix} 3 \\ -5 \end{pmatrix}$ is not the same as $\begin{pmatrix} -5 \\ 3 \end{pmatrix}$

*On a square grid, by a translation vector*

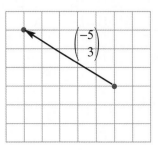

$\begin{pmatrix} -5 \\ 3 \end{pmatrix}$

Here, it is much easier to perform the translation.

Use the diagram on page 295. You will need a piece of tracing paper. Place the tracing paper over the diagram, draw the shape P and mark the position of the origin with a large dot.

1 Draw the mirror line ($x + y = 0$) on the tracing paper. Fold the tracing paper along the mirror line and then place the tracing paper over the diagram, with the mirror line fixing the correct position. Note the position of P.

2 Place a pin (or compass point, or pencil point) at O and rotate the tracing paper 90° in a clockwise direction. Note the position of P.

3 Put the tracing paper back in its position in step 1. Slide the tracing paper until the large dot at the origin is over the point (−5 , −2). Note the position of P.

Describe how you would transform P to A, B and C without using tracing paper.

## Exercise 113

1 On graph paper, draw $x$- and $y$-axes from −6 to 6, and plot triangle P with vertices (4, 1), (4, 2), (2, 1) as shown.

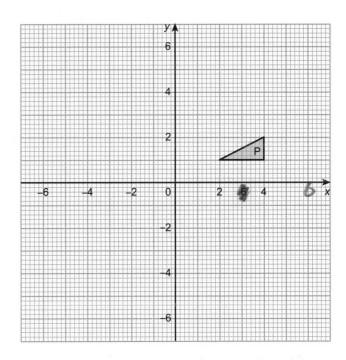

Then do these operations to triangle P, and label the images A, B, C, …, F.

| A | Reflect in $y$-axis | B | Reflect in $x$-axis |
|---|---|---|---|
| C | Rotate 90° anticlockwise about (0, 0) | D | Rotate 180° about (0, 0) |
| E | Translate by the vector $\begin{pmatrix} 2 \\ 4 \end{pmatrix}$ | F | Translate by the vector $\begin{pmatrix} 2 \\ -6 \end{pmatrix}$ |

**2** On graph paper, draw $x$- and $y$-axes from $-6$ to $6$, and plot flag Q with vertices $(-2, 2)$, $(-2, 4)$, $(-1, 3)$ as shown.

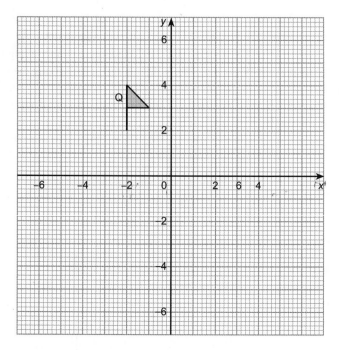

Then do these operations to flag Q, and label the images A, B, C, ..., F.

| A | Reflect in $y$-axis | B | Reflect in $x$-axis |
|---|---|---|---|
| C | Rotate 90° clockwise about (0, 0) | D | Rotate 180° about (0, 0) |
| E | Translate by the vector $\begin{pmatrix} 7 \\ -2 \end{pmatrix}$ | F | Translate by the vector $\begin{pmatrix} -3 \\ -4 \end{pmatrix}$ |

**3** The tables below the figure give details of 12 reflections. Use this figure to copy and complete the tables.

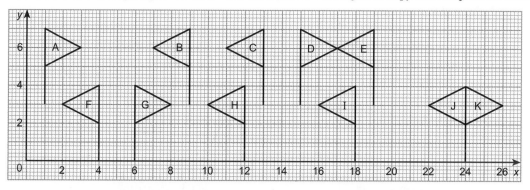

| Object | Reflection in line | Image |
|---|---|---|
| A | $x = 5$ | B |
| F | $x = 5$ | |
| G | | F |
| A | $x = 7$ | |
| D | | B |
| | $x = 9$ | H |

| Object | Reflection in line | Image |
|---|---|---|
| K | | J |
| | $x = 12$ | G |
| | $x = 18$ | |
| | $x = 10$ | A |
| J | | G |
| | $x = 17$ | |

**4** The table below gives details of 12 reflections. Using the figure, copy and complete the table.

| Object | Reflection in line | Image |
|---|---|---|
| B | $y = 0$ | A |
| D | $y = 8$ | |
| B | | E |
| | $y = 4$ | D |
| D | | E |
| I | $y = 9$ | |
| H | | G |
| | $y = 1$ | F |
| F | | J |
| J | $x = 5$ | |
| H | | C |
| | $y = 4$ | B |

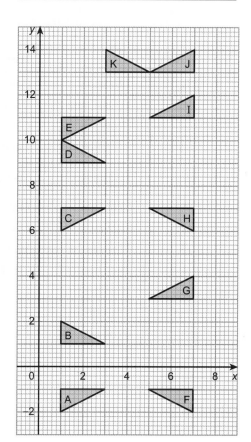

5 Each triangle in this diagram can be transformed onto another one by rotation or translation.

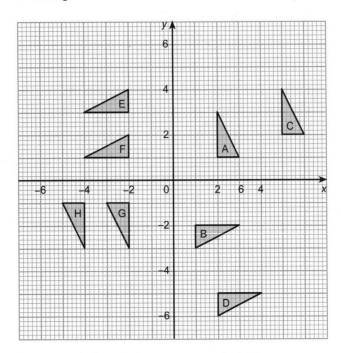

Using the diagram, copy and complete this table.

| Object | Transformation | | | Image |
|--------|----------------|--|--|-------|
| B | Rotation | Centre (0, 0) | angle 90° | A |
| A | Translation | Vector | | C |
| B | Rotation | Centre | angle | C |
| A | Rotation | Centre | angle 90° | F |
| F | Translation | Vector $\begin{pmatrix} 0 \\ 2 \end{pmatrix}$ | | |
| A | Rotation | Centre (−1, 0) | angle 90° | |
| A | Rotation | Centre (0, 0) | angle 180° | |
| G | Translation | Vector | | H |
| A | Rotation | Centre | angle | H |
| B | Rotation | Centre | angle −90° | H |
| H | Translation | Vector | | G |
| B | Rotation | Centre | angle | G |

6   Each flag in this diagram can be transformed onto another one by rotation or translation.

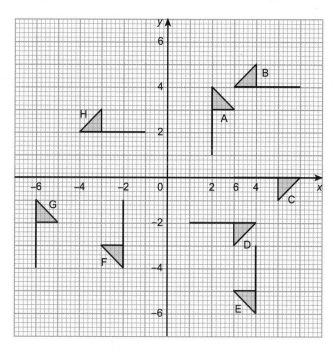

Using the diagram, copy and complete this table.

| Object | Transformation | | | Image |
|---|---|---|---|---|
| A | Rotation | Centre (0, 0) | angle −90° | D |
| D | Translation | Vector | | C |
| A | Rotation | Centre | angle | C |
| H | Rotation | Centre (0, 0) | angle 90° | |
| F | Translation | Vector $\begin{pmatrix} 6 \\ -2 \end{pmatrix}$ | | |
| H | Rotation | Centre (4, 2) | angle 90° | |
| D | Rotation | Centre | angle | F |
| F | Translation | Vector | | E |
| D | Rotation | Centre | angle | E |
| F | Rotation | Centre (0, 0) | angle 180° | |
| | Translation | Vector $\begin{pmatrix} -8 \\ -5 \end{pmatrix}$ | | G |
| F | Rotation | Centre | angle | G |
| H | | | | B |
| C | | | | B |

**7** P(1, 2) is one corner of the shaded triangle in the figure. Point A is the image of P after reflection in the line $y = 3$. Without drawing, find the image of P after a reflection in the line

**a** $y = 0$    **b** $y = 2$    **c** $y = 6$    **d** $y = 10$

Point B is the image of P after a rotation of $-90°$ about O. Without drawing, find the image of P after a rotation of $-90°$ about the point

**e** (2, 0)    **f** (3, 0)    **g** (4, 0)    **h** (10, 0)

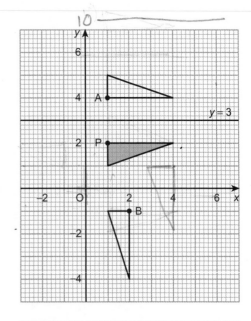

**8** P(6, 2) is the foot of the shaded flag in the figure. Point A is the image of P after reflection in the line $x = 4$. Without drawing, find the image of P after a reflection in the line

**a** $x = 5$    **b** $x = 6$    **c** $x = 7$    **d** $x = 12$

Point B is the image of P after a rotation of $90°$ about (4, 5). Without drawing, find the image of P after a rotation of $90°$ about the point

**e** (4, 4)    **f** (4, 3)    **g** (4, 0)    **h** (4, −10)

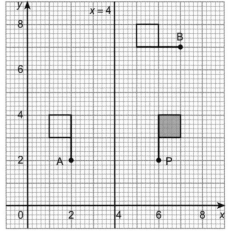

**9** Draw $x$- and $y$-axes from −6 to 6. Draw the line $y = x$.
   Plot triangle P with vertices at (1, 5), (3, 5), (3, 6).
  **a** Reflect P in $y = x$, and label the image A.
  **b** Reflect A in the $x$-axis, and label the image B.
  **c** Describe fully the single transformation which takes P to B.

**10** Draw $x$- and $y$-axes from −6 to 6. Draw the line $y = x$.
   Plot triangle Q with vertices at (3, 1), (4, 2), (6, 1).
  **a** Reflect Q in the $x$-axis, and label the image A.
  **b** Reflect A in $y = x$, and label the image B.
  **c** Describe fully the single transformation which takes Q to B.

**11** Draw $x$- and $y$-axes from −6 to 6. Plot triangle R with vertices at (2, 1), (5, 1), (5, 3).
  **a** Rotate R by $90°$ about O, and label the image A.
  **b** Reflect A in the $y$-axis, and label the image B.
  **c** Describe fully the single transformation which takes R to B.

**12** Draw $x$- and $y$-axes from $-6$ to $6$. Plot triangle S with vertices at $(-2, 1)$, $(-5, 1)$, $(-5, 3)$.

   **a** Reflect S in the $x$-axis, and label the image A.

   **b** Rotate A by $-90°$ about O, and label the image B.

   **c** Describe fully the single transformation which takes S to B.

# Combining transformations

We combine transformations when we apply more than one operation to a shape.

## Activity 38

For this activity, you will need one sheet of A4 paper.
Cut or tear off a strip to make a square.
Copy the diagram onto your paper with a thick felt-tip pen
so that you can see the lines through the paper.
You can do these simple operations.

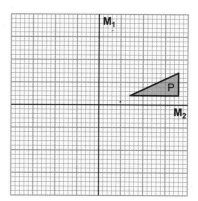

   ♦ 'Do nothing': **I**.

   ♦ Reflect in the vertical (mirror) line: $\mathbf{M}_1$.

   ♦ Reflect in the horizontal (mirror) line: $\mathbf{M}_2$.

   ♦ Rotate $90°$ clockwise: $\mathbf{R}_1$.

   ♦ Rotate $90°$ anticlockwise: $\mathbf{R}_2$.

To reflect, flip the paper round, through the mirror line.

Using one or two of these operations, investigate how
triangle P can be transformed into the other positions.
Record your results.

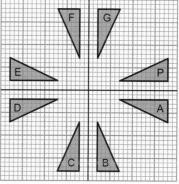

Then draw diagonal lines on the sheet of paper.
You now have three extra operations.

   ♦ Reflect in the diagonal (mirror) line: $\mathbf{M}_3$.

   ♦ Reflect in the diagonal (mirror) line: $\mathbf{M}_4$.

   ♦ Rotate $180°$: $\mathbf{R}_3$.

$\mathbf{M}_1$ followed by $\mathbf{R}_2$ and $\mathbf{M}_2$ followed by $\mathbf{R}_1$ are both
combinations that transform P to C.
Which single operation transforms P to C?

Investigate how the other *single* operations can be formed
by a combination of *two* others from this group.
Record your results in a table.

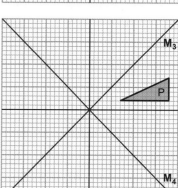

# Exercise 114

Questions 1–8 refer to this diagram. For each one, describe the three transformations, and comment on what you notice.

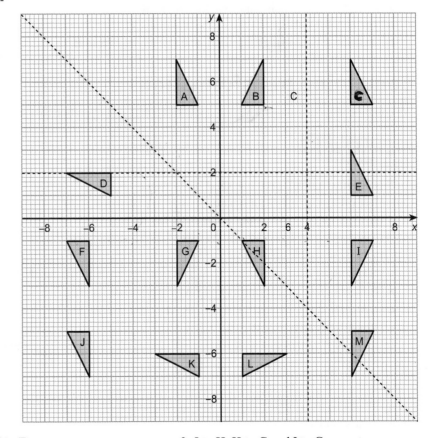

**1** C to I, I to E and C to E

**2** I to H, H to G and I to G

**3** B to A, A to G and B to G

**4** I to H, H to B and I to B

**5** K to L, L to I and K to I

**6** B to A, A to D and B to D

**7** F to G, G to A and F to A

**8** E to M, M to J and E to J

**9** Transformation **A** is a reflection in the line $x = 2$. Transformation **A**, followed by **B** is a 90° rotation about the point (2, 4).
Describe fully the transformation **B**. What single transformation is produced by **B**, then **A**?

**10** Transformation **A** is a rotation about the origin through −90°. Transformation **A**, followed by **B** is a translation of $\begin{pmatrix} -4 \\ 2 \end{pmatrix}$.
Describe fully the transformation **B**. What single transformation is produced by **B**, then **A**?

**11** Transformation **A** is a reflection in the line $x + y = 0$. Transformation **A**, followed by **B** is a reflection in the line $2y = x + 6$.
Describe fully the transformation **B**. What single transformation is produced by **B**, then **A**?

**12** Transformation **A** is a reflection in the line $x + y = 0$. Transformation **A**, followed by **B** is a translation of $\begin{pmatrix} 8 \\ 8 \end{pmatrix}$.
Describe fully the transformation **B**. What single transformation is produced by **B**, then **A**?

# Enlargements

There are many real-world applications of enlargement, from photographic negative enlargement to microscopes and telescopes. This chapter only deals with positive enlargements.

### Positive enlargement

Photographs can be enlarged for displaying in frames. They can also be reduced in size for identity cards.

> ### Remember
>
> An enlargement is defined by the scale factor *and* the centre of enlargement.
> - The *scale factor* defines the size of the image.
> - The image will be a *reduction* of the object when the scale factor is less than 1.
> - The image will be an *enlargement* of the object when the scale factor is greater than 1.
>
> The *centre of enlargement* defines the position of the image.

## Activity 39

This diagram shows triangle T transformed onto triangle $T_1$ and onto triangle $T_2$ by enlargements from centre O.

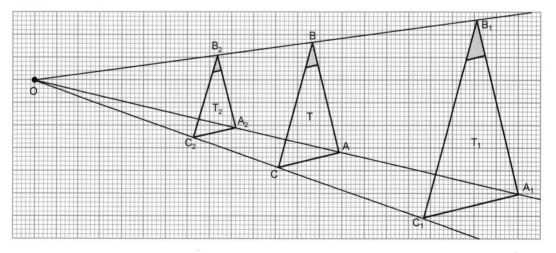

♦ Copy and complete this table, and comment on both sets of ratios.

| T | AB = | AC = | BC = |
|---|---|---|---|
| $T_1$ | $A_1B_1 =$ | $A_1C_1 =$ | $B_1C_1 =$ |
| Ratios | $A_1B_1/AB =$ | $A_1C_1/AC =$ | $B_1C_1/BC =$ |
| $T_2$ | $A_2B_2 =$ | $A_2C_2 =$ | $B_2C_2 =$ |
| Ratios | $A_2B_2/AB =$ | $A_2C_2/AC =$ | $B_2C_2/BC =$ |

♦ What factor changes the lengths of the triangle T to produce $T_1$ and to produce $T_2$?
Your answers are called the 'scale factor of enlargement'.

♦ Investigate the relative areas of the three triangles.

**Example 1**

**a** Enlarge triangle ABC, with scale factor = 3 from the point O = (0, 0).
Label the image $A_1B_1C_1$.

**b** Enlarge triangle $A_1B_1C_1$, with scale factor $= \frac{1}{2}$ from the point P = (−16 , 12).
Label the image $A_2B_2C_2$.

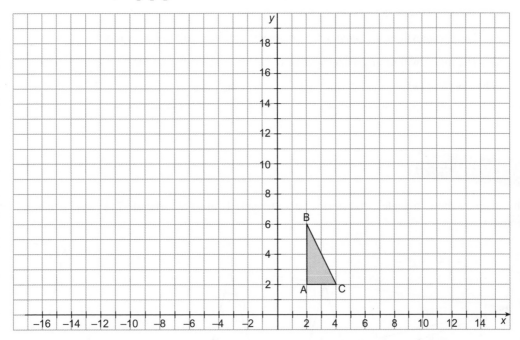

**Solution**

**a**
- Draw 'ray lines' from the point $(0, 0)$ through the points A, B and C.
- On the ray line OA, mark the point $A_1$ such that $OA_1 = 3OA$.
- Repeat, marking $B_1$ on ray line OB and $C_1$ on ray line OC.
- Draw triangle $A_1B_1C_1$.

Note that translation $OA_1 = 3 \times$ translation OA, and so on.

**b**
- Draw 'ray lines' from the point P $(-16, 12)$ through the points $A_1$, $B_1$ and $C_1$.
- On the ray line $PA_1$, mark the point $A_2$ such that $PA_2 = \frac{1}{2}PA_1$.
- Repeat, marking $B_2$ on ray line $PB_1$ and $C_2$ on ray line $PC_1$.
- Draw triangle $A_2B_2C_2$.

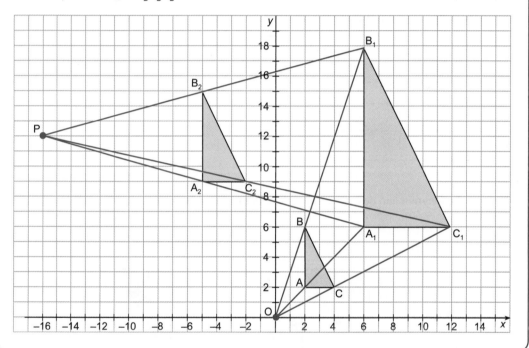

# Exercise 115

**1** Draw *x*- and *y*-axes from −4 to 4. Plot the triangle ABC, for points A(1, 0), B(1, −1) and C(0, −1). Draw the image for each of these enlargements.

|   | Scale factor | Centre of enlargement |
|---|---|---|
| A | 4 | (0, 0) |
| B | 3 | (1, −2) |
| C | $1\frac{1}{2}$ | (4, 2) |
| D | $\frac{1}{2}$ | (4, 2) |

**2** Draw *x*- and *y*-axes from −4 to 4. Plot the triangle DEF, for points D(2, 0), E(1, 0) and F(0, 2). Draw the image for each of these enlargements.

|   | Scale factor | Centre of enlargement |
|---|---|---|
| A | 2 | (4, 0) |
| B | 3 | (2, 2) |
| C | $1\frac{1}{2}$ | (−2, −2) |
| D | $\frac{1}{2}$ | (0, 0) |

**3** Refer to the diagram and describe fully these enlargements: B to A; A to E; D to B; E to D and D to C.

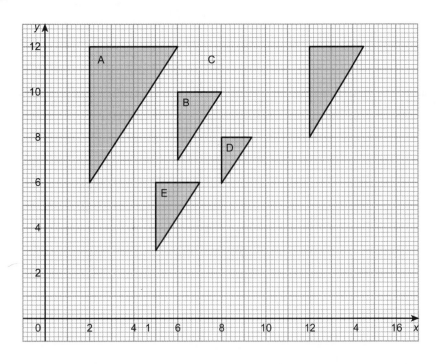

**4** Refer to the diagram, and describe fully these enlargements: A to C; A to D; C to D; C to E; B to C and E to D.

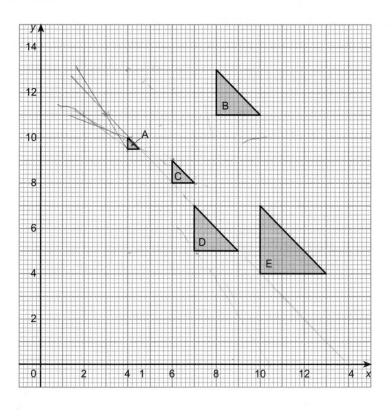

In Europe, the sizes of most paper used in printing and photocopying belong to a standard set of sizes called the A series. The sizes in the series are numbered from A0 to A10.

The area of a sheet of A0 paper is $1\,m^2$.
Most file paper is A4.
One A4 sheet equals two A5 sheets in size,
one A5 sheet equals two A6 sheets in size
and so on.

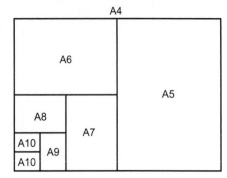

♦ Investigate how many sheets of each type can be obtained from one A1 sheet of paper. Tabulate your results, and comment on the 'area ratio' between paper sizes.

♦ Investigate the width:length ratio of each paper size.

An open-ended cuboid can be made from four A7-sized rectangles.

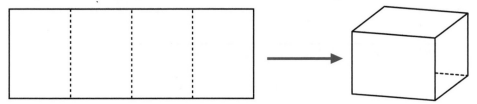

♦ Make a cuboid from A7 rectangles and a *similar* one from A9 rectangles. Find the volume ratio between these two cuboids, and investigate the volume ratio between similar cuboids made from other paper sizes.

♦ Investigate the relationship between the length, area and volume ratios.

# Exercise 116 (Revision)

For Questions 1–10, refer to the diagram and describe fully the transformation.

**1** Q to A      **2** Q to B      **3** Q to C      **4** Q to D      **5** Q to E

**6** Q to F      **7** Q to G      **8** Q to H      **9** Q to I      **10** Q to J

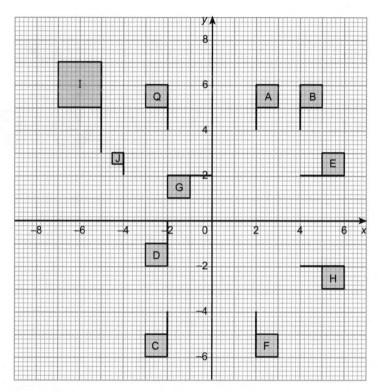

**11** Draw $x$- and $y$-axes from $-8$ to $8$.

Plot triangle P with vertices at $(1, 2)$, $(1, 4)$ and $(2, 4)$.

Do these transformations, and label your images A–E.

**a** Rotate P by $-90°$ about $(2, 0)$.

**b** Translate P along the vector $\begin{pmatrix} 4 \\ 2 \end{pmatrix}$.

**c** Reflect P in the line $x + y = 0$.

**d** Enlarge P from the point $(5, 2)$ with scale factor $2$.

**e** Enlarge P from the point $(2, 0)$ with scale factor $\frac{1}{2}$.

# Exercise 116★ (Revision)

1 What are the coordinates of the *nose* after the clown undergoes these transformations?

   **a** Rotation of 90° about (1, −2)

   **b** Translation along the vector $\begin{pmatrix} 3 \\ -5 \end{pmatrix}$

   **c** Enlargement from the point (5, 5) with scale factor 4

   **d** Reflection in the line $x + 2y = 6$

   **e** Enlargement from the point (6, −5) with scale factor $\frac{2}{3}$

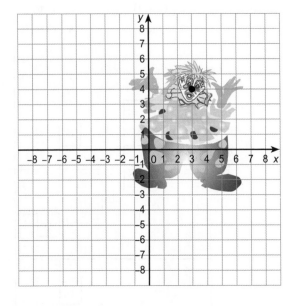

2 Transformation **A** is a reflection in the line $y = x$.
Transformation **B** is a rotation of 90° about the origin.

Transformation **C** is a translation along the vector $\begin{pmatrix} -2 \\ 2 \end{pmatrix}$.

Draw $x$- and $y$-axes, and plot an object near the origin.
Describe fully the single transformation which results from these combined transformations.

   **a** **A** followed by **B**
   **b** **B** followed by **A**
   **c** **A** followed by **C**

   **d** **C** followed by **A**
   **e** **B** followed by **C**
   **f** **C** followed by **B**

# Handling data 5

♦ The average family in England has 2.2 children.
How many families have 2.2 children?
How many families have 4 children?
Who might like to know?

♦ The average shoe size is 9, but how do shoe shops decide how many to stock in each size?

Size 6  Size 7  Size 8  Size 9  Size 10  Size 11  Size 12

♦ The average height of the students in this class photo is 146.3 cm. How useful is this statistic?

# Distributions

For an average (mean, median or mode) to be really useful, we need to know more about the shape and spread of the distribution.

Since the early 19th century, statisticians have investigated and analysed distributions. They have discovered that most sets of data produce predictable shapes, when frequency is plotted against the variable.

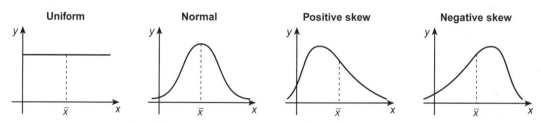

The **mean** $\bar{x}$ 'centres' the shape, but the 'spread' needs to be measured as well. There are simple statistics that can be used to define the **spread** of a distribution.

# Quartiles

The median divides a set of data into two halves. Half of the data is smaller than the median, and half is bigger. The **three quartiles** divide the data into four quarters.

$Q_1$ is the first quartile, or the lower quartile.

$Q_2$ is the second quartile, or the median $m$.

$Q_3$ is the third quartile, or the upper quartile.

In Example 1, the three quartiles divide the other twelve figures exactly into four groups of three. This will not always happen, and two consecutive numbers may have to be considered when calculating a quartile value, as shown in Example 2.

---

**Example 1**

15 children were asked how much money they had in their pockets. Find the quartiles.

45p  30p  10p  40p  35p  1p  22p  35p  1p  5p  20p  10p  £1  7p  50p

First, **sort the data into ascending order**.

Then choose the numbers that divide the data into four equal parts.

1p  1p  5p  | 7p |  10p  10p  20p  | 22p |  30p  35p  35p  | 40p |  45p  50p  £1

           ↑                   ↑                   ↑

         $Q_1$                $Q_2$                $Q_3$

---

**Example 2**

The growth of 20 plants over a 7-day period was measured to the nearest centimetre. The results are shown in ascending order. Estimate the quartiles.

| 1 | 3 | 4 | 5 | 6 | 7 | 7 | 8 | 8 | 8 |
| 9 | 9 | 9 | 10 | 10 | 10 | 11 | 12 | 12 | 17 |

There is no single middle value for the median, so consider the 10th and 11th values. The 10th value is 8 cm and the 11th value is 9 cm.

$$\therefore m = \frac{8\,cm + 9\,cm}{2} = 8.5\,cm$$

Similarly there is no single value for $Q_1$, so consider the 5th and 6th values. The 5th value is 6 cm and the 6th value is 7 cm.

$$\therefore Q_1 = \frac{6\,cm + 7\,cm}{2} = 6.5\,cm$$

For $Q_3$ consider the 15th and 16th values. The 15th and 16th values are both 10 cm.

$$\therefore Q_3 = 10\,cm$$

## Key Points

In general, for $n$ data values, arranged in ascending order:

Lower quartile $= Q_1 = \frac{1}{4}(n+1)$th value

Median $m$ $\quad = Q_2 = \frac{1}{2}(n+1)$th value

Upper quartile $= Q_3 = \frac{3}{4}(n+1)$th value

If this involves two values, use the average of two consecutive data values.

# Measures of spread

The **range** is the difference between the largest value and the smallest value.

This statistic is calculated from just two values from the entire data, so a 'freak' or 'extreme' result at either end of the data set will distort the range and make it unrepresentative of the set as a whole.

A much better statistic is the **interquartile range**: the difference between the upper and lower quartiles. This gives the range of the **middle half** of the data and therefore ignores the 'extreme' values at both ends.

## Key Point

**Interquartile range** (IQR) = upper quartile – lower quartile = $Q_3 - Q_1$

# Cumulative frequency

Cumulative frequency 'accumulates' (or adds) the frequencies as you go along. In many situations, cumulative frequency is a more important statistic than frequency.

- ◆ Examination statistics quote percentage pass rates and percentage fail rates (cumulative frequency), and not the numbers that achieved a particular grade (frequency).
- ◆ For traffic flow along a motorway, the frequency rate is measured, but for cars entering a car park, the cumulative frequency is more relevant.
- ◆ The police at a large stadium would need to consider the rate (frequency per minute) at which the spectators arrive, in case the walkways become dangerously congested. They would also need to be aware of the current size (cumulative frequency) of the crowd so that it remained below the stadium capacity.

Cumulative frequency is calculated with the data in ascending order, so it can also be used to find the **median**, **quartiles** and **interquartile range** of data sets.

To draw **cumulative frequency diagrams for grouped data**:

♦ Cumulative frequency is always plotted on the *vertical* axis.

♦ The cumulative frequencies are plotted against the *exact endpoints* of the groups.

♦ The points that have been plotted are joined by a *smooth curve* to form the cumulative frequency curve. If the points are joined by straight lines they form a cumulative frequency polygon.

♦ The quartiles can be *estimated* from the diagram (from the *vertical axis*, across to the *curve*, and then down to the *horizontal axis*).

♦ As the quartiles can only be estimated, and if *n* is large, then the *n*th values, instead of the (*n* + 1)th values, can be used as starting points.

### Example 3

The frequency table shows the distribution of the crowd arriving at a concert hall for a pop concert. The crowd limit is 2000.

Find the interquartile range of the arrival times.

| Time | Frequency $f$ |
|---|---|
| 6.00–6.15pm | 38 |
| 6.15–6.30pm | 57 |
| 6.30–6.45pm | 102 |
| 6.45–7.00pm | 166 |
| 7.00–7.15pm | 195 |
| 7.15–7.30pm | 549 |
| 7.30–7.45pm | 761 |
| 7.45–8.00pm | 132 |

First, construct a cumulative frequency table. Every group starts at the beginning (6.00pm).

| Time | Frequency $f$ |
|------|------|
| 6.00–6.15pm | 38 |
| 6.15–6.30pm | 57 |
| 6.30–6.45pm | 102 |
| 6.45–7.00pm | 166 |
| 7.00–7.15pm | 195 |
| 7.15–7.30pm | 549 |
| 7.30–7.45pm | 761 |
| 7.45–8.00pm | 132 |

| Time | Cumulative frequency |
|------|------|
| 6.00–6.15pm | $= 38$ |
| 6.00–6.30pm | $38 + 57 = 95$ |
| 6.00–6.45pm | $95 + 102 = 197$ |
| **6.00–7.00pm** | **$197 + 166 = 363$** |
| 6.00–7.15pm | $363 + 195 = 558$ |
| 6.00–7.30pm | $558 + 549 = 1107$ |
| 6.00–7.45pm | $1107 + 761 = 1868$ |
| 6.00–8.00pm | $1868 + 132 = 2000$ |

**363** $(38 + 57 + 102 + 166)$ have arrived by **7.00pm**

Then construct a cumulative frequency curve.

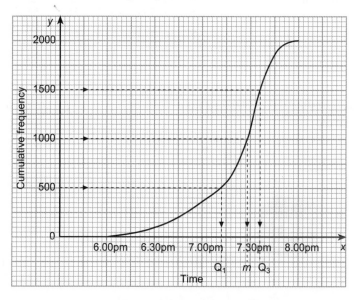

To estimate the median arrival time and the IQR of the arrival times:

Median $= \frac{1}{2}(2000)$th value $= 1000$th value. So, start at 1000 on the vertical axis.

$Q_1 = \frac{1}{4}(2000)$th value $= 500$th value. Start at 500 on the vertical axis.

$Q_3 = \frac{3}{4}(2000)$th value $= 1500$th value. Start at 1500 on the vertical axis.

From the horizontal axis: $m = 7.28$pm, $Q_1 = 7.12$pm, $Q_3 = 7.36$pm.

Therefore IQR $= (Q_3 - Q_1) = 24$ minutes.

# Exercise 117

For Questions 1–4, find the median, quartiles, range and interquartile range of the sets of numbers.

**1**  3    6    8    2    4    3    1

**2**  11    16    14    10    9    7    6    15    21

**3**  2    2    2    4    5    5    9    7    8    4    1

**4**  0.1    0.8    0.2    0.4    0.5    0.3    0.6    0.2

**5**  The table shows the number of emails received by the editor of a magazine on each day in September. Find the median, quartiles, range and interquartile range.

| | | | | | |
|---|---|---|---|---|---|
| 11 | 11 | 12 | 14 | 14 | 16 |
| 18 | 20 | 21 | 22 | 24 | 24 |
| 25 | 25 | 26 | 28 | 29 | 29 |
| 29 | 30 | 31 | 32 | 33 | 34 |
| 35 | 35 | 37 | 41 | 42 | 42 |

**6**  The table shows the weights in grams of 35 packets of sweets. Find the median, quartiles, range and interquartile range of the weights.

| | | | | | | |
|---|---|---|---|---|---|---|
| 54.7 | 54.9 | 55.0 | 55.3 | 55.3 | 55.3 | 55.5 |
| 55.6 | 55.8 | 55.9 | 56.1 | 56.2 | 56.4 | 56.4 |
| 56.5 | 56.6 | 56.7 | 56.7 | 56.8 | 56.9 | 56.9 |
| 57.0 | 57.0 | 57.1 | 57.2 | 57.3 | 57.5 | 57.5 |
| 57.6 | 57.7 | 57.8 | 57.8 | 58.0 | 58.1 | 58.1 |

**7**  This frequency table gives the scores obtained from 40 throws of a biased dice.
Find the median, quartiles, range and interquartile range of the score.

| Score | Frequency |
|---|---|
| 1 | 9 |
| 2 | 10 |
| 3 | 7 |
| 4 | 5 |
| 5 | 4 |
| 6 | 5 |
| | Total = 40 |

**8**  A questionnaire filled in by all the students at a college who had passed their driving test revealed these results.
Find the median, quartiles, range and interquartile range of the number of tests.

| Number of driving tests $t$ | Number of students $f$ |
|---|---|
| 1 | 18 |
| 2 | 27 |
| 3 | 19 |
| 4 | 8 |
| 5 | 3 |

**9** A sample of 60 baking potatoes was taken at a supermarket. Each one was weighed, and the results are shown on this cumulative frequency diagram.

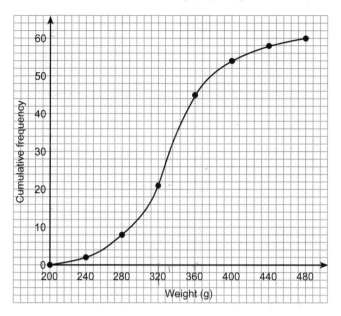

   **a** Estimate the median weight and the quartiles from the graph.
   **b** Work out the interquartile range for the weight of a baking potato.
   **c** Comment on which of the mean and the median is the better average.

**10** This cumulative frequency diagram shows the times of the 48 finishers in a regional cross-country race.

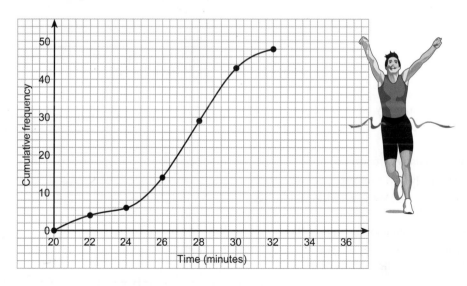

   **a** Estimate the median time and the quartiles from the graph.
   **b** Work out the interquartile range for the finishing time.
   **c** Comment on which of the mean and the median is the better average.

**11** This cumulative frequency curve shows the distribution of teachers' salaries at a small secondary school.

Salary (£'000s)

   **a** Estimate the median and quartiles from the graph.
   **b** Calculate the interquartile range for the salaries.

**12** This cumulative frequency curve shows the distribution of the heights of the players at a basketball club.

   **a** Estimate the median and quartiles from the graph.
   **b** Work out the interquartile range of the heights.

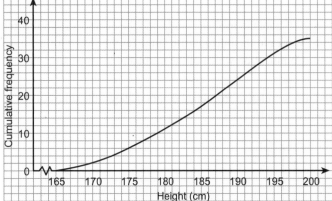

Height (cm)

**13** This cumulative frequency table shows the distribution of the lengths of pencils in the art classroom box.

   **a** Draw the cumulative frequency curve. (Use a horizontal axis scale of 2 cm = 2.5 cm, and a vertical axis scale of 1 cm = 5 pencils.)
   **b** Estimate the median and quartiles of the pencil lengths.
   **c** Give an estimate of the interquartile range.

| Length $l$ (cm) | Cumulative frequency |
|---|---|
| $0 \leqslant l < 2.5$ | 0 |
| $0 \leqslant l < 5$ | 2 |
| $0 \leqslant l < 7.5$ | 8 |
| $0 \leqslant l < 10$ | 21 |
| $0 \leqslant l < 12.5$ | 36 |
| $0 \leqslant l < 15$ | 45 |
| $0 \leqslant l < 17.5$ | 48 |
| $0 \leqslant l < 20$ | 50 |

**14** This cumulative frequency table shows the distribution of the weights of athletes in a club.

  **a** Draw the cumulative frequency curve. (Use a horizontal axis scale of 2 cm = 5 kg, and a vertical axis scale of 1 cm = 5 athletes.)

  **b** Estimate the median and quartiles of the weight.

  **c** Give an estimate of the interquartile range.

| Weight $w$ (kg) | Cumulative frequency |
|---|---|
| $w < 60$ | 0 |
| $60 \leqslant w < 65$ | 5 |
| $60 \leqslant w < 70$ | 16 |
| $60 \leqslant w < 75$ | 29 |
| $60 \leqslant w < 80$ | 42 |
| $60 \leqslant w < 85$ | 47 |
| $60 \leqslant w < 90$ | 50 |

**15** This frequency table gives the weights of the first 55 'millennium' babies born in a city hospital.

  **a** Construct a cumulative frequency table for the data.

  **b** Draw the cumulative frequency curve. (Use a horizontal axis scale of 2 cm = 0.5 kg, and a vertical axis scale of 1 cm = 5 babies.)

  **c** Estimate the median and quartiles for the weights of these babies.

| Weight (kg) | Frequency $f$ |
|---|---|
| $1.0 \leqslant w < 1.5$ | 2 |
| $1.5 \leqslant w < 2.0$ | 7 |
| $2.0 \leqslant w < 2.5$ | 10 |
| $2.5 \leqslant w < 3.0$ | 14 |
| $3.0 \leqslant w < 3.5$ | 9 |
| $3.5 \leqslant w < 4.0$ | 8 |
| $4.0 \leqslant w < 4.5$ | 5 |
| $4.5 \leqslant w$ | 0 |
| | Total = 55 |

**16** The table gives the distribution of the lengths of marriages that ended in divorce in 1996.

  **a** Construct a cumulative frequency table for the data.

  **b** Draw the cumulative frequency curve. (Use a horizontal axis scale of 2 cm = 5 years, and a vertical axis scale of 1 cm = 10 000 marriages.)

  **c** Estimate the median and quartiles for the length of these marriages.

  **d** Comment on which of the mean and the median is the better average.

| Time $t$ (years) | Number $f$ (1000s) |
|---|---|
| $0 \leqslant t < 3$ | 15 |
| $3 \leqslant t < 5$ | 22 |
| $5 \leqslant t < 10$ | 48 |
| $10 \leqslant t < 15$ | 31 |
| $15 \leqslant t < 20$ | 21 |
| $20 \leqslant t < 25$ | 15 |
| $25 \leqslant t < 30$ | 10 |
| $30 \leqslant t$ | 9 |

# Exercise 117★

1 This cumulative frequency diagram gives the distribution of ages for 100 teachers at a college.

    **a** Estimate the median age and the interquartile range.

    **b** Estimate the proportion of teachers who are younger than 40.

    **c** Compare these statistics with those for the teachers at your school.

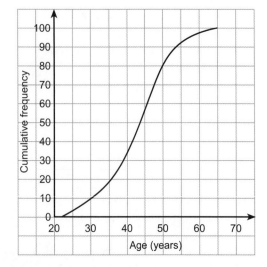

2 The distribution of the ages of the first 100 mothers to give birth in a new hospital maternity unit is shown in this cumulative frequency diagram.

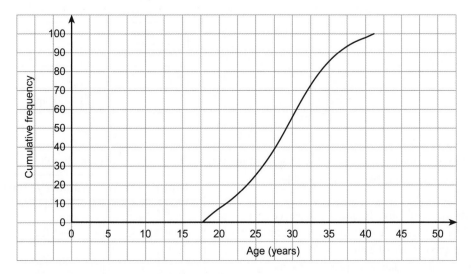

    **a** Estimate the median age and the quartiles from the graph.

    **b** Work out the interquartile range for the age of a mother giving birth.

    **c** Comment on which of the mean or the median is the better average.

3   The marks of 120 pupils in a history examination are shown on this cumulative frequency curve.

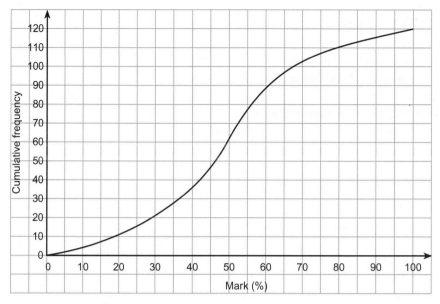

a  Estimate the median and the interquartile range of the marks.
b  Estimate how many candidates obtained a grade A if the boundary mark was 72%.

4   The heights in metres of a random sample of 60 soldiers from a regiment are shown on this
    cumulative frequency diagram.

a  Estimate the median and the
   interquartile range of the heights.
b  What proportion of soldiers are
   eligible to be considered for the
   special forces, for which all soldiers
   have to be taller than 1.85 m?

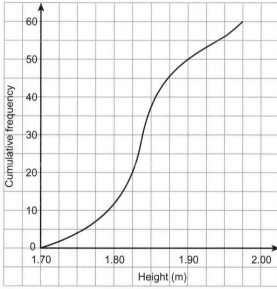

5 A random sample of 60 motorists recorded their weekly mileage to and from their place of work.

a Construct a cumulative frequency table showing the exact group boundaries.

b Draw a cumulative frequency diagram.
(Use a horizontal axis scale of 1 cm = 5 miles, and a vertical axis scale of 1 cm = 5 motorists.)

c Estimate the median and the interquartile range.

d Estimate the number from the sample who drove less than 85 miles.

| Mileage *m* | *f* |
|---|---|
| 0–39 | 0 |
| 40–49 | 6 |
| 50–59 | 11 |
| 60–69 | 16 |
| 70–79 | 14 |
| 80–89 | 8 |
| 90–99 | 5 |

6 The weights of fish caught in an angling competition are shown in this frequency table.

a Construct a cumulative frequency table and then draw a cumulative frequency diagram.
(Use a horizontal axis scale of 1 cm = 50 g, and a vertical axis scale of 1 cm = 5 fish.)

b Estimate the median and IQR of the weights of the fish.

c Estimate the number of fish that weighed more than 0.65 kg.

| Weight *g* | *f* |
|---|---|
| 300– | 5 |
| 400– | 10 |
| 500– | 14 |
| 600– | 7 |
| 700– | 7 |
| 800– | 5 |
| 900– | 4 |

7 A sample of 100 tea-drinkers was surveyed to discover the length of time between waking and their first cup of tea. The results are shown in the table.

a Construct a cumulative frequency table and draw the cumulative frequency curve.
(Use a horizontal axis scale of 1 cm = 10 minutes, and a vertical axis scale of 1 cm = 10%.)

b Use your diagram to estimate the median and IQR of the time between waking and the first cup of tea for these people.

c What proportion of people waited more than 45 minutes in the morning before having a cup of tea?

| Time | *f* |
|---|---|
| Less than 5 minutes | 19 |
| 5–14 minutes | 20 |
| 15–29 minutes | 15 |
| 30–59 minutes | 18 |
| 1–2 hours | 14 |
| More than 2 hours | 14 |

8 This frequency table gives the ages of first-year undergraduates studying for a BA degree in European History at a university.

a Construct a cumulative frequency table and then draw a cumulative frequency diagram.
(Use a horizontal axis scale of 2 cm = 5 years, and a vertical axis scale of 1 cm = 20 students.)

b Estimate the median and IQR of the students' ages.

c Estimate the number of students who were older than 35.

d Comment on which of the mean or the median is the better average.

| Age (years) | *f* |
|---|---|
| 18–19 | 56 |
| 20–21 | 33 |
| 22–25 | 26 |
| 26–29 | 17 |
| 30–39 | 25 |
| 40–49 | 16 |
| 50–59 | 7 |

**9** A schools' inspector recorded the actual time that a group of 15-year-old children were studying during an 80-minute double period. The results are given in this cumulative frequency table.

**a** Draw a cumulative frequency diagram and estimate the median and IQR.
(Use a horizontal axis scale of
1 cm = 5 minutes, and a vertical axis scale of
1 cm = 5 children.)

**b** Calculate the frequencies from the cumulative frequencies and use them in a calculation table to estimate the mean.

**c** Comment on the differences between the mean and the median.

| Time (mins) | Cumulative Frequency |
|---|---|
| 0–10 | 1 |
| –20 | 8 |
| –30 | 28 |
| –40 | 41 |
| –50 | 50 |
| –60 | 54 |
| –70 | 56 |
| –80 | 56 |

**10** A hockey club has 50 male and 50 female players. This table shows the distribution of their ages.

**a** Construct the cumulative frequency table for both sets of data.

**b** Draw both cumulative frequency curves on one diagram.
(Use a horizontal axis scale of
1 cm = 2 years, and a vertical axis scale of
1 cm = 5 players.)

**c** Estimate the median and quartiles for the ages of both sexes.

**d** Comment on the differences.

| Age (years) | Men | Women |
|---|---|---|
| $16 \leqslant t < 19$ | 2 | 5 |
| $19 \leqslant t < 22$ | 3 | 9 |
| $22 \leqslant t < 25$ | 8 | 16 |
| $25 \leqslant t < 28$ | 14 | 9 |
| $28 \leqslant t < 31$ | 11 | 6 |
| $31 \leqslant t < 34$ | 7 | 3 |
| $34 \leqslant t < 37$ | 3 | 2 |
| $37 \leqslant t < 40$ | 2 | 0 |

# Exercise 118 (Revision)

1   The ages of 50 people who passed their driving test at a local test centre are shown on this
    cumulative frequency diagram.

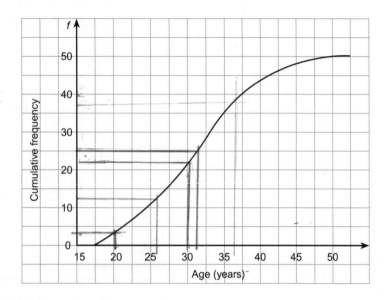

Use the diagram to:
a   Estimate the median and IQR of the ages.
b   Estimate how many people were in their twenties.

2   This cumulative frequency table shows the weights of 40 people who joined a weight reduction
    club.

a   Draw a cumulative frequency diagram.
    (Use a horizontal axis scale of 2 cm = 5 kg, and a vertical
    axis scale of 1 cm = 5 people.)
b   Use the diagram to find estimates of the median and
    interquartile range of the weights.

| Weight (kg) | F |
|---|---|
| $70 \leqslant w < 75$ | 4 |
| $70 \leqslant w < 80$ | 15 |
| $70 \leqslant w < 85$ | 29 |
| $70 \leqslant w < 90$ | 35 |
| $70 \leqslant w < 95$ | 40 |

# Exercise 118★ (Revision)

1 A council surveyed the delays that a sample of 160 motorists suffered on their journeys into the city centre one morning. The results are shown on this cumulative frequency diagram.

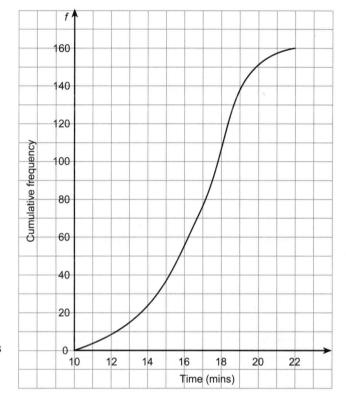

Use the diagram to:

a Estimate the median and IQR of the delay time.

b Estimate the number of motorists who were delayed for less than a quarter of an hour.

2 The table shows the ages of people taking out a mortgage on their house with a particular mortgage company.

| Age (years) | $f$ |
|---|---|
| 18–19 | 2 |
| 20–29 | 35 |
| 30–39 | 45 |
| 40–49 | 31 |
| 50–59 | 12 |
| 60–69 | 6 |
| 70–79 | 1 |
| | Total = 132 |

a Construct a cumulative frequency table, and then draw a cumulative frequency diagram. (Use a horizontal axis scale of 1 cm = 5 years, and a vertical axis scale of 1 cm = 10 mortgages.)

b Estimate the median and IQR of the ages of the clients.

c Estimate the number of people between the ages of 32 and 42 who have mortgages.

## Number

### Proportion

If a passenger jet uses 2700 litres of fuel *every* hour, the number of litres used is **directly proportional** to the time. So, the jet would use half as much fuel in half an hour.

The jet uses 2700 litres in 1 hour.

$$\therefore \quad \text{it uses 1 litre in } \frac{1}{2700} \text{ hours} \qquad \text{(Divide both sides by 2700)}$$

$$1 \text{ litre in } \frac{1}{2700} \times 3600\,\text{s} \qquad \text{(Change to seconds)}$$

$$1 \text{ litre in } 1.\dot{3} \text{ secs}$$

A **density** of $11\,\text{g/cm}^3$ means that $11\,\text{g}$ of a substance has a volume of $1\,\text{cm}^3$. So, $2\,\text{cm}^3$ has a mass of $22\,\text{g}$.

To **find the best buy**, find, for each item, either the cost of 1 unit (say in $ per kg), or the weight you can buy for $1 (in kg per $).

### Positive integer powers of number

$$a^m \times a^n = a^{m+n} \qquad a^m \div a^n = a^{m-n} \qquad (a^m)^n = a^{m \times n}$$

| Question | Working | Rule | Calculator | Answer |
|---|---|---|---|---|
| $4^5 \times 4^4$ | $4^{5+4} = 4^9$ | Add indices | 4 $\boxed{\wedge}$ 9 $\boxed{=}$ | 262 144 |
| $3.1^7 \div 3.1^3$ | $3.1^{7-3} = 3.1^4$ | Subtract indices | 3.1 $\boxed{\wedge}$ 4 $\boxed{=}$ | 92.3521 |
| $(1.05^4)^6$ | $1.05^{24}$ | Multiply indices | 1.05 $\boxed{\wedge}$ 24 $\boxed{=}$ | 3.23 to 3 s.f. |

### Simple recurring decimals

◆ $\frac{1}{7} = 1 \div 7 = 0.142\,857\,142\,8571 \ldots$ and this is written as $0.\dot{1}42\,85\dot{7}$

◆ Fractions that produce terminating decimals have 2 and 5 as the only factors in their denominator, for example $\frac{3}{5}, \frac{3}{8}, \frac{7}{10}, \frac{3}{20}$. Fractions with other prime numbers as factors in their denominator recur, e.g. $\frac{1}{6}, \frac{1}{15}$.

◆ Express $0.32\dot{3}\dot{2}$ as an exact fraction.

$$x = 0.32\dot{3}\dot{2}$$

$$100x = 32.32\dot{3}\dot{2}$$

$$99x = 32$$

$$x = \frac{32}{99}$$

## Algebra

### Expanding brackets

$$(x + 3)(x - 1) = \begin{array}{cccc} \text{First} & \text{Outside} & \text{Inside} & \text{Last} \\ x^2 & - \; x & + \; 3x & - \; 3 \end{array} = x^2 + 2x - 3$$

### Factorising quadratics

This is the process of working backwards from $x^2 + 2x - 3$ to give $(x + 3)(x - 1)$.

The factors of $x^2$ are $x$ and $x$, and the factors of $-3$ are $+3$ and $-1$ or $-3$ and $+1$.

Always check your answer by expanding the brackets.

### Solving quadratics by factors

$x^2 + 2x - 3 = 0$ (first factorise) $(x + 3)(x - 1) = 0$.

$\therefore$ *either* $(x + 3) = 0$, so $x = -3$, *or* $(x - 1) = 0$, so $x = 1$.

### Factorising the difference of two squares

$x^2 - 9 = (x + 3)(x - 3)$

$x^2 - y^2 = (x + y)(x - y)$

## Sequences

### To continue a sequence

Find the difference between each pair of terms.

| Sequence | | 2 | | 7 | | 12 | | 17 | ... |
|---|---|---|---|---|---|---|---|---|---|
| Differences | | | 5 | | 5 | | 5 | | |

The difference is 5, so the next term is 22.

### To find the rule if the first row of differences is constant

If the differences are equal to $a$, the formula for the $n$th term will be $an \pm b$, where $b$ is another constant.

| Sequence | | −5 | | −2 | | 1 | | 4 | ... |
|---|---|---|---|---|---|---|---|---|---|
| Differences | | | +3 | | +3 | | +3 | | |

Therefore the rule is $+3n + b$. When $n = 1$, the formula must give the first term as $-5$.

Thus $3 \times 1 + b = -5$, giving $b = -8$. Therefore the rule for the $n$th term is $3n - 8$.

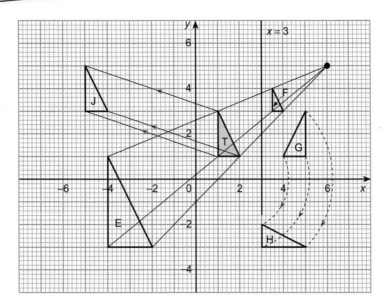

- ♦ △G is a **reflection** of △T in the line $x = 3$.

- ♦ △H is a **rotation** of △G 90° clockwise (−90°), centre (2, 0).

- ♦ △J is a **translation** of △T by the vector $\begin{pmatrix} -6 \\ +2 \end{pmatrix}$.

- ♦ △E is an **enlargement** of △T by a scale factor 2, centre (6, 5). The area of △E is four times the area of △T.

- ♦ △F is an **enlargement** of △T by a scale factor $\frac{1}{2}$, centre (6, 5). The area of △F is one-quarter of the area of △T.

## Cumulative frequency

Cumulative frequency is always plotted on the vertical axis against the **end-points** of each group. Points are joined by a smooth curve or straight line segments. This cumulative frequency curve can be used to *estimate* quartiles.

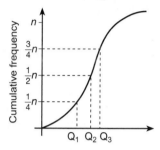

| Quartiles | Cumulative frequency | Percentile (%) |
|---|---|---|
| Lower quartile $Q_1$ | $\frac{1}{4}n$ | 25 |
| Median $m$ | $\frac{1}{2}n$ | 50 |
| Upper quartile $Q_3$ | $\frac{3}{4}n$ | 75 |

## Measures of spread

Range = largest value − smallest value

Interquartile range = upper quartile − lower quartile = $Q_3 - Q_1$

# Examination practice 5

1 Write these as a single power, and then calculate the answer correct to 3 significant figures.
   **a** $1.6^4 \times 1.6^5$       **b** $16.7^{10} \div 16.7^5$
   **c** $(1.065^3)^3$

2 Without doing any working, write down which of these fractions produce terminating decimals.
   $$\frac{1}{3}, \frac{4}{25}, \frac{5}{14}, \frac{7}{20}, \frac{3}{8}$$

3 Expand and simplify
   **a** $(x-1)(2x+5)$     **b** $(4-3x)(2-x)$

4 The weights of 15 packets of sweets are
   55.0 g, 55.3 g, 55.3 g, 55.3 g, 55.5 g, 55.6 g,
   55.8 g, 55.9 g, 56.1 g, 56.2 g, 56.2 g, 56.4 g,
   56.7 g, 56.7 g, 56.8 g.
   **a** Write down the values of the median and quartiles.
   **b** Calculate the interquartile range.

5 Find a formula for the $n$th term and hence evaluate the 100th term for these sequences.
   **a** 4, 11, 18, 25, …     **b** 32, 23, 14, 5, …

6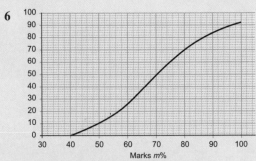
   Marks $m\%$

   The diagram shows the marks of 92 pupils from a geography examination. Work out
   **a** the median mark
   **b** the interquartile range of the marks
   **c** how many pupils scored an A grade if an A grade was awarded for 75%.

7 A flea jumps 40 mm in 0.0025 s. Calculate the average speed of the flea in metres per second.

8 Solve by factorising
   **a** $x^2 - 2x = 80$     **b** $x^2 = 10x + 24$

9 Use the method of differences to find the next two terms in the sequence:
   1     6     14     25     39

10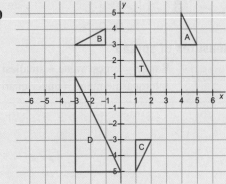

   Define fully the transformations that transform T to
   **a** A     **b** B     **c** C     **d** D

11 Using factors or otherwise, calculate the exact value of $913\,827^2 - 913\,830 \times 913\,824$.

12 How many terms of the sequence 3, 10, 17, 24 are there between 100 and 200?

13 The frequency table gives some results of a short multiple-choice test.
   **a** State the median of these scores.
   **b** Work out the interquartile range of these scores.

| No. of correct answers, $x$ | Frequency, $f$ |
|:---:|:---:|
| 4 | 5 |
| 5 | 9 |
| 6 | 16 |
| 7 | 13 |
| 8 | 11 |
| 9 | 6 |

A sheet of graph paper is needed for Questions 14 and 15. Draw x- and y- axes from −8 to +8 using a scale of 1 cm to 1 unit on both axes. Draw and label triangle T with vertices at (1, 1), (4, 1) and (4, 2).

**14 a** A translation of $\binom{1}{3}$ transforms T to A. Draw A on the diagram.

   **b** A clockwise rotation of 90° about the origin transforms T to B. Draw B on the diagram.

   **c** Describe the transformation that transforms A to B.

**15 a** Reflect T in the line $x + 2 = 0$. Label the image C.

   **b** Reflect T in the line $x + y = 0$. Label the image D.

   **c** Describe the transformation that transforms C to D.

**16** An athlete trains by jumping over low hurdles placed between two mats. The diagram shows the two ways of crossing with two hurdles.

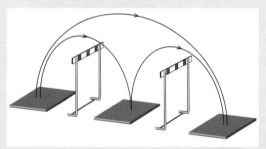

   **a** Draw a diagram to show that there are four different ways in which the athlete can jump over three hurdles.

   **b** Work out a rule to find the number of different ways in which the athlete can jump over $n$ hurdles.

   **c** In theory, there are 512 different ways of jumping over a given number of hurdles. How many hurdles are there?

**17** Arrange these in order with the best buy first.
500 g of rice for 124p
400 g of rice for 99p
300 g of rice for 76p

**18** Solve for $x$.
   **a** $x^2 - 4x + 3 = 0$
   **b** $x^2 + 5x + 6 = 0$

# NUMERACY PRACTICE

## Skills practice 1

### Number

Calculate these.

| | | | |
|---|---|---|---|
| **1** $45 \times 6$ | **2** $4 \times 76$ | **3** $438 \div 6$ | **4** $658 \div 7$ |
| **5** $56.7 + 67.8$ | **6** $409.8 + 69.7$ | **7** $64.6 - 28.8$ | **8** $64.3 - 8.9$ |
| **9** $456 \times 0.1$ | **10** $4.5 \times 0.1$ | **11** $843 \div 0.1$ | **12** $34 \div 0.1$ |
| **13** $78.9 + 4.84$ | **14** $23.6 + 127.9$ | **15** $414 - 99.7$ | **16** $307 - 7.9$ |
| **17** $0.3 \times 0.8$ | **18** $(0.5)^2$ | **19** $0.9 \div 0.3$ | **20** $0.3 \div 0.8$ |

Simplify these, and give each answer as a decimal number.

| | | | |
|---|---|---|---|
| **21** $\frac{4}{5}$ | **22** $\frac{2}{5}$ | **23** $\frac{42}{5}$ | **24** $\frac{23}{5}$ |
| **25** $\frac{3}{8}$ of 56 | **26** $\frac{4}{7}$ of 49 | **27** $\frac{3}{20}$ | **28** $\frac{3}{40}$ |
| **29** $\frac{2}{9}$ of 8.1 | **30** $\frac{2}{7}$ of 8.4 | | |

### Algebra

Simplify these.

| | | | |
|---|---|---|---|
| **1** $2x + 3x$ | **2** $7y + y$ | **3** $6x - 4x$ | **4** $3y - y$ |
| **5** $2x \times 3$ | **6** $4 \times 2x$ | **7** $6x \div 3$ | **8** $8y \div 4$ |
| **9** $3y \times 4y$ | **10** $5y \times 5y$ | | |

Solve these for $x$.

| | | | |
|---|---|---|---|
| **11** $11 = x + 2$ | **12** $3 + x = 19$ | **13** $\frac{x}{2} = 13$ | **14** $36 = \frac{x}{2}$ |
| **15** $13 = x - 12$ | **16** $x - 9 = 9$ | **17** $4x = 68$ | **18** $7x = 105$ |
| **19** $\frac{20}{x} = 5$ | **20** $3 = \frac{39}{x}$ | | |

Substitute $x = 4$ and $y = 3$ to find the value of these.

| | | | |
|---|---|---|---|
| **21** $3x + y$ | **22** $\frac{6x}{y}$ | **23** $\frac{6 + 2y}{x}$ | **24** $\frac{2y + x}{x}$ |
| **25** $2x^2$ | **26** $x^2 - y^2$ | **27** $xy^2$ | **28** $(xy)^2$ |
| **29** $(x - y)^2$ | **30** $(y - x)^2$ | | |

# Skills practice 2

## Number

Calculate these.

| | | | |
|---|---|---|---|
| **1** $87 \times 7$ | **2** $6 \times 79$ | **3** $344 \div 8$ | **4** $414 \div 9$ |
| **5** $2.34 + 0.234$ | **6** $2.34 - 0.234$ | **7** $20.5 \times 6$ | **8** $3.04 \times 7$ |
| **9** $13.8 \div 0.3$ | **10** $14.8 \div 0.4$ | **11** $10^2 \times 0.67$ | **12** $10^3 \times 0.049$ |
| **13** $69.5 \div 10^2$ | **14** $145.9 \div 10^3$ | **15** $10^{-2} \times 78$ | **16** $10^{-3} \times 91$ |
| **17** $0.65 + 10^{-1}$ | **18** $0.58 + 10^{-2}$ | **19** $0.65 - 10^{-1}$ | **20** $0.58 - 10^{-2}$ |

Change each of these to a mixed number.

**21** $\frac{13}{3}$  **22** $\frac{12}{7}$

Change each of these to an improper fraction.

**23** $3\frac{2}{3}$  **24** $4\frac{3}{4}$

Simplify these and give each answer as a fraction.

| | | | |
|---|---|---|---|
| **25** $\frac{1}{6} + \frac{1}{3}$ | **26** $\frac{5}{12} - \frac{1}{3}$ | **27** $\frac{4}{5} \times \frac{5}{12}$ | **28** $\frac{3}{8} \times \frac{16}{21}$ |
| **29** $\frac{7}{9} \div \frac{1}{3}$ | **30** $\frac{7}{9} \div \frac{14}{27}$ | | |

## Algebra

Simplify these.

| | | | |
|---|---|---|---|
| **1** $2x^2 + 3x^2$ | **2** $3y^3 - y^3$ | **3** $2a + 2b - a$ | **4** $5a - 3b - 2a$ |
| **5** $3x^2 \times x$ | **6** $4x \times x^2$ | **7** $2x^2 \times 3$ | **8** $4 \times 2x^2$ |
| **9** $6x^2 \div 2$ | **10** $6x^2 \div x$ | | |

Solve these for $x$.

| | | | |
|---|---|---|---|
| **11** $9.6 = x + 4.68$ | **12** $4.7 + x = 5.61$ | **13** $\frac{x}{6} = 5.8$ | **14** $\frac{x}{9} = 7.6$ |
| **15** $6.8 - x = 4.9$ | **16** $8.1 - x = 2.7$ | **17** $\frac{39}{x} = 0.5$ | **18** $0.6 = \frac{44.4}{x}$ |
| **19** $2(x + 2) = 3(8 - x)$ | **20** $3(2x - 3) = 7(13 - 2x)$ | | |

Substitute $x = -2$ and $y = 3$ to find the value of these.

| | | | |
|---|---|---|---|
| **21** $2y + x$ | **22** $y + 2x$ | **23** $3y - x$ | **24** $y - 3x$ |
| **25** $yx^2$ | **26** $xy^2$ | **27** $3x^2$ | **28** $(3x)^2$ |
| **29** $y - x^2$ | **30** $(y - x)^2$ | | |

# Skills practice 3

## Number

Calculate these.

| | | | |
|---|---|---|---|
| 1 $35 \times 7$ | 2 $7 \times 17$ | 3 $27 \times 9$ | 4 $315 \div 7$ |
| 5 $216 \div 6$ | 6 $22.4 \div 7$ | 7 $122.2 \div 13$ | 8 $68.9 \div 1.3$ |
| 9 $432.1 - 123.4$ | 10 $132 - 41.6$ | 11 $176.8 - 84$ | 12 $70 \times 0.1$ |
| 13 $70 \div 0.1$ | 14 $70 \times (0.1)^2$ | 15 $67.5 \times 0.1$ | 16 $67.5 \div 0.1$ |
| 17 $24 \div 0.4$ | 18 $25 \div (0.5)^2$ | 19 $10^7 \div 10^{-3}$ | 20 $(2 \times 2 \times 2 \times 2)^2$ |

Change each of these to a mixed number.

21 $\frac{7}{3}$　　　　　　　　22 $\frac{15}{6}$

Change each of these to an improper fraction.

23 $4\frac{1}{5}$　　　　　　　　24 $5\frac{1}{3}$

Simplify these, and give each answer as a fraction.

| | | | |
|---|---|---|---|
| 25 $\frac{1}{3} + \frac{1}{4}$ | 26 $\frac{2}{3} - \frac{1}{4}$ | 27 $\frac{3}{5} \times \frac{5}{18}$ | 28 $\frac{5}{7} \times \frac{21}{25}$ |
| 29 $\frac{5}{12} \div \frac{1}{4}$ | 30 $\frac{3}{5} \div \frac{6}{7}$ | | |

Estimate each of these.

31 $24 \times 807$　　32 $7982 \div 194$　　33 $\dfrac{24 \times 396}{96}$　　34 $\dfrac{8.97 \times 0.47}{1.89 \times 0.08}$

## Algebra

Simplify these.

| | | | |
|---|---|---|---|
| 1 $3x + x$ | 2 $3x - x$ | 3 $3x \div x$ | 4 $(3x)^2 \div x$ |
| 5 $6x - 5y + 2x$ | 6 $xy \times yx$ | 7 $3x^2 - 2y + (2x)^2$ | 8 $xyz \times (zyx)^2$ |
| 9 $3x^2y \div xy^2$ | 10 $(x + x)^3$ | | |

Solve each of these for $x$.

11 $12 = x - 3$　　12 $123 = 76 - x$　　13 $12 + x = 3x - 48$　　14 $3 = \dfrac{267}{x}$

15 $2(x - 1) = 48$　　16 $8 = \dfrac{104}{(x - 1)}$

Substitute $x = 6$ and $y = -5$ in each of these to find their value.

| | | | |
|---|---|---|---|
| 17 $x + 2y$ | 18 $x^2 + y$ | 19 $x + y^2$ | 20 $x^2 + y^2$ |
| 21 $(xy)^2$ | 22 $x^3$ | 23 $(x + y)^2$ | 24 $(x - y)^2$ |
| 25 $x^2 - y^2$ | 26 $(x + y)^x$ | | |

# Skills practice 4

## Number

Calculate these.

**1** $23 \times 9$      **2** $27 \times 7$      **3** $27 \times 8$      **4** $1440 \div 9$

**5** $196 \div 7$      **6** $312 \div 8$      **7** $123.6 + 23.6$      **8** $727 + 72.7$

**9** $987.6 - 678.9$      **10** $999 - 99.9$      **11** $727 - 72.7$      **12** $\frac{5}{100} + 0.09$

**13** $84 \times 0.2$      **14** $84 \div 0.2$      **15** $84 \times (0.2)^2$      **16** $2\frac{1}{3} \times 1\frac{5}{6}$

**17** $2\frac{1}{3} \div 1\frac{5}{6}$      **18** $2\frac{1}{3} + 1\frac{5}{6}$      **19** $2\frac{1}{3} - 1\frac{5}{6}$      **20** $\sqrt{256}$

Change these to a mixed number.

**21** $\frac{5}{4}$      **22** $\frac{17}{5}$

Change these to an improper fraction.

**23** $3\frac{1}{7}$      **24** $2\frac{3}{5}$

Calculate these, giving each answer as a fraction in its simplest terms.

**25** $\frac{2}{5} + \frac{1}{7}$      **26** $\frac{5}{9} - \frac{2}{11}$      **27** $\frac{2}{7} \times \frac{7}{12}$      **28** $\frac{3}{28} \times \frac{7}{9}$

**29** $\frac{2}{5} \div \frac{3}{10}$      **30** $\frac{5}{8} \div \frac{1}{4}$

Estimate these.

**31** $32 \times 694$      **32** $5987 \div 1492$      **33** $\dfrac{18 \times 597}{37}$      **34** $\dfrac{11.7 \times 10.9}{2.8 \times 5.6}$

## Algebra

Simplify these.

**1** $2 - 3(4 - 5x)$      **2** $2xy \times 3yx$      **3** $2x^2y + 6yx^2$      **4** $2x^2y - 6yx^2$

**5** $2x^2y \times 6yx^2$      **6** $2x^2y \div 6yx^2$      **7** $(2x + 3x)^2$      **8** $\dfrac{x + x + x}{3}$

**9** $\dfrac{(x + x + x)^2}{3x}$      **10** $2(3x)^2$

Solve each of these for $x$.

**11** $10 = 2x - 6$      **12** $3(x - 2) = 45$      **13** $3(x - 2) = 2(x - 1)$      **14** $\dfrac{114}{x} = 6$

**15** $\dfrac{x - 2}{3} = x - 1$      **16** $\dfrac{3}{x + 1} = 2$

Substitute $x = 4$ and $y = -1$ to find the value of each of these.

**17** $x - y$      **18** $(x - y)^2$      **19** $x^2 + y^2$      **20** $(xy)^2$

**21** $y^3$      **22** $x - 6y$      **23** $(2x - y)^2$      **24** $x^2y$

**25** $(x + 2y)^x$      **26** $y^x$

# Skills practice 5

## Number

Calculate these.

**1** $4.8 \times 9$  **2** $5.9 \times 8$  **3** $4.5 + 45 + 40.5$  **4** $100 - 9.9 - 0.09$

**5** $0.9 - 0.06 + 0.087$  **6** $78 - 5.9 + 6.4$  **7** $(1.3)^2$  **8** $(0.6)^3$

**9** $\sqrt{169}$  **10** $\sqrt[3]{27}$  **11** $98.8 \div 13$  **12** $83.3 \div 17$

**13** $24 \times 6.7$  **14** $3.8 \times 69$  **15** $23.2 \div 0.4$  **16** $5.22 \div 0.6$

**17** $4.1 + 5.2 \times 6.3$  **18** $17.2 - 12 \div 30$  **19** $\dfrac{4.2 \times 16}{3.5 \times 6}$  **20** $\dfrac{39 \times 2.1}{7 \times 7.8}$

**21** $\dfrac{3}{8} + \dfrac{3}{4}$  **22** $\dfrac{3}{8} \div \dfrac{3}{4}$  **23** $2 - \dfrac{5}{9}$  **24** $2\dfrac{1}{3} \times \dfrac{3}{13}$

**25** Change 0.67 to a fraction.  **26** Change $\dfrac{3}{8}$ to a decimal.

**27** Change $\dfrac{1}{7}$ to a recurring decimal.  **28** Change $0.\dot{3}$ to a fraction.

Estimate these.

**29** $48 \times 631$  **30** $7853 \div 15.8$  **31** $753.99 \times 0.0418$  **32** $0.694 \div 0.00486$

## Algebra

Simplify these.

**1** $x^2(3x)^2$  **2** $x^2 + (3x)^2$  **3** $(3x)^2 - x^2$  **4** $(3x)^2 \div x^2$

**5** $\dfrac{2x + x^2}{x}$  **6** $\dfrac{3y - 6y^3}{3y}$  **7** $(x - 2)^2$  **8** $(a - b)(a + b)$

**9** $\dfrac{x}{3} + \dfrac{2x}{5}$  **10** $\dfrac{x}{3} \div \dfrac{2x}{5}$

Factorise these.

**11** $3x^2y - 6xy$  **12** $x^2 - 6x + 8$  **13** $x^2 - y^2$

Solve for $x$.

**14** $\dfrac{306}{x} = 45$  **15** $(2x)^2 + 1 = 26$  **16** $(x - 4)(x - 2) = 0$

Make $x$ the subject of these.

**17** $c = \dfrac{x}{d} - a$  **18** $ad - bx = c$  **19** $cx = bx + a$

Solve these inequalities, writing your answer in the form $x \geqslant, >, \leqslant$ or $< \ldots$.

**20** $3 - x > 6$  **21** $3(x - 7) \geqslant 8x + 19$

Substitute $x = -2$ and $y = -3$ to find the value of these.

**22** $(x + y)^2$  **23** $(x - y)^2$  **24** $x^2y$  **25** $(xy)^2$

**26** $2x^2 + y$  **27** $(2x)^2 - y$  **28** $x^y$

# CHALLENGES

1   The table lists some giant numbers.

| Million | $1 \times 10^6$ |
|---|---|
| Quintillion | $1 \times 10^{30}$ |
| Vigintillion | $1 \times 10^{120}$ |
| Centillion | $1 \times 10^{600}$ |

Write down in standard form the value of
**a** a million multiplied by a centillion
**b** a vigintillion divided by a quintillion
**c** a centillion squared

2   In the universe, the distances are so vast that light years are used as units of length.
A light year is the distance travelled by light in one Earth year of 365 days.
Given that the speed of light is $3 \times 10^5$ km/s, find out how many cubic millimetres there are in a
volume of a cubic light year to three significant figures in standard form.

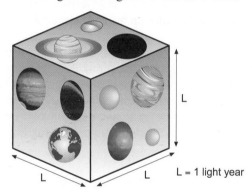

L

L

L

L = 1 light year

3   What is the value of $\sqrt{2 + \sqrt{2 + \sqrt{2 + \sqrt{2 + \sqrt{2 + \ldots}}}}}$?

4   If $A$, $B$, $C$ and $D$ are all positive whole numbers, each of value between 1 and 9 inclusive, solve
this equation.
$$A^B \times C^D = ABCD$$

5   The diagram shows a sheet of 12 postage stamps.
In how many different ways can four stamps be torn
off so that all four of them remain joined together?
Draw diagrams to illustrate your answer.

6   How many positive whole numbers have square roots that are between 100 and 101?

7 Can you complete the seating plan for a committee meeting
   around a table to meet these conditions?

   ♦ The chairman, Mr Grim, will sit at the head of the table.

   ♦ Anyone may sit next to him, except for Mrs Pain, who can sit
     next to Mr Nice, who can have Mrs Chatty on his other side.

   ♦ Mr Woof can sit on Mrs Hack's left.

   ♦ Mrs Smart must sit next to at least one male, but not the
     chairman.

8 The height of each tower of the Humber
   Bridge is 162.5 m. The towers are 1.41 km
   apart at the base but 36 mm further apart
   at the top. Estimate the radius of the Earth.

162.5 m

9 Unfortunately, the final table of the local ice hockey league table is lost on its way to the sports
   editor of the local paper. However, the report does arrive at her office, and it gives enough
   information to enable the editor to draw up the final league table.

> Each team played every other team twice, and the final order was alphabetical.
> Only two teams won more than half their matches, but only one team lost
> more than half its matches. No two teams won the same number of matches.
> No team was unbeaten, and only Barnwell failed to draw at least one match.
> Ashton won the same number of matches as Elton lost. Deene lost half its
> matches, and Cotterstock got 8 points. No two teams ended up with the same
> number of points.

(A win counts as 2 points and a draw counts as 1 point.)
Copy and complete this table, using the information given in the report.

| Team | Played | Won | Drew | Lost | Points |
|------|--------|-----|------|------|--------|
| Ashton | | | | | |
| Barnwell | | | | | |
| Cotterstock | | | | | |
| Deene | | | | | |
| Elton | | | | | |

10 a What is the last digit of $4^{50}$?
    b What is the last digit of $3^{100}$?
    c Is $3^{444} + 4^{333}$ divisible by 5?

**11** Which is a better fit, a round peg that just fits in a square hole or a square peg that just fits in a round hole?

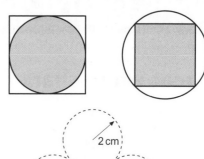

**12** This vase shape is made from three-quarters of the circumference of a circle with radius 2 cm and three separate quarter-circle arcs with radius 2 cm. Find the shaded area.

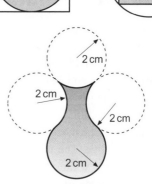

**13** There are ten beads in a box. Some are red and some are blue. The probability of picking at random a red bead and a blue bead, in either order, is $\frac{8}{15}$. How many red beads are in the box?

**14** Popeye is buying his supper in the local supermarket. He finds he has the exact amount of money to buy either 5 cans of spinach and 6 cans of baked beans, or 7 cans of spinach and 3 cans of baked beans. In the end he buys as many cans of spinach as possible. How many cans of spinach does he buy, and does he receive any change?

**15** Mr Hony, Mr Potts and Mr Turner spend the evening gambling at home. At the start of the evening the amount of money they have between them is in the ratio 5 : 4 : 3 respectively, while at the end of the evening the amount of money is in the ratio 4 : 3 : 2 respectively. One of them loses £1. How much money did each have at the start of the evening?

**16** The diagram shows two circles, both of radius 2 cm, inside a semicircle. Find the shaded area.

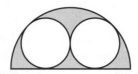

**17** The diagram shows a square surrounded by a circle which in turn is surrounded by a square. The area of the inner square is 100 cm². Find the area of the outer square.

**18** Make the following numbers, using all the numbers 3, 4, 5, 6 each exactly once and the symbols +, −, ×, ÷ and brackets (as many of each as you like).
For example, 29 could be formed as $29 = 6 \times 5 - 4 + 3$.
**a** 42   **b** 38   **c** 47   **d** 39   **e** 67

**19** [*n*] means the largest integer less than or equal to *n*. For example, [3.2] = 3, [7] = 7, [5.9] = 5 etc.
Solve the equation $x + \left[\frac{x}{100}\right] = 2004$.

**20** A 4 cm × 4 cm square is crossed by lines from each corner to a point three-quarters of the way along the opposite side, as shown. Find the area of the shaded shape.

## Great white shark

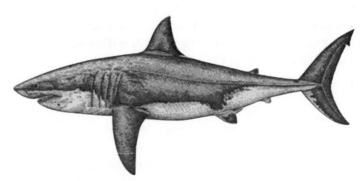

Of the **350** species of shark, the great white shark is the most well known and feared, being large, aggressive and fast. It has extraordinary senses, and is one of the most efficient hunters alive. The territory of the great white shark can span thousands of miles. They are ferocious predators, and have about **3000** razor-sharp teeth that are up to **7.5 cm** in length. Their body length averages **3–5 m**, although the biggest great white ever recorded was a massive **7 m** long, and it weighed **3200 kg**. Even this is relatively small when compared with the great white's ancient, extinct ancestor, the *Carcharodon megalodon*, which was **18 m** long and had teeth up to **20 cm** long.

Great whites are propelled through the water by their powerful tails. Their average speed is **2 miles/hour**, but they swim for short bursts at **15 miles/hour**. Because of the speed of their prey, they rely heavily on stealth and surprise, often attacking from below. They can smell one drop of blood in **100 litres** of water, and sense the tiny electrical charges generated by all animals by using a series of jelly-filled canals in their head called the ampullae of Lorenzi. A shark's lateral line can 'feel' vibrations up to **180 m** away. Great white sharks eat a lot less than you would imagine, consuming about **3–5%** of their body weight at each feed. They do not chew their food. Their teeth rip prey into mouth-sized pieces which are swallowed whole. A large meal can satisfy the shark for up to **2 months**.

Sharks have survived for over **400 million years**; but within a span of 20 years, man has drastically reduced the population by killing them for sport and food, and overfishing the marine life they feed on. **100 million** sharks are killed each year. After the release of the film *Jaws*, the general public perceived these creatures as maneaters. In fact, great whites rarely attack humans. By comparison with the number of people exposed to them (knowingly or unknowingly), the ratio is very small. Of the **100 annual shark attacks**, great whites account for about **one-third**, resulting in **10 deaths** worldwide. When attacks do occur, it is often a case of mistaken identity. A person on a surfboard looks much like a sealion from below. However, the first bite is often deadly, because of the shark's extraordinary bite strength. Humans can exert a bite of **11 kg/cm$^2$**. A great white's jaws exert **310 kg/cm$^2$**.

The average life span of the great white is about **20–30 years**, but it is believed that they can live to be about **100 years** old. Despite the number of pups born to each female (**2–14** fully formed babies that can be up to **1.5 m** in length), great whites are now an endangered species, because of ignorance and over-hunting by man. They are a protected species along the coasts of the USA, Australia and South Africa. Their greatest threat is mankind.

# Exercise

Where appropriate, give your answers correct to 3 significant figures.

**1** How many sharks are killed in the world on average per second?

**2** If all the teeth of the great white shark were laid end to end, approximately how long would the greatest length of this line be in metres?

**3** Thirty species of shark are responsible for attacks on humans. What percentage of shark species could be termed maneaters?

**4** Express the length of the largest ever recorded great white shark compared with your own height as a ratio of $1:x$. Find $x$.

**5** Calculate the top speed of the great white shark in metres per second (1 mile $\simeq 1609\,\mathrm{m}$).

**6** Can the great white sense the vibrations of the paddling surfboarder below?

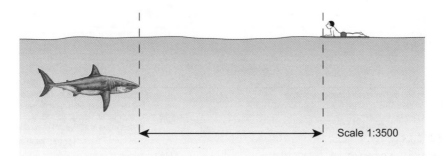

Scale 1:3500

**7** The length of the extinct *Carcharodon megalodon* is $x$% of the biggest great white ever recorded. Find $x$.

**8** The mako shark is the fastest of the species at 97 km/h. What percentage of this speed can the great white shark reach?

**9** The weight ratio of the largest ever great white to the *Carcharodon megalodon* is 1:17. What was the weight of the *Carcharodon megalodon* in kilograms?

**10** Express the bite strength of humans as a percentage of that of the great white shark.

**11** Calculate the bite strength of the great white shark in tonnes per square metre.

**12** Estimate the greatest gross weight of prey in tonnes consumed by a 1400 kg great white shark in a lifetime of 30 years. Give your answer in standard form.

# This troubled planet

During the last decade of the twentieth century, our planet underwent dramatic changes in terms of pollution, population and the pressures that humanity puts on its natural resources.

The Earth Summit in Rio de Janeiro spotlighted ten trends challenging world leaders as the dawn of the twenty-first century approached.

**1990**

**2000**

### POPULATION
The Earth's population was **5.31 billion** people, of whom **73%** lived in developing countries.

### POPULATION
It is now **6.17 billion** people, of whom **79%** live in developing countries, and the population is growing faster than ever before. Each year the number rises by **95 million**, which was the planet's entire population in around 1000 BC.

### WAR AND REFUGEES
Nations spent **$788 billion** on arms and armed forces. The number of refugees fleeing war was estimated to be about **13.8 million**.

### WAR AND REFUGEES
Global military spending is expected to be just under **$908 billion**. Refugee numbers are put at **22.7 million**.

### NUCLEAR POWER
There were just over **368 nuclear reactors**.

### NUCLEAR POWER
There are **428 nuclear reactors** in **31 countries**.

### TRANSPORT
There were **565 million motor vehicles**, including **452 million cars**. Their pollution was confined almost entirely to developed countries.

### TRANSPORT
There are just over **740 million motor vehicles**, including **592 million cars**. Developed nations still have the great majority, but vehicles help to make the air dangerous in many Third World cities.

### GLOBAL WARMING
**22 billion tonnes of carbon dioxide**, chief of the climate-changing greenhouse gases, were released into the air from the burning of fossil fuels and cement manufacture. The concentration in the atmosphere stood at **353 parts/million**.

### GLOBAL WARMING
**26 billion tonnes of carbon dioxide** are released, and the concentration is now **368 parts/million**. Forest burning is adding more.

### OZONE LAYER
Ozone-layer-destroying chlorine had caused a hole over the Antarctic. Chlorine concentration was **2.8 parts/billion**.

### OZONE LAYER
Chlorine concentration is now **3.5 parts/billion**, which is enough to open a hole in the ozone layer each Antarctic spring.

### MEGA-CITIES
There were **12 cities** with over **10 million people**. **45%** of the world's population lived in towns and cities.

### MEGA-CITIES
There are **17 cities** with over **10 million people**. **50%** of the world's population is urban.

### RAINFORESTS
Up to one-third of the Earth's girdle of tropical rainforest had been destroyed. Some **163 000 km²** were being lost each year, which is an area the size of Iceland.

### RAINFORESTS
The deforestation rate is now thought to have risen to around **190 000 km²/year**.

### FISHERIES
**89 million tonnes of fish** were taken from the Earth's oceans as nations expanded deep-sea fleets and sought protein from the sea.

### FISHERIES
The catch has risen to about **126 million tonnes/year**. Fish stocks are increasingly at risk of collapse.

### SPECIES
There were just under **750 000 African elephants** left. This was one of thousands of species known to have become endangered by humanity. A wave of extinctions unprecedented since the last Ice Age had begun.

### SPECIES
There are now only about **525 000 African elephants** left, mainly because of ivory poaching.

**FACT FINDER**

# Exercise

Give your answers to these questions correct to 3 significant figures.

1 How many people lived in developing countries in 2000?

2 Calculate how much was spent in US dollars per person in the world on military spending in
  **a** 1990                 **b** 2000

3 Find the mean increase in cars per hour between 1990 and 2000.

4 Calculate the mean number of kilograms of fish caught per person in
  **a** 1990                 **b** 2000

5 Using the given deforestation rate, calculate how many areas equivalent to a football pitch are disappearing per minute, given that the dimensions of a football pitch are 75 m × 110 m, in
  **a** 1990                 **b** 2000

6 Find the mean number of African elephants which disappeared per day between 1990 and 2000.

7 Calculate the mean rate of population increase in people per second for the period 1990–2000.

8 Calculate the population density of the Earth in people per square kilometre in
  **a** 1990                 **b** 2000
  (Assume that the Earth is a sphere of diameter 12 700 km. The area of a sphere is $4\pi r^2$, where $r$ is the radius, and 30% of the Earth's surface is land.)

9 Assuming that the rate of deforestation remains unchanged from 2000, use the graph to estimate
  **a** the total area of rainforest in 2010
  **b** when the area of rainforest will be zero.

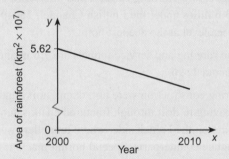

10 Assume that the African elephant population will continue to decline at the same rate. Calculate when this elephant will become extinct.

11 Given that the volume in cubic kilometres $V$ of the Earth's atmosphere is given by the formula
  $$V = \frac{4\pi(R^3 - r^3)}{3}$$
  where $R = 6407$ km and $r = 6400$ km, calculate the density of carbon dioxide in the atmosphere in kilograms per cubic kilometre in 2000.

12 The total length of all the roads in the USA in 2000 was 6.9 million kilometres.
  Given that the average motorvehicle is 5 m long, calculate the average spacing between these vehicles if they were all to be simultaneously on US roads in 2000.

# The Channel Tunnel

The building of the Channel Tunnel between England and France is the largest civil engineering project ever undertaken in Europe. The first proposal for such a connection was in **1802**, but was ill-timed due to war resuming between Britain and France. The construction was completed in **1994**.

The tunnel itself is in fact three: **two running tunnels** and **one service tunnel**. The tunnels are **30.7 miles** in length with **23.6 miles** under the English Channel, **25–40 m** below the sea bed, reaching **100 m** below sea level at the deepest point.

The internal diameters of the running and service tunnels are **7.6 m** and **4.8 m** respectively and their maximum gradient is about **1:90**.

Conditions underground during construction were hot, damp, noisy and dangerous. The main engineering challenge was having to drill through fractured chalk saturated with water. This problem was particularly acute on the French side, where tunnellers were expected to progress at half the speed of their English counterparts. Special boring machines were fitted with a seal behind the cutting head to prevent water gushing in as boring progressed; in fact water had to be pumped away at **41 litres per second**! The British workers advanced at **300 m** per week (**100 m** more than planned) while the French also exceeded their target of **100 m per week**.

The total cost of the construction was about **£8 billion**, and there were **15 000 workers** on site at its busiest.

The crossing by ferry takes **90–100 minutes**, but the train crossing takes **35–40 minutes**. The high-speed trains (HSTs) are designed to travel at **80 mph** in the tunnel and **100 mph** once outside, reducing the travel time between London and Paris from **5 hours and 12 minutes** to just **3 hours**, not much longer than by air! Each train carries an estimated **800 passengers** in **24** specially designed wagons. In **1994**, there were **315 000** passengers, only four years later this rose to **12 799 000**. As more people become accustomed to this service, it is expected that these figures will continue to rise.

# Exercise

1 After how many years since its first proposal was the Channel Tunnel completed?

2 What percentage of the tunnel length is actually underneath the English Channel?

3 Calculate the total volume of air in the three tunnels in m³ expressed in standard form to 3 significant figures.

(Assume the tunnels are empty and are perfectly straight cylinders. 1 mile $\approx$ 1600 m.)

4 All three tunnels were coated with a sealant costing £1.40 per square metre. Find the cost of this coating.

5 On average, how many people used the tunnel per hour in 1994?

6 Use the photograph to show that the diameter of the drill-head is about 10 m.

7 The whole construction started at the beginning of 1987 and was completed at the end of 1994. Calculate the mean cost of the scheme per second.

8 According to initial estimates, how long after tunnelling began were the British tunnellers expected to meet the French?

9 How many litres of water were pumped away per day by one boring machine? (Answer in standard form to 3 significant figures.)

10 Calculate the percentage increase in the number of trains required from 1994 to 1998.

11 A particular tunnel crossing takes 35 minutes. For the first third of the journey, the train travels at $x$ mph. The final two-thirds is completed at three times this speed. Find $x$.

12 If $p \times 10^3$ is the number of passengers using the tunnel, $y$ years after 1994, and assuming that this passenger increase is constant up to 1998, show that $p = ay + b$, where $a$ and $b$ are constants, and find the values of $a$ and $b$.

# Recycling

Recycling is important because it conserves natural resources, saves energy and landfill space and reduces air and water pollution.

**Aluminium,** a light metal with a density of **2.7 g cm⁻³**, is one of the major recycled materials in the United Kingdom. It can be recycled indefinitely, as reprocessing does not damage its structure.

The UK's latest used-aluminium recycling plant produces **8 m** long aluminium ingots weighing **26 tonnes**, each of which makes **1.6 million** cans. If you were to drink one can per day, it would take over **4000 years** to use up the cans produced by one ingot.

Production of aluminium is costly, demanding large quantities of energy. Recycling **one** aluminium can takes about **5%** of the energy needed to produce the can from raw materials.

In 1998 about $4 \times 10^9$ cans were used in the UK. The average can has a height of **14.5 cm** and a diameter of **6.5 cm**. **36%** were recycled and the rest went into landfills. If all the recycled cans were laid end to end, they would stretch from Land's End to John O'Groats **160 times**.

A bin liner holds about **200** loose cans, while **one tonne** of crushed cans takes up about **4 m³**. In October 2000 you would have received **£650 per tonne** for loose cans.

**Glass** is another of the major recycled materials, with the first bottle banks appearing in 1977.

At end of 1999 there were about **22 020** bottles banks in the UK, one for every **2700** people, and **499 000** tonnes of glass were recycled. Bottle banks are about **1.5 m** high with a diameter of **1.25 m**, and can hold up to **3000** bottles or jars before they need emptying.

On average every household in the UK consumes around **500** glass bottles and jars per year.

The density of glass is the same as aluminium and there are approximately **3000** bottles or jars to a **tonne**. The average bottle or jar has a height of **19 cm** and diameter of **7 cm** and contains **22%** recycled glass. The largest glass furnaces can produce about **400 tonnes** of glass every day.

# Exercise

1 Estimate the mass of an aluminium can in grams.

2 Find the density of aluminium in tonnes per m$^3$.

3 What is the cross-sectional area in m$^2$ of an aluminium ingot produced by the latest recycling plant?

4 It is claimed that if you were to drink one can per day it would take over 4000 years to use one ingot. How many years would it actually take?

5 If all the cans used in the UK in 1998 were put one on top of each other, how high would the pile be?

6 Estimate the mass of a glass bottle in grams.

7 Estimate the volume of a bottle bank.

8 Estimate the volume of an average bottle.

9 Show that a bottle bank with 3000 bottles in it must contain some broken bottles.

10 If all the bottles recycled in 1999 were laid end to end, how many times would they go round the Earth's equator? (Assume the Earth is a sphere with radius 6400 km.)

11 How many recycled cans can be made using the same energy as is needed to make a new can?

12 Estimate the population of the United Kingdom in 1999, giving your answer in standard form.

13 If all the glass that is recycled comes from bottles, how many bottles were recycled in 1999?

14 What is the distance from Land's End to John O'Groats?

15 What would be a fair selling price for a bin liner of loose cans?

16 In 1998, if all the cans that went to landfill sites were crushed, what volume did they take up?

17 When aluminium cans are crushed, approximately what percentage is air?

18 Estimate how many times each bottle bank is emptied each year.

19 Assume that all the recycled glass from 1999 together with new glass is used to make new bottles.
   a How many bottles can be made?
   b How many furnaces would be required?
   c What volume (in m$^3$) of glass is involved?

20 Estimate the average number of people in a family in the UK.

# The Solar System

Our Solar System consists mainly of nine planets, some with moons, orbiting the Sun, together with rocky lumps called the asteroids. Here are some facts about the Solar System.

| Body | Diameter (km) | Distance from the Sun (km) |
|---|---|---|
| Sun | 1 390 000 | |
| Earth | 12 800 | $1.5 \times 10^8$ |
| Mars | 6790 | $2.28 \times 10^8$ |
| Jupiter | 143 000 | $7.78 \times 10^8$ |
| Pluto | 3000 | $5.95 \times 10^9$ |

Our Solar System, with the Sun at its centre, is only one small part of our galaxy, which is called the Milky Way. A galaxy is a system of stars, and the Milky Way, which is shaped like a thin disc, has over **100 000 million** of them. The largest known star has a diameter of **$2.5 \times 10^8$ km**.

Because the distances between stars are so large, they are measured in light years, which is the distance light travels in one year. Light travels at **300 000 km/s**; in contrast, a fighter jet travels at around **2000 km/h**. The Milky Way is **120 000 light years in diameter**, and our Sun is **30 000 light years** from the centre. The nearest star to our Sun is **4.2 light years** away, while the Andromeda Galaxy, the remotest object in our skies that is (just) visible with the naked eye, is **$2.2 \times 10^6$ light years** away.

# Exercise

1 Imagine making a scale model of the Solar System, with the Sun represented by an orange 8 cm in diameter. Show that the Earth could be represented by a ball of plasticine 0.7 mm in diameter placed 8.6 m away from the orange.

2 For the model, work out the diameter and distance from the orange of the other planets given in the table.

3 What is the diameter of the largest star in our model?

4 Work out how long it would take a fighter jet to fly from the Earth to the Sun (assume this is possible!)

5 How long does it take light to travel from the Sun to the Earth?

6 How many km are there in a light year?

7 Imagine the nearest star is represented by another orange in the scale model. How far away is the nearest star in this model?

8 How long would it take a fighter jet to fly to the nearest star?

9 What is the diameter of the Milky Way in km and how long would it take a fighter jet to fly across the Milky Way?

10 How far away is the Andromeda Galaxy in km?

Investigations are open-ended – you are not told what to do, so you need to plan your own path through the problem to reach a solution. The aim of any investigation is to use and apply mathematical knowledge and skills in three areas:

- Decision making to solve problems
- Communicating mathematically
- Mathematical reasoning

## Decision making to solve problems

You need a strategy. Think how you might tackle the problem, and plan what to do before you begin.

- A general formula is often asked for. Starting with small numbers, look for patterns and then use algebra to simplify the problem.
- Widening your collection of information may result in clearer ideas – particularly true of any statistical investigation. So, gather information and tabulate your results.
- As you make progress, use your initiative. Be curious and ask yourself further questions so that you explore the whole problem, before arriving at what you think is the solution.

## Communicating mathematically

- Explain everything you do and justify your results. Even if you find yourself down a 'blind alley', you should comment on this and then move on, having learnt from your experience.
- Diagrams, graphs and tables should help you to compare your results and to spot patterns.
- Use algebra to make generalisations. Notation will enable you to express yourself in a clear, mathematical statement.
- Observations should be recorded and perhaps pursued further.
- Suggestions should be made: 'I wonder what would happen if...?'
- ICT should be used where appropriate, showing all relevant methods and working.

## Mathematical reasoning

- Generalise your results where possible into a rule or formula, test this formula and decide if it has any limitations.
- Formal proof is very desirable and is often the hardest part of a mathematical investigation.
  A general proof illustrates that the rule is correct for *all* values in a given range.
  Proving it for a particular value only demonstrates that a rule works for that single value!
  However, a single counter-example can be used to disprove a statement.

You might need to use a variety of mathematical skills – such as algebra, graph plotting, geometry and statistics – as well as ICT. The success of your investigation will hinge on how you combine your mathematical knowledge into a coherent answer and a sensible solution.

These four examples give tips on strategies you might use when tackling investigations.

## Example 1

**Investigate whether $n^2 - n + 41$ is prime for all integers $n$.**

The strategy for this investigation seems to be to try lots of different values of $n$.

♦ You might start with $n = 1$. Having worked out that $n$ is prime for all values up to 20, say, you might begin to believe that the result is always prime, but how can you prove this?

♦ Alternatively, you might try $n = 41$, because then $-n + 41 = 0$ and the remaining term ($n^2$) is not prime. A single – carefully chosen – counter-example disproves the statement.

## Example 2

**Investigate the possible areas obtained by a rectangle of perimeter 100 m.**

For this investigation, you might try to find the dimensions of the largest possible area.

♦ You might use algebra and introduce a variable to represent the length of one side.
Using the fact that the perimeter is 100 m, you could find an expression for the other side. From this, you could produce a formula for the area.

♦ You might decide to plot a graph of area against one length, and use the graph to make sensible conclusions.

## Example 3

**Investigate the number of matches required in a tennis club tournament of 20 members if every member plays each of the other members once.**

To solve this problem for 20 members might prove difficult and time-consuming, and it might be more useful to try to find a general rule for any number of people.

♦ You might start with a much smaller number, and try to draw a diagram or list the matches.

♦ You might then try other numbers and draw more diagrams, to see if you can spot a pattern.

♦ You might tabulate your results and introduce a variable to represent the number of people.

♦ Having found the pattern, and expressed it using a formula, you might use trial and error to check that your formula works. Then, substituting 20 into your formula solves the original problem.

## Example 4

**Investigate whether there is any connection between people's ability in Mathematics and in Science.**

This investigation will involve gathering some data. Your sample could be a group of, say, 30 people – perhaps the people in your class. You will need to think about lots of things:

♦ *How to collect the data*
This might involve designing a questionnaire, or setting tests in Mathematics and Science, or relying on previous examination results. Data from the Internet could also be used.

♦ *How to avoid bias in your sample*
You may want to extend your investigation to other age groups, other areas of the country – even to other countries. You might want to consider current data against data 20 years ago. You might also consider taking a stratified sample to make allowance for bias.

♦ *What statistics to collect, and how to display your results*
It is important to decide what data to collect and then to display this data sensibly, using a bar chart, pie chart, histogram, scatter graph or cumulative frequency graph.

♦ *Are there any exceptions in the data?*
You will need to explain any exceptions, and to think carefully before making wide-sweeping generalisations from your results from a relatively 'small' sample.

♦ *Will your survey be constrained by costs, lack of time and/or personnel?*
Any increase in sample size increases the amount of evidence and may produce more reliable results, so if your survey could be improved, suggest how this could be done.

# Answers to activities and exercises

# Unit 1

## Number 1

### EXERCISE 1 (page 1)

**1** $\frac{3}{4}$, 75%  **3** $\frac{1}{4}$, 0.25  **5** 0.15, 15%

**7** $\frac{7}{20}$, 35%  **9** $2\frac{2}{3}$  **11** $3\frac{2}{5}$

**13** $\frac{7}{3}$  **15** $\frac{11}{6}$  **17** $\frac{2}{7}$  **19** $\frac{1}{6}$

**21** $\frac{1}{2}$  **23** $\frac{3}{40}$  **25** $\frac{5}{6}$  **27** $\frac{2}{3}$

**29** $\frac{4}{9}$  **31** $\frac{7}{40}$  **33** 680  **35** 0.765

**37** 39  **39** 24  **41** 26.25  **43** 41.86

### EXERCISE 2 (page 2)

**1** 5  **3** 13  **5** 13  **7** −3
**9** −12  **11** −6  **13** −2  **15** 3
**17** 10  **19** −2  **21** 23  **23** 18
**25** 18  **27** 12  **29** 12  **31** 36
**33** 49  **35** 144

### EXERCISE 3 (page 4)

**1** 1.8  **3** 5.52 m  **5** 0.684
**7** £31.80  **9** £45.50  **11** 15%
**13** 10%  **15** 5%  **17** 6%
**19** €6880

### EXERCISE 3★ (page 4)

**1** 0.75  **3** $171  **5** 52.5
**7** 68.005 km  **9** $80.04  **11** 15%
**13** 12.5%  **15** 0.0402%  **17** 250%
**19** There are 500 million telephones in the world, 60 000 in Ethiopia. Population of Ethiopia is 30 million.

### ACTIVITY 1 (page 6)

Neurons: $10^{11}$
Connections: $10^{15}$
Connections per second = $10^{15} \div$
$(60 \times 60 \times 24 \times 365 \times 75) \approx$ half a million

### EXERCISE 4 (page 6)

**1** $10^5$  **3** $10^{10}$  **5** $10^3$
**7** $10^{13}$  **9** $10^2$  **11** $10^3$
**13** $10^3$  **15** $10^1$  **17** $4.56 \times 10^2$
**19** $1.2345 \times 10^2$  **21** $5.68 \times 10^2$  **23** $7.0605 \times 10^2$
**25** 4000  **27** 4 090 000  **29** 560
**31** 7 970 000  **33** 1000  **35** $8.4 \times 10^9$
**37** $5 \times 10$

### EXERCISE 4★ (page 6)

**1** $4.5089 \times 10^4$  **3** $2.983 \times 10^7$
**5** $10^3$  **7** $10^5$
**9** $10^{21}$  **11** $10^0$ or 1

**13** $6.16 \times 10^6$  **15** $4 \times 10$
**17** $9.1125 \times 10^{16}$  **19** $2.5 \times 10^4$
**21** $9.653 \times 10^8$  **23** $10^{10}$
**25** Saturn 10 cm, Andromeda Galaxy 1 million km, Quasar 1000 million km

### EXERCISE 5 (page 8)

**1** 800  **3** 10  **5** 3740
**7** 45.7  **9** 0.44  **11** 0.069
**13** 0.506  **15** 0.0495  **17** 34.78
**19** 9.00  **21** 3.0  **23** 7.0
**25** $1 \times 10^5$  **27** $1.06 \times 10^5$  **29** $1 \times 10^5$
**31** $9.88 \times 10^4$

### REVISION EXERCISE 6 (page 8)

**1** $10^5$  **3** 4570  **5** 3700
**7** 0.48  **9** 2.625  **11** $\frac{1}{30}$
**13** $\frac{1}{16}$  **15** 420  **17** 14.4
**19** 0.046  **21** $9.2 \times 10^8$  **23** 2.04
**25** 5%

### REVISION EXERCISE 6★ (page 9)

**1** $\frac{1}{4}$  **3** $\frac{5}{12}$  **5** $8.10 \times 10^3$
**7** 45 600  **9** $10^2$  **11** $1.586 \times 10^8$
**13** $8 \times 10^2$  **15** 18%  **17** $42 928.40

## Algebra 1

### ACTIVITY 2 (page 10)

**3** 6, half number added.

### INVESTIGATE (page 10)

Substituting any value of $x$ gives the same values for both expressions because $x^3 + x^2 + x + 1 = (x + 1)(x^2 + 1)$.

### EXERCISE 7 (page 11)

**1** $4ab$  **3** $7xy$  **5** $-3pq$
**7** $y - xy$  **9** $2 - 6x$  **11** $2cd$
**13** $-4xy$  **15** $2ab + 5bc$  **17** 0
**19** $2gh - 5jk + 7$  **21** $-3p^2 - 2p$  **23** $5x^2y - 3xy^2$

### EXERCISE 7★ (page 11)

**1** $-xy$  **3** $4ab - b$  **5** $6ab$
**7** 0  **9** $3ab + 3bc$  **11** $3q^2$
**13** $x + 1$  **15** $a^3 + 2a^2 + a$
**17** $h^3 + h^2 + 3h + 4$  **19** $7a^2b - 3ab$
**21** $a^2b^3c - 0.6a^3b^2c + 0.3$  **23** $4pq^2r^5 - 2pq^2r^4$

### EXERCISE 8 (page 12)

**1** $6a$  **3** $2x^2$  **5** $3x^3$  **7** $15a^5$
**9** $6st$  **11** $4rs^2$  **13** $2a^2b^2$  **15** $4y^3$
**17** $12x^3$  **19** $20a^3$

### EXERCISE 8★ (page 12)

**1** $8a^3$  **3** $15x^4y^2$  **5** $6a^7$
**7** $18y^3$  **9** $36x^5y^3$  **11** $30a^3b^3c^5$

**13** $56xy^4$  **15** $10x^3y^3$  **17** $3x^3y^4 - 2x^3y^2$
**19** $14a^4b^6$

## EXERCISE 9 (page 13)

**1** $10 + 15a$  **3** $2b - 8c$  **5** $-6a - 24$
**7** $4x - 12$  **9** $2b - a$  **11** $5a + 4b$
**13** $3t - 18$  **15** $6x + y$  **17** $1.4x + 0.3y$
**19** $2.1a - 11.7$

## EXERCISE 9★ (page 13)

**1** $12m - 8$  **3** $2x - 2y + 2z$
**5** $15a + 5b - 20c$  **7** $2x - 3y + 4$
**9** $3y - x$  **11** $-1.4x - 2.2$
**13** $-6x - 3y$  **15** $4.6x - 6.2y - 0.4z$
**17** $-0.6a - 4.2b + 0.7$  **19** $-0.44x^2 - 3.8xy - 1.2y^2$

## EXERCISE 10 (page 14)

**1** 4  **3** 15  **5** 25  **7** 100
**9** 12  **11** 15  **13** 2.4  **15** 13.5
**17** 26.6  **19** 1.4  **21** 0.985  **23** 6.8
**25** 99.9  **27** 5.13  **29** 40.7  **31** 580
**33** 8.49  **35** 38.8

## EXERCISE 11 (page 15)

**1** $x = 3$  **3** $x = -1$  **5** $x = -2$
**7** $x = 2$  **9** $x = 8$  **11** $x = 1$
**13** $x = -6$  **15** $x = 1$  **17** $x = -2$
**19** $x = \frac{2}{3}$  **21** $x = \frac{5}{9}$  **23** $x = \frac{4}{3}$
**25** $x = -1$  **27** 238, 239  **29** $x = 10$; 40, 80, 60
**31** $b = 4$  **33** $-15$

## EXERCISE 11★ (page 16)

**1** $x = 4$  **3** $x = 11$  **5** $x = -2$
**7** $x = -5$  **9** $x = -4$  **11** $x = 5$
**13** $x = 0$  **15** 72, 74, 76  **17** 11, 44, 67 kg.
**19** $x = -50$  **21** 45 m (2sf)  **23** $1\frac{2}{3}$ km

## EXERCISE 12 (page 17)

**1** $x = 1$  **3** $x = 2$  **5** $x = 4$
**7** $x = -3$  **9** $x = 1$  **11** $x = -1$
**13** $x = 0$  **15** $x = -\frac{1}{2}$  **17** $x = 2, 38$
**19** 9

## EXERCISE 12★ (page 18)

**1** $x = 4$  **3** $x = -2$  **5** $x = 1\frac{1}{2}$
**7** $x = \frac{4}{5}$  **9** $x = \frac{7}{9}$  **11** $x = 3$
**13** $x = 5$  **15** $x = -9$  **17** $x = 0.576$ (3 sig figs)
**19** $x = 1.28$ (3 sig figs)  **21** $x = 30$

## EXERCISE 13 (page 18)

**1** $x = 13$  **3** $x = 3$  **5** $x = 2$
**7** $x = 4$  **9** $x = \frac{5}{2}$

## EXERCISE 13★ (page 19)

**1** $x = 8$  **3** $x = 5$  **5** $x = 4$
**7** $x = 9$  **9** $x = \frac{3}{4}$  **11** 6 hits
**13** 4

## REVISION EXERCISE 14 (page 19)

**1** $3x - 2$  **3** $6a$  **5** $a^3$
**7** $4a^4$  **9** $x + 7y$  **11** $x = 4.8$
**13** 145, 146, 147

## REVISION EXERCISE 14★ (page 20)

**1** $4xy^2 - 3x^2y$  **3** 1
**5** $x = 20$  **7** $x = -6$
**9** $x = 4$  **11** 11 years old
**13** $98

# Graphs 1

## INVESTIGATE (page 22)

Gradient of AB = 0.1
Closer to zero. Gradient AC = 0
Larger and larger. Gradient BC is infinite.

## EXERCISE 15 (page 22)

**1** 1  **3** 0.5  **5** 3  **7** $\frac{1}{4}$
**9** $-\frac{1}{4}$  **11** 10 m  **13** 1.5 m  **15** 2.325 m
**17 a** 14 m  **b** $\frac{1}{30}$

## EXERCISE 15★ (page 23)

**1** $\frac{3}{8}$  **3** $-\frac{3}{4}$  **5** 6  **7** Yes
**9** Yes  **11 a** $\frac{1}{6}$ cm  **b** 0.1 cm  **13** $p = -2$

## ACTIVITY 3 (page 24)

| $x$ | $-2$ | 0 | 2 | Gradient | $y$ intercept |
| --- | --- | --- | --- | --- | --- |
| $y = x + 1$ | $-1$ | 1 | 3 | 1 | 1 |
| $y = -x + 1$ | 3 | 1 | $-1$ | $-1$ | 1 |
| $y = 2x - 1$ | $-5$ | $-1$ | 3 | 2 | $-1$ |
| $y = -2x + 1$ | 5 | 1 | $-3$ | $-2$ | 1 |
| $y = 3x - 1$ | $-7$ | $-1$ | 5 | 3 | $-1$ |
| $y = \frac{1}{2}x + 2$ | 1 | 2 | 3 | $\frac{1}{2}$ | 2 |
| $y = mx + c$ |  |  |  | $m$ | $c$ |

## EXERCISE 16 (page 25)

**1** 3, 5  **3** 1, −7  **5** $\frac{1}{3}$, 2
**7** $-\frac{1}{2}$, 5  **9** $-\frac{1}{3}$, −2  **11** −2, 4
**13** 0, −2  **15** $y = 3x - 2$  **17** $y = \frac{x}{3} + 10$
**19** $y = 2x + 4$  **21** $y = -5x - 1$

ANSWERS ◆ Unit 1

**23** For example

   **a** $y = x - 1$    **b** $y = -\frac{1}{2}x + 2$    **c** $y = 1$

## EXERCISE 16★ (page 26)

**1** $5, \frac{1}{2}$        **3** $0, -\frac{3}{4}$        **5** $-3, 2.5$

**7** $6, -\frac{3}{2}$      **9** $\infty$         **11** $-\frac{2}{3}, -\frac{5}{3}$

**13** $y = 2.5x - 2.3$        **15** $y = \frac{x}{4} + 1$

**17** $y = 2.5x - 3.5$        **19 a** $7x + 6y = 84$

**21** For example

   **a** $x = 2$     **b** $2x + y = 2$     **c** $y = \frac{1}{2}x - 1$

## ACTIVITY 4 (page 27)

A baked bean tin or a cardboard tube is ideal.

$\pi = \frac{22}{7} = 3.142\ldots$

## EXERCISE 17 (page 27)

**1** $(3, 0), (0, 6)$       **3** $(4, 0), (0, 6)$

**5** $(5, 0), (0, 4)$       **7** $(4, 0), (0, -2)$

**9** $(-8, 0), (0, 6)$     **11 b** £10    **c** 42 000

                    **d** £0; no, not a sensible value

## EXERCISE 17★ (page 28)

**1** $(6, 0), (0, 12)$       **3** $(3.5, 0), (0, 5.25)$

**5** $(7.5, 0), (0, -6)$      **7** $(-10.5, 0), (0, 3)$

**9** $(-3.5, 0), (0, 3)$

**11 b** ~23    **c** $-\frac{3}{25}$, 23; about $25H + 3W = 575$

      **d** 92 weeks; no, longer, unlikely to continue linear

## ACTIVITY 5 (page 28)

Gradient of AB is 0.5. AB extended intercepts the $y$ axis at $(0, 2.5)$.

The equation of the line through AB is $y = \frac{1}{2}x + 2.5$.

    $y = 3x + 7$; $y = -\frac{1}{3}x + 3$; $y = \frac{2}{3}x + \frac{1}{3}$

## REVISION EXERCISE 18 (page 28)

**1 a** 2                     **b** $-1$

**3 a** $3, -2$             **b** $-2, 5$

**5 a** Gradient 2, intercept $-3$

   **b** Gradient $-1$, intercept 4

   **c** Passes through $(5, 0)$ and $(0, 2)$

**7 b** 27 °C; −30 °C; 77 °F     **c** −40

**9** $y = x, y = -x, x = 3, x = -3, y = 3, y = -3$

## REVISION EXERCISE 18★ (page 29)

**1 a** $-\frac{1}{3}$                 **b** 2

**3 a** Gradient 3, intercept $-2$

   **b** Gradient $-2$, intercept 3

---

   **c** Gradient $-\frac{1}{2}$, intercept 2.5

   **d** Passes through $(2, 0)$ and $(0, \frac{10}{3})$

**5** $b = \pm 1.5$

**7 a** $F = 20P + 20$; $S = 35P + 35$

   **c** 3 hrs 48 mins; 2 hrs 40 mins

   **d** 10.25 pounds          **e** about 1.15 pm

## Shape and space 1

### ACTIVITY 6 (page 32)

| | □ | ▭ | ▱ | ▱ | △ | ⬯ | ◇ |
|---|---|---|---|---|---|---|---|
| Diags equal | Y | Y | N | N | N | N | N |
| Diags bisect each other | Y | Y | Y | Y | N | N | N |
| Diags perpendicular | Y | N | Y | N | Y | N | Y |
| Diags bisect angles | Y | N | Y | N | N | N | N |
| Opp. angles equal | Y | Y | Y | Y | N | N | N |

### EXERCISE 19 (page 33)

**1** $a = 102°, b = 78°$      **3** $a = 65°$

**5** $a = 73°, b = 34°$      **7** $a = 57°, b = 123°$

**9** $a = 31°, b = 31°$      **11** $a = 124°, b = 56°$

**13** $a = 58°, b = 32°$     **15** $45°, 135°, 1080°$

**17** 9 sides            **19** $x = 74°$; $74°$ and $148$

### EXERCISE 19★ (page 34)

**1** $a = 137°, b = 43°$      **3** $a = 36°$

**5** $a = 17°$              **7** $180° - 2x$

**9** $x = 50°$            **11** $a = 56°, b = 34°$

**13** $a = 40°, b = 113°$     **15** 20 sides

### EXERCISE 20 (page 38)

**1** 5.86 cm             **3** 4.2 cm, 14.7 cm²

**5 c** PS = QS = RS = 4.7 cm; all equal, on a circumcircle

**7**

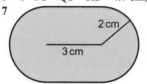

Scale 1 cm : 10 m

**9 a**

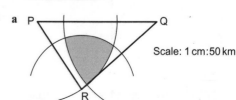

Scale: 1 cm : 50 km

   **b** 180 km

ANSWERS ◆ Unit 1

**11**  577 m

**13**

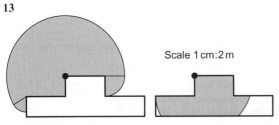

Scale 1 cm:2 m

## EXERCISE 20★                                    (page 40)

**1**  5.4 cm                              **3**  6.8 cm

**5**

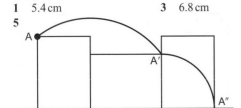

**7**

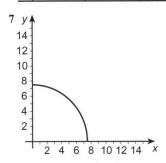

**9**

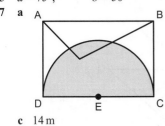

← wind

## REVISION EXERCISE 21                     (page 41)

**1**  **a** = 52°          **b** = 128°          **c** = 76°
**3**  **a**  12 sides      **b**  30°
**5**  $a = 75°$,          $b = 50°$
**7**  **a**

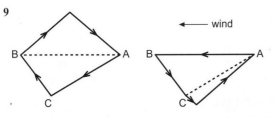

 **c**  14 m

## REVISION EXERCISE 21★                    (page 42)

**1**  $a = 40°$, $b = 100°$, $c = 40°$, $d = 260°$
**3**  **b** 6.5 cm          **c**  30 cm$^2$

**5**  ∠LKJ = 15°, ∠HKL = 45°

**7**  **c**  5.2 cm

## EXERCISE 22                                    (page 44)

**1**  **a**  Any two vegetables
  **b**  Any two colours
  **c**  Any two letters
  **d**  Any two odd numbers
**3**  **a**  {Sunday, Monday, Tuesday, Wednesday,
       Thursday, Friday, Saturday}
  **b**  {1, 4, 9, 16, 25, 36, 49, 64, 81, 100}
  **c**  For example, {Mathematics, Science, English, …}
  **d**  {2, 3, 5, 7, 11, 13, 17, 19}
**5**  **a**  {the first four letters of the alphabet}
  **b**  {days of the week beginning with T}
  **c**  {first four square numbers}
  **d**  {even numbers}
**7**  **a**  False
  **b**  False
  **c**  False
  **d**  True
**9**  **b** and **c**

## EXERCISE 22★                                   (page 45)

**1**  **a**  Any two planets
  **b**  Any two polygons
  **c**  Any two elements
  **d**  Any two square numbers
**3**  **a**  {2, 3, 4}
  **b**  {1, 4, 6}
  **c**  {1, 5, 7, 35}
  **d**  {1, 10, 100, 1000, 10 000, 100 000}
**5**  **a**  {seasons of the year}
  **b**  {conic sections}
  **c**  {first five powers of 2}
  **d**  {Pythagorean triples}
**7**  **a**  True
  **b**  False
  **c**  True
  **d**  False
**9**  **a**, **c** and **d**

## EXERCISE 23                                    (page 48)

**1**  **a**  16
  **b**  $n(T) = 14$; 14 pupils like toffee
  **c**  $n(C \cap T) = 12$; 12 pupils like both chocolate
       and toffee
  **d**  21

**3 a**

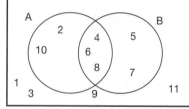

**b** {c, e}, 2
**c** Yes
**d** No, d∈B but d∉A, for example
**5 a** Pink Rolls-Royce cars
**b** There are no pink Rolls-Royce cars in the world.
**7 a** 35      **b** 3      **c** 11
**d** 2      **e** 64

## EXERCISE 23★                    (page 49)

**1 a**

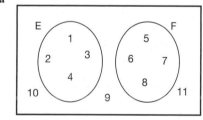

**b** {4, 6, 8}, 3
**c** Yes
**d** {1, 2, 3, 5, 7, 9, 10, 11}
**e** Yes

**3 a**

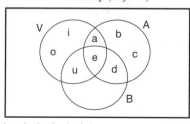

**b** { } or ∅
**c** The sets don't overlap (*disjoint*).

**5 a**

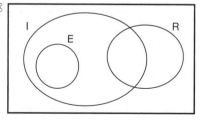

**b** {a, e}, {a, i, o}, {u}
**c** {e}

**7 a**

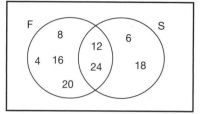

**b** {12, 24}
**c** 12
**d** LCM is 12. It is the smallest member of F∩S.
**e** **i)** 24      **ii)** 40
**9** $2^n$

## EXERCISE 24                    (page 51)

**1 a** ℰ

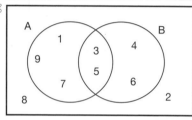

**b** {1, 3, 4, 5, 6, 7, 9}, 7
**c** Yes
**d** {2, 8}
**e** No

**3 a** Diagram not unique

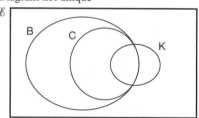

**b** All cards that are black or a king or both
**c** All cards that are black or a king or both
**d** All cards that are red or a king or both

**5 a** ℰ

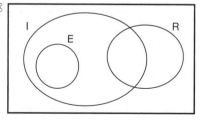

**b** An isosceles right-angled-triangle
**c** Isosceles triangles, triangles that are isosceles
or right-angled or both
**d** Equilateral triangles, ∅

## EXERCISE 24★ (page 52)

**1 a**

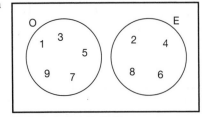

**b** {1, 2, 3, 4, 5, 6, 7, 8, 9}

**c** E∩O = ∅

**d** E∪O = ℰ

**3** $n(A∪B∪C) = 28$

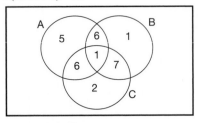

**5** B⊂A or B = ∅

## REVISION EXERCISE 25 (page 52)

**1 a** Any spice      **b** Any pet or animal

   **c** Any fruit      **d** Any colour

**3 a** {first four prime numbers}

   **b** {even numbers between 31 and 39}

   **c** {days of the week beginning with S} or {days of the weekend}

   **d** {vowels}

**5 a** {3, 6, 9, 12}

   **b** ℰ

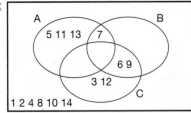

   **c** {5, 6, 7, 9, 11, 13}      **d** {6, 9}

   **e** ∅

## REVISION EXERCISE 25★ (page 53)

**1 a** {4, 8, 12, 16}

   **b** {red, orange, yellow, green, blue, indigo, violet}

   **c** {CAT, CTA, ACT, ATC, TCA, TAC}

   **d** {2, 3, 6}

**3 a** 9

   **b** ℰ

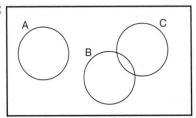

**5 a** Isosceles right-angled triangles

   **b** E              **c** ∅

   **d** ℰ

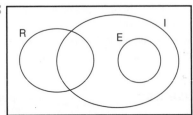

## Examination practice 1 (page 57)

**1 a** $7a^2 + 4a^3$    **b** $2a^3$    **c** $12x - 15$

**3 a** $10^{14}$    **b** $1 \times 10^{14}$

**5 a** 92.78    **b** 182.8    **c** 70.75

**7 a** 38    **b** −8    **c** 64

**9 a** $y = 3x + 2$ and $y = 3x - 3$; the $x$ co-efficents are the same.

**11 a** 0.2    **b** 68    **c** 0.2

**13 a** $3a^2b + a$    **b** $4a^3$    **c** $4c + 4$

**15** 1.3 by 11.1%

**17 a** E′ = {1, 3, 5, 7, 9, 11}, E∩T = {6, 12}, F∩T = {12}

   **b** E′ = {odd integers less than 13}, E∩T = {multiples of 6}, F∩T = {multiples of 12}.

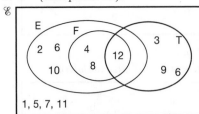

**19 a** $y = -\frac{2}{3}x + 2$ or $2x + 3y = 6$

   **b** $y = -\frac{2}{3}x - 1$

**21 a** $20x - 10(30 - x) = 180$    **b** 16

# Unit 2

## Number 2

### ACTIVITY 7 (page 59)

| Decimal form | Multiples of 10 or fraction form | Standard form |
|---|---|---|
| 100 | $10 \times 10$ | $10^2$ |
| 10 | 10 | $10^1$ |
| 1 | $\frac{1}{1}$ | $10^0$ |
| 0.1 | $\frac{1}{10}$ | $10^{-1}$ |
| **0.01** | $\frac{1}{100} = \frac{1}{10^2}$ | $\mathbf{10^{-2}}$ |
| 0.001 | $\frac{1}{1000} = \frac{1}{10^3}$ | $\mathbf{10^{-3}}$ |
| 0.0001 | $\frac{1}{10\,000} = \frac{1}{10^4}$ | $\mathbf{10^{-4}}$ |
| **0.000 001** | $\frac{1}{1000\,000} = \frac{1}{10^6}$ | $10^{-6}$ |

### ACTIVITY 8 (page 60)

Mouse = $10\,\text{g} = 10^1\,\text{g}$; pigmy shrew = $1\,\text{g} = 10^0\,\text{g}$; grain of sand = $0.0001\,\text{g} = 10^{-4}\,\text{g}$

10; 10 000; house mouse; grain of sand; grain of sand, staphylococcus; 10, staphylococcus

$<; >; <$

### EXERCISE 26 (page 61)

1 $10^{-1}$    3 $10^{-3}$    5 $10^{-3}$
7 $10^1$    9 $0.001$    11 $0.0012$
13 $0.000001$    15 $0.0467$    17 $5.43 \times 10^{-1}$
19 $7 \times 10^{-3}$    21 $6.7 \times 10^{-1}$    23 $1 \times 10^2$
25 $100$    27 $10\,000$    29 $128$
31 $30$    33 $0.018$

### EXERCISE 26★ (page 61)

1 $10$    3 $0.011$    5 $0.01$
7 $0.0011$    9 $10^3$    11 $10^1$
13 $10^4$    15 $10^2$    17 $6.25 \times 10^{-6}$
19 $6.9 \times 10^7$    21 $4 \times 10^{-6}$    23 $4.8 \times 10^5$
25 $5000$ viruses    27 $3 \times 10^{10}$
29 $(3.4 \times 10^{23}) + (0.34 \times 10^{23}) = 3.74 \times 10^{23}$

### EXERCISE 27 (page 64)

1 $\frac{6}{7}$    3 $\frac{2}{5}$    5 $1\frac{2}{9}$    7 $\frac{1}{2}$
9 $\frac{23}{24}$    11 $4\frac{5}{12}$    13 $\frac{5}{18}$    15 $\frac{6}{7}$
17 $\frac{3}{5}$    19 $\frac{3}{25}$    21 $3$    23 $1\frac{1}{2}$
25 $4\frac{7}{12}$    27 $2\frac{3}{4}$    29 $4\frac{4}{5}$    31 $\frac{3}{10}$

### EXERCISE 27★ (page 64)

1 $\frac{3}{4}$    3 $\frac{3}{5}$    5 $\frac{19}{20}$    7 $\frac{3}{4}$
9 $\frac{4}{45}$    11 $\frac{7}{10}$    13 $\frac{2}{9}$    15 $\frac{3}{8}$
17 $7\frac{2}{3}$    19 $6\frac{1}{2}$    21 $6\frac{7}{9}$    23 $1\frac{7}{15}$
25 $2\frac{2}{3}$    27 $\frac{5}{6}$    29 $4\frac{1}{2}$

### EXERCISE 28 (page 65)

1 $168, $224    3 94 kg, 658 kg
5 574, 410    7 8.1, 5.4

### EXERCISE 28★ (page 65)

1 $45 : $75    3 111 ml    5 117 : 234 : 351
7 $32

### EXERCISE 29 (page 67)

1 Yen 180    3 Aus$2.38    5 NZ$2.75
7 a £8.40    b £75.60
9 a 105 mm/hr    b 1.75 mm/min
11 a 0.9 s    b 66.7 $l$

### EXERCISE 29★ (page 67)

1 a $15    b 2667 yen
3 2.3 million
5 $\approx$ 12 miles
7 a 0.0825 s    b 12.1 litres/s
9 a 94.4 m/s    b 0.00106 s

### REVISION EXERCISE 30 (page 68)

1 $1 \times 10^{-3}$    3 100    5 0.11
7 $\frac{2}{3}$    9 $110, $165    11 1000 gallons/s

### REVISION EXERCISE 30★ (page 68)

1 $1 \times 10^{-6}$    3 $3\frac{11}{12}$    5 $14\frac{7}{12}$
7 $208, $312, $416
9 Midges/m$^2$ = 2000. The area/midge = 5 cm$^2$. However, these answers are misinformative because not all the midges will be at ground level.
11 11.4 tonnes    13 2.65 s

## Algebra 2

### EXERCISE 31 (page 70)

1 4    3 $3y$    5 2    7 $\frac{3a}{b}$
9 $\frac{b}{2}$    11 $\frac{3a}{b}$    13 $4c$    15 $\frac{a}{2}$
17 $\frac{4}{x}$    19 $2b$    21 $\frac{1}{5b^2}$    23 $\frac{a}{4}$

### EXERCISE 31★ (page 71)

1 $\frac{1}{2}$    3 $\frac{2}{b}$    5 $\frac{x}{4}$    7 $\frac{a}{2}$

**9** $\dfrac{2}{b}$  **11** $\dfrac{6}{b^2}$  **13** $\dfrac{a}{2b}$  **15** $\dfrac{3}{abc}$

**17** $\dfrac{3a}{4b^2}$  **19** $\dfrac{1}{a^2b^2}$  **21** $\dfrac{3a}{8b}$  **23** $\dfrac{3z^2}{10x^2}$

## EXERCISE 32 (page 71)

**1** $\dfrac{5x^2}{4}$  **3** $\dfrac{x^3}{y}$  **5** $1$  **7** $\dfrac{4c^2}{5}$

**9** $6$  **11** $\dfrac{ab}{2}$  **13** $\dfrac{b}{6}$  **15** $1$

**17** $\dfrac{2}{y}$  **19** $\dfrac{b}{2c}$

## EXERCISE 32★ (page 72)

**1** $2a^3$  **3** $\dfrac{3x}{z}$  **5** $\dfrac{3pq^2}{2}$

**7** $\dfrac{1}{2x}$  **9** $\dfrac{5xy^3}{z^2}$  **11** $y$

**13** $\dfrac{3x^4}{8y}$  **15** $\dfrac{1}{2a^2b^2}$

## EXERCISE 33 (page 72)

**1** $\dfrac{7x}{12}$  **3** $\dfrac{a}{12}$  **5** $\dfrac{4a+3b}{12}$

**7** $\dfrac{5x}{12}$  **9** $\dfrac{a}{2}$  **11** $\dfrac{3a+4b}{12}$

**13** $\dfrac{a}{6}$  **15** $\dfrac{3a+8b}{12}$

## EXERCISE 33★ (page 73)

**1** $\dfrac{7x}{18}$  **3** $\dfrac{5a}{21}$  **5** $\dfrac{14x+20y}{35}$

**7** $\dfrac{a}{4}$  **9** $\dfrac{17}{6b}$  **11** $\dfrac{2d+3}{d^2}$

**13** $\dfrac{7-3x}{10}$  **15** $\dfrac{y-2}{30}$  **17** $\dfrac{3x+5}{12}$

**19** $\dfrac{2a-1}{a(a-1)}$

## EXERCISE 34 (page 74)

**1** $x = \pm 3$  **3** $x = \pm 6$  **5** $x = \pm 4$  **7** $x = \pm 8$
**9** $x = \pm 3$  **11** $x = \pm 1$  **13** $x = 13$  **15** $x = \pm 4$
**17** $x = 16$  **19** $x = 81$

## EXERCISE 34★ (page 74)

**1** $x = \pm 5$  **3** $x = \pm 7$  **5** $x = \pm 9$  **7** $x = 81$
**9** $x = 7$  **11** $x = \pm 4$  **13** $x = \pm 5$
**15** $x = 10$ or $x = -16$  **17** $x = \pm 3$  **19** $x = 1$

## CHALLENGE (page 75)

Parallelogram: draw a diagonal. Area of each triangle
$= \frac{1}{2}bh$. So area of parallelogram $= bh$.

Trapezium: draw a diagonal. Area of trapezium
$= \dfrac{ah}{2} + \dfrac{bh}{2} = \dfrac{h}{2}(a+b)$

## EXERCISE 35 (page 75)

**1** $4.5\,\text{cm}$  **3** $44\,\text{cm}$; $154\,\text{cm}^2$
**5** $h = 8\,\text{cm}$  **7** $YZ = 4\,\text{cm}$
**9 a** $9.42 \times 10^8\,\text{km}$  **b** $110\,000\,\text{km/hr}$
**11** $h = 4.7\,\text{cm}$  **13** $11.7\,\text{km}$
**15** $1$ second

## EXERCISE 35★ (page 77)

**1** $14\,\text{cm}$  **3** $AC = 12\,\text{cm}$; $4.2\,\text{cm}$
**5** $h = 5.5\,\text{cm}$  **7** $\sqrt{9^2 + 12^2} > 14$ ∴ no
**9** $5.30\,\text{cm}^2$  **11** $10.5\,\text{cm}^2$ for each
**13** $15.9\,\text{km}$  **15** Obtuse
**17** $8.37\,\text{cm}$ x
**19 a** £1560  **b** £2130  **c** £2890

## EXERCISE 36 (page 79)

**1** $2^{10} = 1024$  **3** $4^7 = 16384$
**5** $2^6 = 64$  **7** $7^3 = 343$
**9** $2^{12} = 4096$  **11** $6^8 = 1679616$
**13** $a^5$  **15** $c^4$  **17** $e^6$  **19** $a^9$
**21** $c^5$  **23** $12a^6$  **25** $6a^5$  **27** $2e^8$

## EXERCISE 36★ (page 80)

**1** $6^{12} = 2.18 \times 10^9$  **3** $7^6 = 1.18 \times 10^5$
**5** $8^{12} = 6.87 \times 10^{10}$  **7** $4^{17} = 1.72 \times 10^{10}$
**9** $a^{12}$  **11** $3c^6$  **13** $2e^8$  **15** $8g^{12}$
**17** $48j^{12}$  **19** $24m^7$  **21** $27a^6$  **23** $2$
**25** $8b^4$  **27** $6$

## EXERCISE 37 (page 82)

**1** $<$  **3** $<$  **5** $>$  **7** $<$
**9** $x \leqslant 0, x > 2$  **11** $-3 < x \leqslant 3$
For Questions 13 to 28 the result should be shown on a
number line.
**13** $x > 5$  **15** $x \leqslant 4$  **17** $x < 3$  **19** $x \geqslant 3$
**21** $x \geqslant 9$  **23** $x < 4$  **25** $x < 0$  **27** $x > 3$
**29** $x < -2$  **31** $x \geqslant -5$  **33** $x < -1$  **35** $x \leqslant -2$
**37** $x \geqslant -\frac{2}{3}$  **39** $x \geqslant -7$  **41** $\{5,6\}$  **43** $\{3,4\}$
**45** $\{0,1\}$  **47** $\{1,2,3,4\}$  **49** $\{2,3\}$

## EXERCISE 37★ (page 83)

**1** $x \leqslant 0$ or $x > 3$; $0 \geqslant x > 3 \Rightarrow 0 \geqslant 3$!
For Questions 3 to 14 the result should also be shown on
a number line.
**3** $x \leqslant 2.5$  **5** $x < 5\frac{1}{3}$  **7** $x < 1.5$
**9** $x < -3\frac{1}{5}$  **11** $x \leqslant 2$  **13** $-1 < x \leqslant 3$

**15** $x \leqslant 7$ {7,6,5,4}

**17** 23       **19** {1,2,3}

## REVISION EXERCISE 38 (page 84)

**1** 3      **3** $3x$      **5** $a$

**7** $\dfrac{9y}{20}$      **9** $\dfrac{4a+b}{10}$      **11** $\pm6$

**13** $a^{10}$      **15** $c^{12}$      **17** 12 cm

**19** $<$      **21** $=$      **23** $x > 5$

**25** $x \geqslant 2$

**27** {3,4}      **29** 11.34 km

## REVISION EXERCISE 38★ (page 85)

**1** $\dfrac{4a}{b}$      **3** $\dfrac{b}{4a}$      **5** $\dfrac{5}{xy}$

**7** $\dfrac{8a}{5}$      **9** $\dfrac{2x+6}{21}$      **11** 2

**13** $a^4$      **15** $81c^7$      **17** $-3 < x \leqslant 0$; $-2$

**19** $x > -4$

**21** 37      **23** 50.1 cm      **25** $4.07 \times 10^6$ km/hour

## Graphs 2

## ACTIVITY 9 (page 86)

Pineapple

| Minutes online ($t$) | 0 | 500 | 1000 |
|---|---|---|---|
| Cost in cents ($C$) | 999 | 1549 | 2099 |

Banana

| Minutes on line ($t$) | 0 | 500 | 1000 |
|---|---|---|---|
| Cost in pence ($C$) | 495 | 1395 | 2295 |

720 minutes per month results in the same charge of £17.91.

## EXERCISE 39 (page 86)

**1**

| $x$ | 0 | 2 | 4 |
|---|---|---|---|
| $y = x + 1$ | 1 | 3 | 5 |

| $x$ | 0 | 2 | 4 |
|---|---|---|---|
| $y = 2x - 2$ | −2 | 2 | 6 |

Intersection point is (3, 4).

**3** (2, 5)      **5** (3, 8)      **7** (1.5, 9)

**9** (2, 2)

**11 a** $d = 35t$

    **b** $d = 5 + 15t$; after 15 mins at distance 8.75 km.

## EXERCISE 39★ (page 87)

**1** (6, 13)      **3** (1.75, 4.25)

**5** (2.57, 0.29)      **7** (0.53, −0.9)

**9** (698, 2930), (7358, 2264)

## EXERCISE 40 (page 90)

**1** $x \leqslant 2$      **3** $y > 4$      **5** $y \leqslant -2$

**7** $x + y \geqslant 6$

**9** 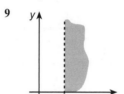      **11**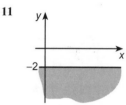

**13** 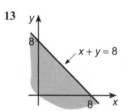      **15**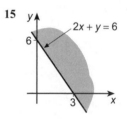

## EXERCISE 40★ (page 91)

**1** $y > -2$      **3** $2x + y \geqslant 6$

**5** $y - x < 4$      **7** $2y + x \leqslant 4$

**9** 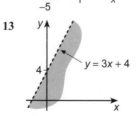      **11**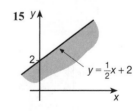

**13** 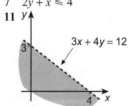      **15**

## INVESTIGATE (page 91)

The resultant graph would be very confusing.

## EXERCISE 41 (page 92)

**1** $2 < x < 5$      **3** $-2 < y \leqslant 3$

**5** $x \geqslant 4$ or $x \leqslant -3$      **7** $y \geqslant 9$ or $y < 3$

**9** $x + y > 3$ and $x - y \leqslant 2$

**11** $y < x + 3$, $2y + x \leqslant 6$ and $y \geqslant 0$

**13**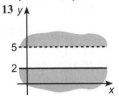

**15**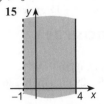

**17**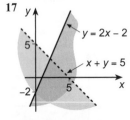

**19**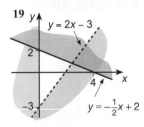

## EXERCISE 41★ (page 93)

**1** $-3 \leqslant x < 4$

**3** $2y + x \geqslant 10$ or $2y + x \leqslant 4$

**5** $4x + 3y \leqslant 12$, $y \geqslant 0$ and $y < 2x + 4$

**7** $x \geqslant 0$, $y \geqslant 0$, $y < \dfrac{-3x}{2} + 9$ and $y \leqslant \dfrac{-2x}{3} + 6$

**9**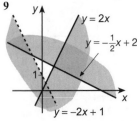

**11**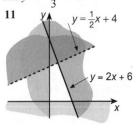

**13 b** $y < x + 2$, $y < -2x + 2$ and $2y + x > -2$
 **c** $y = -1$

**15**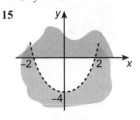

**17**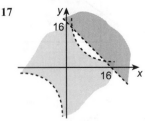

4,10; 4,11; 5,8; 5,9; 5,10; 6,7; 6,8; 6,9; 7,7; 7,8 and all negative pairs with $xy \leqslant 36$

## REVISION EXERCISE 42 (page 94)

**1** (2, 1)  **3** (−1, 3)

**5** A: $y \geqslant 2x$, $2y + x \geqslant 6$  B: $y \leqslant 2x$, $2y + x \geqslant 6$
 C: $y \leqslant 2x$, $2y + x \leqslant 6$  D: $y \geqslant 2x$, $2y + x \leqslant 6$

**7**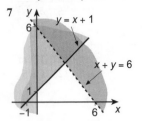

## REVISION EXERCISE 42★ (page 94)

**1** (2.29, 2.14)  **3** (−1.6, 1.2)

**5** $y < 3x + 2$, $y < 7 - 2x$ and $2y > x - 1$

**7** $\{y \geqslant 3$ and $2x + y \leqslant 6\}$

# Shape and space 2

## ACTIVITY 10 (page 95)

$\dfrac{o}{a} \approx 0.6$ for $x$, $y$, $z$

| $x$ | 0° | 15° | 30° | 45° | 60° | 75° | 89° | 90° |
|---|---|---|---|---|---|---|---|---|
| $\tan x$ | 0 | 0.268 | 0.577 | 1 | 1.73 | 3.73 | 57.3 | ∞ |

$\tan 89°$ is large because the opposite side becomes much larger compared with the adjacent side as the angle approaches 90°.

## EXERCISE 43 (page 97)

**1** $x$: hyp, $y$: opp, $z$: adj  **3** $x$: opp, $y$: adj, $z$: hyp

**5** $\dfrac{4}{3}$  **7** 5.8  **9** 87

**11** 100  **13** 6.66 cm  **15** 8.20 cm

**17** 11.3 cm  **19** 87.5 m  **21** 100 m²

## EXERCISE 43★ (page 98)

**1** 14.4 cm  **3** $x = 200$  **5** 8.45 m

**7** 100 m  **9** 22.4 m

**11** $x = 10.9$ cm, 6.40 cm

**13** $x = 27.5$ cm, $y = 9.24$ cm

**15** BX = 2.66 m  BC = 4.00 m

## EXERCISE 44 (page 100)

**1** 45°  **3** 15°  **5** 70.0°

**7** 45°  **9** 75°  **11** 37.9°

**13** 28.2°  **15** 27.1°  **17** 23.4°

## EXERCISE 44★ (page 101)

**1** a = 69°, b = 139°  **3** 60°  **5** 160°

**7 a** 080.5°  **b** 260.5°  **c** 108.4°  **d** 236.3°

**9** 101°

## REVISION EXERCISE 45 (page 102)

| | | | | | |
|---|---|---|---|---|---|
| **1** | 7.00 | **3** | 6.99 | **5** | 8.57 |
| **7** | 59.0° | **9** | 58.0° | **11** | 30° |

## REVISION EXERCISE 45★ (page 103)

**1** 549 m

**3 a** 1.01 m

   **b** Undesirable to have too large a blind distance.

# Handling data 2

## EXERCISE 46 (page 106)

**1**

| Score | | Frequency |
|---|---|---|
| 1 | ‖‖‖‖ ‖‖‖‖ | 9 |
| 2 | ‖‖‖‖ ‖‖‖‖ | 10 |
| 3 | ‖‖‖‖ ‖‖ | 7 |
| 4 | ‖‖‖‖ | 5 |
| 5 | ‖‖‖‖ | 4 |
| 6 | ‖‖‖‖ | 5 |
| | | Total = 40 |

**3**

| Weight in kg | | Frequency |
|---|---|---|
| $1.0 \leqslant w < 1.5$ | ‖ | 2 |
| $1.5 \leqslant w < 2.0$ | ‖‖‖‖ | 5 |
| $2.0 \leqslant w < 2.5$ | ‖‖‖‖ ‖‖‖‖ | 9 |
| $2.5 \leqslant w < 3.0$ | ‖‖‖‖ ‖‖‖‖ ‖‖ | 12 |
| $3.0 \leqslant w < 3.5$ | ‖‖‖‖ ‖‖‖‖ | 9 |
| $3.5 \leqslant w < 4.0$ | ‖‖‖‖ ‖‖‖ | 8 |
| $4.0 \leqslant w < 4.5$ | ‖‖‖‖ | 4 |
| $4.5 \leqslant w < 5.0$ | ‖ | 1 |
| | | Total = 50 |

$\frac{14}{25}$ of the babies weigh less than 3 kg.

| | mean | median | mode |
|---|---|---|---|
| **5** | 10 | 12 | 14 |
| **7** | 15 | 16 | 4 |

**9** Mean = 48.9 sec, median = 48 sec.

**11** Mean 92, median 91, mode 91; therefore either median or mode.

**13** 10.5 years

**15** For example: 1, 2, 3, 4, 5, 16; mean = 5.2

## EXERCISE 47 (page 109)

**1** 96°, 60°, 48°, 48°, 108°

**3** Frequencies: 9, 10, 7, 5, 4, 5

**7** 7; 6.1

**9** 13.8, 13, 17        **11** 720 pupils

**13** 14 minutes 41 s

## REVISION EXERCISE 48 (page 112)

**1** Frequencies are 5, 9, 11, 21, 14

**5** 43.5

## REVISION EXERCISE 48★ (page 112)

**1** 30       **3** 15 min 55 secs, 14 mins

**5** Sector angles = 84°, 72°, 36°, 72°, 36°, 60°

# Examination practice 2 (page 117)

**1** £1

**3 a** $k^5$    **b** $p^4$    **c** 3 m    **d** $y^6$

**5** 112°, 29°

**7 a** $1.28 \times 10^9$ km    **b** 71 mins

**9 a** $5.24 \times 10^{12}$ m    **b** $9.5 \times 10^{15}$ m

   **c** $5.6 \times 10^{-4}$ light years

**11 a** 15    **b** 9       **c** 10

**13 a** 4000    **b** 120°

**15** 73.7°

**17 a** 1.28    **b** 1       **c** 0.9

# Unit 3

## Number 3

### ACTIVITY 11 (page 120)

£120 × 1.08 = £129.60; £120 × 0.92 = £110.40

| To increase by (%) | 100% → | Multiply by |
|---|---|---|
| 15 | 115% | 1.15 |
| 70 | 170% | 1.70 |
| 56 | 156% | 1.56 |
| 2 | 102% | 1.02 |
| 8 | 108% | 1.08 |
| 80 | 180% | 1.80 |

| To decrease by (%) | 100% → | Multiply by |
|---|---|---|
| 15 | 85% | 0.85 |
| 70 | 30% | 0.30 |
| 20% | 80% | 0.8 |
| 2% | 98% | 0.98 |
| 8% | 92% | 0.92 |
| 80% | 20% | 0.20 |

### ACTIVITY 12 (page 121)

**Bread:**

| Time (yrs) | 10% | 20% | 30% | 40% |
|---|---|---|---|---|
| 3 | 1.1 | 1.5 | 1.9 | 2.3 |
| 6 | 1.5 | 2.5 | 4.1 | 6.4 |
| 9 | 2.0 | 4.4 | 9.0 | 18 |
| 12 | 2.7 | 7.6 | 20 | 48 |
| 15 | 3.6 | 13 | 44 | 130 |

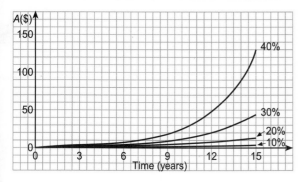

Five-fold increase: 40% ≈ 4.8 years (2 s.f.),
30% ≈ 6.1 years (2 s.f.), 20% ≈ 8.8 years (2 s.f.),
10% ≈ 17 years (2 s.f.)
**Car:** 13.5 years

### INVESTIGATE (page 122)

**Peru:** When $R = 6000$, the multiplying factor becomes 61. Therefore all prices are likely to increase by a factor of 61 times in a year. Using this figure, the cost of an item, after a chosen number of years, could be worked out.

If $M\%$ is the inflation rate per month and $Y\%$ is the inflation rate per year, then

$$Y = 100\left[\left(1 + \frac{M}{100}\right)^{12} - 1\right]$$

If the rate of inflation in country A is higher than the rate of inflation in country B then exports from country A to country B will become more expensive and therefore less competitive. Conversely, exports from country B to country A will become cheaper and therefore more competitive. Therefore, in an inflationary period there is a tendency for imports to increase and unemployment to rise as the country become less competitive. Pupils should illustrate this scenario by taking two different inflation rates, one in each country, and investigate what happens to the cost of imports and exports after a number of years. Graphs should be drawn to illustrate their findings.

### EXERCISE 49 (page 122)

1   **a**   1.1     **b**   1.2     **c**   1.30
    **d**   1.01    **e**   1.15    **f**   1.25
3   $440      5   $520      7   230 km
9   £270     11 £255     13 £435
15 "… by 1.03 × 1.04"    17 660 pupils
19 "… by $(1.04)^5$"      21 €675
23 $98 400, $804 000

### ACTIVITY 13 (page 123)

$R = 0\%$: multiplying factor = 1 (no change)
$R = 100\%$: multiplying factor = 2 (double)
$R = 200\%$: multiplying factor = 3 (treble)
$2^{30} \approx 1$ billion, which is about 1000 times the population of the UK at that time!

### EXERCISE 49★ (page 124)

1   **a**   1.125     **b**   1.04     **c**   1.05
3   90 km     5   $437
7   Result is a multiplying factor of 1.2 × 0.8 = 0.96, that is 4% decrease. You do not get back to the original quantity because the 80% reduction is applied to the increased value.
9   €808
11 €10 100, €20 200, €40 700
13 $4.39 million      15 SF22 800
17 $P = \$614$       19 $R = 12.7\%$
21 $n = 14.2$ years

**23**

| No. times done | No. of sheets in pile | Height of pile |
|---|---|---|
| 3 | 8 | 0.8 mm |
| 5 | 32 | 3.2 mm |
| 10 | 1024 | 0.102 m |
| 50 | $1.13 \times 10^{15}$ | 1.13 billion km |

42 times

## EXERCISE 50 (page 126)

**1** 7, 14, 21, 28, 35, 42   **3** 6, 12, 18, 24, 30, 36
**5** 1, 2, 3, 4, 6, 12   **7** 1, 2, 3, 5, 6, 15, 30
**9** $2 \times 2 \times 7$   **11** $2 \times 2 \times 3 \times 5$
**13** Yes   **15** $2 \times 3 \times 5 \times 7$
**17** $2 \times 2 \times 2 \times 11$   **19** Divisible by 7

## EXERCISE 50★ (page 126)

**1** No   **3** 3, 7, 19
**5** 3, 7, 11   **7** 11
**9** 1, 3, 5, 15, 25, 75
**11** 1, 2, 3, 6, 9, 18, 27, 54
**13** $3 \times 5 \times 11$   **15** $3 \times 7 \times 19$
**17** 59, 61   **19** $2^3 \times 3^2 \times 7$
**21** $2^4 \times 3^2 \times 7$

## EXERCISE 51 (page 128)

**1** 2   **3** 5   **5** 22   **7** 6
**9** 30   **11** 30   **13** $2x$   **15** $4y^2$
**17** $6ab$   **19** $\frac{3}{4}$   **21** $\frac{4}{7}$   **23** $\frac{1}{2}$
**25** $\frac{3}{16}$   **27** $\frac{1}{5}$   **29** $\frac{7}{24}$   **31** $\frac{17}{140}$
**33** $\frac{11}{30}$   **35** $\frac{7}{12}$

## EXERCISE 51★ (page 128)

**1** HCF = 6   **3** HCF = 15   **5** HCF = $y$
LCM = 36   LCM = 210   LCM = $6xyz$
**7** HCF = $2xy$   **9** HCF = $xy$   **11** HCF = $xy$
LCM = $12xy$   LCM = $xyz$   LCM = $x^3y^4$
**13** HCF = $3xyz$
LCM = $18x^2y^2z^2$
**15** $\frac{13}{24}$   **17** $\frac{13}{36}$   **19** $\frac{93}{140}$
**21** $\frac{29}{72}$

## ACTIVITY 14 (page 130)

$abc$   $7 \times 11 \times 13 = abcabc$ (because $7 \times 11 \times 13 = 1001$)

## ACTIVITY 15 (page 130)

710.77345, SHELL OIL; 710.0553, ESSO OIL

## EXERCISE 52 (page 130)

**1** 6.96   **3** 3.35   **5** 6.96
**7** 3.35   **9** 134   **11** 0.384

**13** 12.9   **15** 16.1   **17** 2.58
**19** 14.6   **21** 2.69   **23** 7.16
**25** 11.3   **27** 1.22   **29** 625
**31** 191   **33** 191   **35** 245 000
**37** $1.75 \times 10^{10}$

## EXERCISE 52★ (page 131)

**1** 3.43   **3** 0.005 80   **5** −1.01
**7** 12.4   **9** −0.956   **11** 15.2
**13** 0.103   **15** 0.0454   **17** 3.60
**19** $a^2 + b^2 \neq (a + b)^2$

## REVISION EXERCISE 53 (page 131)

**1** 1.15   **3** $35.20
**5** 2, 3, 5, 7, 11, 13, 17, 19, 23, 29, 31, 37, 41, 43, 47
**7** 2, 5, 7   **9** $168 = 7 \times 24$
**11** 2   **13** 10   **15** 12
**17** 42.6   **19** 9.56   **21** 1.09
**23** £960, £1400, £2060   **25** 50%

## REVISION EXERCISE 53★ (page 132)

**1** 406 kg   **3** 47, 53, 59, 61, 67
**5** $2^4 \times 7$   **7** 777   **9** 28
**11** $3a^2$   **13** $12x^2y$   **15** 0.914
**17** −0.0852   **19** 1.02   **21** £2590

# Algebra 3

## EXERCISE 54 (page 133)

**1** $x(x + 3)$   **3** $x(x - 4)$
**5** $5(a - 2b)$   **7** $x(y - z)$
**9** $2x(x + 2)$   **11** $3x(x - 6)$
**13** $ax(x - a)$   **15** $3xy(2x - 7)$
**17** $3pq(3p + 2)$   **19** $a(p + q - r)$

## EXERCISE 54★ (page 134)

**1** $5x^3(1 + 3x)$   **3** $3x^2(x - 6)$
**5** $3x^2y^2(3x - 4y^2)$   **7** $x(x^2 - 3x - 3)$
**9** $ab(c^2 - b + ac)$   **11** $4pq(pqr^2 - 3r + 4q)$
**13** $3x(10x^2 + 4y - 7z)$   **15** $0.1h(2h + g - 3g^2h)$
**17** $\frac{xy(2x^2 - 4y + xy)}{16}$   **19** $\pi r(r + 2h)$
**21** $4pqr(4pr^2 - 7 - 5p^2q)$   **23** $(a + b)(x + y)$
**25** $(x - y)^2(1 - x + y)$

## EXERCISE 55 (page 134)

**1** $x + 1$   **3** $\frac{(x + y)}{z}$   **5** 2
**7** $\frac{(a - b)}{b}$   **9** $\frac{t}{r}$   **11** $\frac{x}{z}$

## EXERCISE 55★ (page 135)

**1** $x + y$   **3** $\frac{1}{z + 1}$   **5** $2 + 3x^2$
**7** $\frac{2}{3}\left(x - 3y^2\right)$   **9** $y$   **11** $\frac{2x}{z}$

**13** 1          **15** $\dfrac{b}{a}$          **17** 5

**19** $-x$

## EXERCISE 56 (page 136)

**1** $x = 8$          **3** $x = -10$          **5** $x = 2$
**7** $x = 0$          **9** $x = -6$          **11** $x = 5$
**13** $x = -4$          **15** $x = 6$          **17** $x = 14$
**19** $x = 3$          **21** $x = 0$          **23** $x = 10$
**25** 6 km

## EXERCISE 56★ (page 136)

**1** $x = 9$          **3** $x = \dfrac{3}{5}$          **5** $x = 9$
**7** $x = -6$          **9** $x = 0$          **11** $x = \dfrac{1}{9}$
**13** $x = 3$          **15** $x = -1$          **17** $x = -\dfrac{5}{13}$
**19** $x = 7$          **21** 84 years

## EXERCISE 57 (page 137)

**1** $x = 2$          **3** $x = -3$          **5** $x = \dfrac{3}{5}$
**7** $x = -8$          **9** $x = 10$          **11** $x = -2.4$
**13** $x = 50$          **15** $x = -25$          **17** $x = \dfrac{5}{3}$
**19** $x = \pm 3$

## EXERCISE 57★ (page 138)

**1** $x = 4$          **3** $x = -8$          **5** $x = \dfrac{1}{6}$
**7** $x = -64$          **9** $x = 4$          **11** $x = \pm 8$
**13** $x = \pm 2$          **15** $x = 0.32$          **17** $x = \dfrac{5}{6}$
**19** $x = \dfrac{(a + b)}{ab}$

## EXERCISE 58 (page 139)

**1** (8, 3)          **3** (4, 5)          **5** (1, 5)
**7** (0, −2)          **9** (−1, 5)

## EXERCISE 59 (page 140)

**1** (2, 5)          **3** (5, 1)          **5** (1, 3)
**7** (5, −1)          **9** (2, 1)

## EXERCISE 59★ (page 140)

**1** (3, −1)          **3** (1, 2)          **5** (−0.4, 2.6)
**7** (7, 3)          **9** (0.5, 0.75)          **11** (4, 6)
**13** (−0.6, −0.8)          **15** (0.4, 0.5)

## EXERCISE 60 (page 141)

**1** $x = 3, y = 1$          **3** $x = 1, y = 4$
**5** $x = 1, y = 6$          **7** $x = -1 , y = 2$
**9** $x = 3, y = -1$

## EXERCISE 60★ (page 142)

**1** $x = 1, y = 2$          **3** $x = 4, y = 1$
**5** $x = 2, y = 1$          **7** $x = 1 , y = -2$
**9** $x = -3, y = \dfrac{1}{2}$

## EXERCISE 61 (page 143)

**1** 29, 83          **3** 12, 16          **5** 9, 4
**7** $x = 2, y = 3$, area = 180
**9** Burger 99p, cola 49p
**11** 27 @ 20p, 12 @ 50p
**13** 11

## EXERCISE 61★ (page 144)

**1** (2, 3)          **3** $m = 2, c = -1$
**5** $\dfrac{12}{17}$          **7** $1.5\,\text{m s}^{-1}$          **9** 7.5 km
**11** 37          **13** 50 m

## REVISION EXERCISE 62 (page 145)

**1** $x(x-8)$          **3** $6xy(y - 5x)$          **5** $x - 1$
**7** $x = 4$          **9** −4          **11** 24
**13** (0, 3)          **15** (1, 3)
**17** 19 @ 10p, 11 @ 20p

## REVISION EXERCISE 62★ (page 146)

**1** $3x^3 (x - 4)$          **3** $6x^2y(4xy - 3)$
**5** $\dfrac{x}{y}$          **7** $x = \dfrac{1}{3}$
**9** $x = 6$          **11** 70 years
**13** (4, 1)          **15** $\left(3\dfrac{1}{3}, 2\right)$
**17** Mike is 38, Ben is 14

## Graphs 3

## EXERCISE 63 (page 147)

**1** **a** 65 km/h          **b** 50 km/h          **c** 12:00
   **d** 72.5 km          **e** 11:08 approx

**3** **a**

   **b** 14:00

## EXERCISE 63★ (page 149)

**1** **a**          **b**

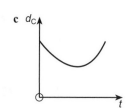

**c** $d_C$

**d** $d_D$

**3** **a** **(i)** B & C joint 1st, A 2nd
**(ii)** C 1st, B 2nd, A 3rd
**(iii)** A 1st, B 2nd, C 3rd
**b** 28.5 s          **c** B
**d** **(i)** A    **(ii)** C

## EXERCISE 64 (page 151)

**1** **a** $2\,\text{m/s}^2$   **b** $4\,\text{m/s}^2$   **c** $150\,\text{m}$   **d** $10\,\text{m/s}$
**3** **a** $2\,\text{m/s}^2$   **b** $1\,\text{m/s}^2$   **c** $8000\,\text{m}$   **d** $50\,\text{m/s}$

## EXERCISE 64★ (page 152)

**1** **a**

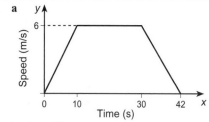

**b** $0.6\,\text{m/s}^2$
**c** $-0.5\,\text{m/s}^2$
**d** $4.43\,\text{m/s}$ (3 sf)
**3** **a** $t = 10\,\text{s}$, so distance $= 1900\,\text{m}$
**b** $-3\,\text{m/s}^2$          **c** $47.5\,\text{m/s}$
**5** Bee cannot have two speeds nor be at two different distances at any given time.

## REVISION EXERCISE 65 (page 153)

**1** **a** 20 mins   **b** 10:00   **c** 10 km/h   **d** $3\frac{1}{3}\,\text{km}$
**3**

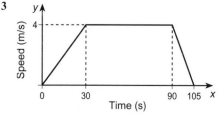

**a** $\frac{2}{15}\,\text{m/s}^2$   **b** $0\,\text{m/s}^2$   **c** $\frac{4}{15}\,\text{m/s}^2$   **d** $3\frac{1}{7}\,\text{m/s}$

## REVISION EXERCISE 65★ (page 154)

**1** **b** Elisa home at 12:00, Albert home at 12:00
**c** Elisa 1.48 m/s, Albert 2.22 m/s
**3** **a** False; it is constant at $\frac{2}{3}\,\text{m/s}^2$     **b** True
**c** True
**d** False; it is 72 km/h

## ACTIVITY 16 (page 155)

| Triangle | Opposite | Adjacent | Hypotenuse |
|---|---|---|---|
| P | 15 mm | 27 mm | 31 mm |
| Q | 25 mm | 43 mm | 50 mm |
| R | 19 mm | 33 mm | 38 mm |

| $d$ | 0° | 15° | 30° | 45° | 60° | 75° | 90° |
|---|---|---|---|---|---|---|---|
| $\sin d$ | 0 | 0.259 | 0.5 | 0.707 | 0.866 | 0.966 | 1 |
| $\cos d$ | 1 | 0.966 | 0.866 | 0.707 | 0.5 | 0.259 | 0 |

## EXERCISE 66 (page 157)

**1** 2.46          **3** 5.07          **5** 8.09
**7** 9.44 cm       **9** 8.76 cm       **11** 10.1 m
**13** 1.61 m

## EXERCISE 66★ (page 158)

**1** 6.57 cm
**3** **a** 2.5 m   **b** 4.33 m   **c** $20.6\,\text{m}^2$   **d** 31.7°
**5** **a** 107 m   **b** 79.7 m
**7** 5.88 cm

## ACTIVITY 17 (page 159)

| $\sin d$ | 0 | 0.259 | 0.5 | 0.866 | 0.966 | 1 |
|---|---|---|---|---|---|---|
| $d$ | 0° | 15° | 30° | 60° | 75° | 90° |
| $\cos b$ | 1 | 0.966 | 0.866 | 0.5 | 0.259 | 0 |
| $b$ | 0° | 15° | 30° | 60° | 75° | 90° |

## EXERCISE 67 (page 161)

**1** 48.6°          **3** 46.1°          **5** 19.5°
**7** 78.9°          **9** 1.72°

## EXERCISE 67★ (page 161)

**1** 37.8°          **3** 58.6°          **5** 57.3°
**7** 014.9°; 194.9°   **9** 33.6°

## INVESTIGATE (page 162)

$(\sin f)^2 + (\cos f)^2 = 1$

$\left(\dfrac{b}{a}\right)^2 + \left(\dfrac{c}{a}\right)^2 = 1$

$\dfrac{b^2 + c^2}{a^2} = 1$

$b^2 + c^2 = a^2$ (Pythagoras' theorem)

$\sin f = \dfrac{b}{a}, \; \cos f = \dfrac{c}{a}$

$\dfrac{\sin f}{\cos f} = \dfrac{b}{a} \div \dfrac{c}{a} = \dfrac{b}{\cancel{a}} \times \dfrac{\cancel{a}}{c} = \dfrac{b}{c} \Rightarrow \dfrac{\sin f}{\cos f} = \tan f.$

$\sin 30° = \dfrac{1}{2} \quad \cos 30° = \dfrac{\sqrt{3}}{2} \quad \tan 30° = \dfrac{1}{\sqrt{3}}$

$\sin 60° = \dfrac{\sqrt{3}}{2} \quad \cos 60° = \dfrac{1}{2} \quad \tan 60° = \sqrt{3}$

## EXERCISE 68 (page 162)

| | | | | | |
|---|---|---|---|---|---|
| **1** | $x = 5$ | **3** | $x = 8.7$ | **5** | $x = 5.2$ |
| **7** | $a = 30°$ | **9** | $a = 60°$ | **11** | 9.24 |
| **13** | 32.2° | **15** | 49.7° | **17** | 1.38 cm |
| **19** | 41.6° | **21** | 62.3° | **23** | 250 m |
| **25** | 5.47 km | | | | |

## EXERCISE 68★ (page 164)

| | | | | | |
|---|---|---|---|---|---|
| **1** | $x = 18$ | **3** | $x = 100$ | **5** | $x = 20$ |
| **7** | $a = 45°$ | **9** | $a = 30°$ | | |
| **11** | **a** 4.66 km N | | **b** 17.4 km W | | |
| **13** | $H = 22.2$ m | | | | |
| **15** | 7.99 km | | | | |

## INVESTIGATE (page 166)

Prove that $h = x(1 - \cos a)$ and superimpose the three graphs for $x = 2$, $x = 3$ and $x = 10$ (metres). Range for $a$ is $0 - 90°$.

The three parabolas all go through the origin and exist for $0° \leqslant a \leqslant 90°$. The steepest is for $x = 10$, second steepest for $x = 3$ and third steepest for $x = 2$. This does make sense when compared with the motion of a swing.

## REVISION EXERCISE 69 (page 166)

| | | | |
|---|---|---|---|
| **1** | $x = 14.1$ cm, $d = 70.5°$ | **3** | $x = 16.7$ km, $d = 39.9°$ |
| **5** | $x = 2.38$ m, $d = 4.62°$ | **7** | 43.3 cm$^2$ |
| **9** | **a** 0.5 | **b** $f = 30°$ |

## REVISION EXERCISE 69★ (page 167)

**1** Ascends in 3 mins 52 s, so reaches surface with 8 seconds to spare.

**3** 3.56 m

**5** $p = 25$

## Handling data 3

### ACTIVITY 18 (page 171)

Swimmers' times: exact group boundaries $55 \leqslant t < 60$ etc.; width = 5; mid-points are 57.5, 62.5, 67.5, 72.5

Earthworm lengths: exact group boundaries $2.5 \leqslant l < 5.5$ etc.; width = 3; mid-points are 4, 7, 10, 13

Noon temperatures: exact group boundaries $0 \leqslant t < 8$ etc.; width = 8; mid-points are 4, 12, 20, 28

Babies' weights: exact group boundaries $1 \leqslant w < 2.5$ etc.; width = 1; mid-points are 1.5, 2.5, 3.5, 4.5

Children's ages: exact group boundaries $2 \leqslant y < 4$ etc.; width = 2; mid-points are 3, 5, 7, 9

### EXERCISE 70 (page 171)

| | | | | | |
|---|---|---|---|---|---|
| **1** | **a** f = 7, 8, 5, 4, 3, 3 | **b** 2.83 | | **c** 2.5 | |
| **3** | **a** 50 | **b** 1.44 | | **c** 1 | |
| **5** | **a** 30 | **b** 1.3 | | **c** 2 | |
| **7** | **a** 35 | **b** 167.5 | | **c** 184.5 cm | |

## EXERCISE 70★ (page 173)

| | | | |
|---|---|---|---|
| **1** | 2.9 kg | **3** | 16 mins |
| **5** | 28.15 yrs = 28 yrs 2 months | | |
| **7** | 26.1 | | |

## REVISION EXERCISE 71 (page 175)

**1**   **a**   f = 3, 4, 4, 5, 10, 2
     **b**   $11.85
     **c**   $12

## REVISION EXERCISE 71★ (page 176)

**1**   1670 cc

## Examination practice 3 (page 181)

| | | | | | | | |
|---|---|---|---|---|---|---|---|
| **1** | **a** 3375 | **b** $2 \times 10^5$ | | **c** 2 | | **d** 2 |
| **3** | **a** (i) $x = 6$ | | (ii) $x = 12$ | | | |
| | **b** (i) $x(x + 7)$ | | (ii) $xy(x + 2y + 3)$ | | | |

**5**   177 cm

**7**   **a**   Gain of £275      **b**   29.9%
     **c**   £2.36 approx

**9**   **a**   3358 ft      **b**   3.7%

**11**   **a**   $x = 2, y = 5$
      **b**   $x = 3, y = -1$

**13**   **a** 8.7 cm   **b** 3.7 cm   **c** 5 cm   **d** 12.5 cm$^2$

**15**   **a**   (i)   4880 km      (ii)   4090 km
      (iii)   38 200 km     (iv)   123°

      **b**   Tangent is perpendicular to radius. Therefore, the smallest possible value of angle ODS is 90°, since a smaller value would mean that the signal would pass back into Earth before reaching Delhi.

      **c**   77 000 km

# Unit 4

## Number 4

### EXERCISE 72 (page 184)

| | | | | | | | |
|---|---|---|---|---|---|---|---|
| **1** | $442 | **3** | 100% | **5** | $40 | **7** | $60 |
| **9** | 13.3% | **11** | €76.20 | **13** | €74.10 | **15** | $2426 |
| **17** | 7% | | | | | | |

### EXERCISE 72★ (page 184)

| | | | | | |
|---|---|---|---|---|---|
| **1** | $44 | **3** | $180 000 | **5** | $60 |
| **7** | 11.1% | **9** | €73 000 | **11** | 34.4% |
| **13** | £463 | **15** | 66.7% | **17** | 14.5% |
| **19** | 3400 | | | | |

### EXERCISE 73 (page 188)

**1** 230 m (to 10 m): 235 m; 225 m; (230 ± 5) m
**3** 74°F (to 1°F): 74.5°F; 73.5°F; (74 ± 0.5)°F

| | | | |
|---|---|---|---|
| **5** | 3.4 and 3.6 | **7** | 5.59 and 5.61 |
| **9** | 4.7 and 5.1 | **11** | 6.49 and 6.51 |
| **13** | 2.6 ± 0.1 | **15** | 21.35 ± 0.05 |
| **17** | 2.2 ± 0.1 | **19** | 46.35 ± 0.05 |

### EXERCISE 73★ (page 188)

**1** 5.5 and 6.5; 16.5 and 17.5; 122.5 and 123.5
**3** 2.25 and 2.75; 14.25 and 14.75;
  145.75 and 146.25
**5** 0.1 and 0.3; 7.5 and 7.7; 12.3 and 12.5
**7** 42.5 kg and 43.5 kg
**9** Max length = 18.3 cm, min length = 17.9 cm
  Min difference = 6.7 cm
**11** $p(\text{max}) = 1.82$, $p(\text{min}) = 1.40$
**13** Radius = 1.54 cm, circumference = 9.64 cm

### EXERCISE 74 (page 191)

| | | | | | |
|---|---|---|---|---|---|
| **1** | 150 | **3** | 3 | **5** | 300 |
| **7** | 8 | **9** | 200 | **11** | $6 \times 10^7$ |
| **13** | $4 \times 10^3$ | **15** | $2.3 \times 10^4$ | **17** | $8.7 \times 10^4$ |
| **19** | $5 \times 10^2$ | **21** | $2 \times 10^6$ | **23** | $1 \times 10^2$ |
| **25** | $7 \times 10^6$ | **27** | $4 \times 10^6$ | **29** | $8 \times 10^2$ |

### EXERCISE 74★ (page 191)

| | | | | | |
|---|---|---|---|---|---|
| **1** | 4 | **3** | 8 | **5** | 600cm² |
| **7** | 100cm² | **9** | $1.2 \times 10^9$ | **11** | $2 \times 10^3$ |
| **13** | $7.06 \times 10^8$ | **15** | 50 000 | **17** | 0.2 |
| **19** | 0.06 | **21** | $6 000 000 | **23** | $2 \times 10^6$ |
| **25** | $3 \times 10^{-1}$ | **27** | $2 \times 10^7$ | **29** | $1 \times 10^{-3}$ |

### REVISION EXERCISE 75 (page 192)

| | | | | | |
|---|---|---|---|---|---|
| **1** | 1.15 | **3** | $150 | **5** | 4 |
| **7** | $6 \times 10^{10}$ | **9** | $4 \times 10^5$ | **11** | 33.5 and 34.5 |
| **13** | Max = 17.25; min = 17.15 | | | | |

### REVISION EXERCISE 75★ (page 193)

| | | | | | |
|---|---|---|---|---|---|
| **1** | 406 kg | **3** | $1125 | **5** | 17.2% |
| **7** | $33\frac{1}{3}$ | **9** | $5 \times 10^3$ | **11** | $2 \times 10^{-2}$ |

**13** 3.3 ± 0.05
**15** Max radius = 3.83 cm, min circumference = 23.8 cm

## Algebra 4

### EXERCISE 76 (page 194)

| | | | | | |
|---|---|---|---|---|---|
| **1** | $a - 2$ | **3** | $5 + p$ | **5** | $c - a$ |
| **7** | $\dfrac{b}{5}$ | **9** | $\dfrac{(b-a)}{3}$ | **11** | $\dfrac{(t-s)}{2}$ |
| **13** | $\dfrac{(4-b)}{a}$ | **15** | $\dfrac{(f+g)}{e}$ | **17** | $\dfrac{c}{(a+b)}$ |
| **19** | $\dfrac{(d-8b)}{c}$ | **21** | $\dfrac{(a-3b)}{3}$ | **23** | $\dfrac{(c-a)}{a}$ |
| **25** | $ab$ | **27** | $\dfrac{qr}{p}$ | **29** | $r(p+q)$ |
| **31** | $cd + b$ | | | | |

### EXERCISE 76★ (page 195)

| | | | | | |
|---|---|---|---|---|---|
| **1** | $x = \dfrac{c-b}{a}$ | **3** | $x = cd + b$ | **5** | $x = \dfrac{cd}{b}$ |
| **7** | $x = \dfrac{e}{a} - c$ | **9** | $x = \dfrac{P - b^2}{\pi}$ | **11** | $x = \dfrac{Td^2}{b}$ |
| **13** | $x = \pi - b$ | **15** | $x = \dfrac{ab - c}{d}$ | **17** | $\dfrac{a}{b}$ |
| **19** | $\dfrac{(a+b)}{c}$ | **21** | $\dfrac{s}{(p-q)}$ | **23** | $\dfrac{a}{2\pi}$ |
| **25** | $\dfrac{3V}{\pi r^2}$ | **27** | $\dfrac{(y-c)}{m}$ | **29** | $\dfrac{(v^2 - u^2)}{2a}$ |
| **31** | $2m - b$ | **33** | $\dfrac{S(1-r)}{(1-r^n)}$ | **35** | $\dfrac{S}{n} - \dfrac{(n-1)d}{2}$ |

### EXERCISE 77 (page 196)

| | | | | | |
|---|---|---|---|---|---|
| **1** | $x = \sqrt{\dfrac{b}{a}}$ | **3** | $x = \sqrt{ab}$ | **5** | $x = \sqrt{2D - C}$ |
| **7** | $x = \sqrt{a(c-b)}$ | **9** | $x = \sqrt{\dfrac{c - 2b}{a}}$ | **11** | $x = \dfrac{t}{a+d}$ |
| **13** | $x = \dfrac{ab}{a-1}$ | **15** | $x = \dfrac{2b - a}{a - b}$ | **17** | $\sqrt{\dfrac{A}{4\pi}}$ |
| **19** | $\sqrt{ar}$ | **21** | $\sqrt[3]{\dfrac{3V}{4\pi}}$ | **23** | $\left(\dfrac{T}{2\pi}\right)^2$ |

### EXERCISE 77★ (page 196)

| | | | | | |
|---|---|---|---|---|---|
| **1** | $x = \sqrt{\dfrac{S}{R}}$ | **3** | $x = \sqrt{\dfrac{g-a}{c}}$ | **5** | $x = \dfrac{c}{b-a}$ |
| **7** | $x = \dfrac{c - f}{e + d}$ | **9** | $x = \dfrac{\tan b + ac}{1 - a}$ | **11** | $x = t(p^2 - s)$ |
| **13** | $x = \sqrt{Ab - Da}$ | | | **15** | $\sqrt{\dfrac{3V}{\pi h}}$ |
| **17** | $\sqrt{2gh}$ | | | **19** | $\sqrt{\dfrac{1}{y} - a^2}$ |
| **21** | $b - \sqrt{12s}$ | **23** | $\dfrac{1}{P}\left(\dfrac{S}{r}\right)^2$ | **25** | $\left(\dfrac{k}{F}\right)^3$ |
| **27** | $1 - \dfrac{1}{y^2}$ | **29** | $\dfrac{p(y+1)}{y-1}$ | | |

## ACTIVITY 20 (page 197)

63.4 seconds; 0.25 m, 25 m, 890 m, $1.86 \times 10^9$ m, $1.9 \times 10^9$ m

## ACTIVITY 21 (page 198)

Base area $= \pi r^2$ so volume $= \pi r^2 h$
Circumference of base $= 2\pi r$ so curved surface area $= 2\pi r h$

## EXERCISE 78 (page 199)

1   **a**   155 mins    **b**   2 kg
3   **a**   18    **b**   10
5   **a**   0.15    **b**   200
7   **a**   2350    **b**   25 750
9   $A = \frac{\pi r^2}{4}, P = r\left(2 + \frac{\pi}{2}\right)$
   **a**   $A = 19.6$ cm$^2$, $P = 17.9$ cm
   **b**   7.98    **c**   19.0    **d**   4.55

## EXERCISE 78★ (page 200)

1   **a**   22    **b**   400
3   **a**   354    **b**   3.16
5   **a**   209    **b**   3     **c**   8
7   **a**   27.3    **b**   12    **c**   6.63
9   $A = r^2\left(1 + \frac{\pi}{4}\right), P = r\left(4 + \frac{\pi}{2}\right)$
   **a**   $A = 28.6$ cm$^2$, $P = 22.3$ cm
   **b**   4.10    **c**   12.6    **d**   3.12

## REVISION EXERCISE 79 (page 201)

1   $x = \dfrac{b}{a}$     3   $x = \dfrac{a-c}{b}$     5   $y = \dfrac{b^2}{a}$
7   $y = \dfrac{bc}{c-1}$
9   **a**   26
   **b**   100
   **c**   $1800

## REVISION EXERCISE 79★ (page 201)

1   $x = \dfrac{c-b}{a}$     3   $x = \dfrac{ab - \tan c}{a}$
5   $y = \dfrac{ac - d}{a - b}$
7   **a**   $\frac{2}{3}$   **b**   3 sec   **c**   2.45 m
9   1 : 3

## Graphs 4

## ACTIVITY 22 (page 203)

Distance travelled by a falling stone v time …
Speed of cyclist v time …
Sunshine intensity v time in a day … numerous other examples.

## EXERCISE 80 (page 207)

1
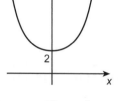

3
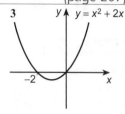

5
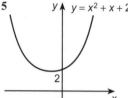

7
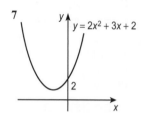

9   **b**

| $x$ | 0 | 0.4 | 0.8 | 1.2 | 1.6 | 2.0 |
|---|---|---|---|---|---|---|
| $v$ | 0 | 0.32 | 1.28 | 2.88 | 5.12 | 8 |

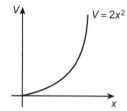

   **c**   1.41 m $\times$ 1.41 m
   **d**   0.72 m$^3$
   **e**   $1.22 \leqslant x \leqslant 1.8$

## EXERCISE 80★ (page 208)

1
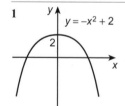

3
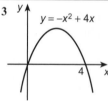

5
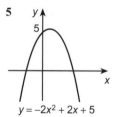

7   **a**   $k = 1.75 \Rightarrow P = 1.75t^2 + t + 1$

| $t$ | 2 | 4 | 6 | 8 | 10 | 12 |
|---|---|---|---|---|---|---|
| $P$ | 10 | 33 | 70 | 121 | 186 | 265 |

**b**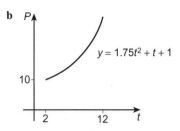

$y = 1.75t^2 + t + 1$

**c** Accurate answer is 49 750 000. Only approximate answers will be available from the graph.

**d** 7.2 days approx.

**9 a** $A = (10 - 2x)^2 + 4x(10 - 2x) = 100 - 4x^2$

**b**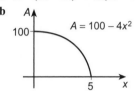

$A = 100 - 4x^2$

**c** $2.5 \leqslant x \leqslant 3.5$

## INVESTIGATE (page 210)

1708 m²; 6831 m²
Try to steer the pupils to producing a general formula for various angles:
$A = \pi(\tan x)^2 \times H^2$ and draw a family of curves for $x = 20°, 30°, 40° \dots 90°$

## EXERCISE 81 (page 210)

**1** $x = 3$ or $x = 2$      **3** $x = -1$ or $x = 3$
**5** $x = 0$ or $x = 3$

## EXERCISE 81★ (page 211)

**1** $x = \frac{1}{2}$ or $x = 2$      **3** $x = \frac{2}{3}$ or $x = 2 -$
**5** $y = x^2 - 6x + 5$      **7** $y = x^2 - 6x + 9$

**9**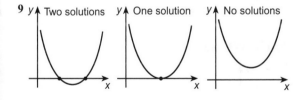
Two solutions   One solution   No solutions

## ACTIVITY 23 (page 211)

| Year interval | Fox numbers | Rabbit numbers | Reason |
|---|---|---|---|
| A–B | Decreasing | Increasing | Fewer foxes to eat rabbits |
| B–C | Increasing | Increasing | More rabbits attract more foxes into the forest |

| C–D | Increasing | Decreasing | More foxes to eat rabbits so rabbit numbers decrease |
|---|---|---|---|
| D–A | Decreasing | Decreasing | Fewer rabbits to be eaten by foxes so fox numbers decrease |

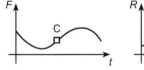

## REVISION EXERCISE 82 (page 212)

**1**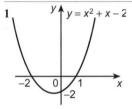

$y = x^2 + x - 2$

**3** $x \approx 0.6$ or $3.4$
**5 a** 4.9 cm²    **b** 4.5 cm    **c** 12.9 cm²

## REVISION EXERCISE 82★ (page 213)

**1**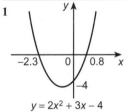

$y = 2x^2 + 3x - 4$

**3** $x \approx -2.6$ or $1.1$
**5 b** 28.3 m

## Shape and Space 4

## ACTIVITY 24 (page 214)

| | BAO | ABO | AOB | BOC | OCB + OBC | OBC | ABC |
|---|---|---|---|---|---|---|---|
| $C_1$ | 20° | 20° | 140° | 40° | 140° | 70° | 90° |
| $C_2$ | 30° | 30° | 120° | 60° | 120° | 60° | 90° |
| $C_3$ | $x$ | $x$ | $180° - 2x$ | $2x$ | $180° - 2x$ | $90° - x$ | 90° |

## EXERCISE 83 (page 216)

**1** $a = 230°$, $b = 25°$     **3** $a = 280°$, $b = 50°$
**5** $a = 100°$, $b = 260°$    **7** $a = 40°$, $b = 320°$
**9** $a = 110°$, $b = 35°$    **11** $a = 60°$, $b = 60°$
**13** $a = 108°$, $b = 72°$, $c = 54°$
**15** $a = 62°$, $b = 62°$, $c = 28°$

17 $a = 55°$
19 $a = 45°$
21 $a = 90°, b = 32°, c = 58°$
23 $a = 50°$
25 a $90°$    b $32°$    c $132°$    d $80°$

## EXERCISE 83★       (page 218)

1 a $37°$    b $53°$
3 a $x$    b $2x$
5 a $22.5°$    b $45°$
7 $\angle OUT = x; \angle TOU = 180° − 2x; \angle TUV = 90° − x$
9 $\angle OBA = 90° − x; \angle AOB = 2x; \angle COB = 180° − 2x$
11 a $\sqrt{r^2 − x^2}$    b $\sqrt{r^2 − x^2}$
13 a $12\,cm$    b $54\,cm^2$    c $7.2\,cm$
15 a $9\,cm$    b $108\,cm^2$

## ACTIVITY 25       (page 220)

| | $\angle ORP$ | $\angle ORQ$ | $\angle RPO$ | $\angle POS$ |
|---|---|---|---|---|
| $C_1$ | $25°$ | $30°$ | $25°$ | $50°$ |
| $C_2$ | $80°$ | $25°$ | $80°$ | $160°$ |
| $C_3$ | $x$ | $y$ | $x$ | $2x$ |

| | $\angle RQO$ | $\angle QOS$ | $\angle PRQ$ | $\angle POQ$ |
|---|---|---|---|---|
| $C_1$ | $30°$ | $60°$ | $55°$ | $110°$ |
| $C_2$ | $25°$ | $50°$ | $105°$ | $210°$ |
| $C_3$ | $y$ | $2y$ | $x + y$ | $2x + 2y$ |

## EXERCISE 84       (page 222)

1 $a = 124°$    3 $a = 56°$    5 $a = 102°$
7 $a = 78°$    9 $a = 56°$
11 $a = 36°, b = 36°$     13 $a = 50°, b = 130°$
15 $a = 67°, b = 113°$     17 $a = 100°, b = 160°$
19 $a = 40°, b = 140°$     21 $a = 67°, b = 85°$
23 $a = 105°$

## EXERCISE 84★       (page 224)

5 $x = 126°$          9 $x = 116°, y = 64°$
11 $x = 27°$

## ACTIVITY 27       (page 227)

XRY = BRA (angle of incidence = angle of reflection)
XYR = BAR (90°)   ∴ RXY = ABR
∴ $\triangle$s $\dfrac{ABR}{YXR}$ are similar. $\dfrac{AB}{XY} = \dfrac{AR}{YR} = \dfrac{BR}{XR}$

## EXERCISE 85       (page 229)

1 $T_1, T_2$     3 $T_2, T_3$     5 $T_1, T_3$
7 $T_1, T_3$     9 $x = 6$
11 $x = 5, y = 8$

13 a $3\,cm$       b $14\,cm$
15 $12.5\,m$

## EXERCISE 85★       (page 231)

1 E(4, 2), F(4, 4)
3 a (3, 5)    b P(1, 4)    c Q(5, 6)
5 b $16\,cm$    c $15\,cm$
7 a $5.7\,cm$    b $\dfrac{AC}{EF} = \dfrac{BC}{FG}; 6.32\,cm$
9 PQS, RPS, RPQ
   a $PS = \dfrac{20}{3}, RS = \dfrac{16}{3}$    b $\dfrac{50}{3}$
11 $18\,m$     13 $55$ cubits     15 $2.5\,m$
17 $239\,000\,cm^3$

## EXERCISE 86       (page 235)

1 $10.3\,cm$       3 $8.94\,cm$
5 $11.8\,m$       7 $70.7\,cm$
9 $3.16\,m$
11 a $\sqrt{y^2 − r^2}$    b $\sqrt{y^2 − r^2}$

## EXERCISE 86★       (page 236)

1 $12.4\,cm$       3 $8.77\,m$
5 $13.9$       7 $17:28:20$
9 $27.5\,m$
11 a $48\,m$    b $24\,m$

## REVISION EXERCISE 87       (page 237)

1 $a = 52°, b = 128°, c = 76°$
3 $a = 80°, b = 50°, c = 130°$
5 $a = 18°$
7 PQ = 3.6 cm; AC = 6 cm
9 $44.7\,cm$

## REVISION EXERCISE 87★       (page 238)

1 $a = 40°, b = 100°, c = 40°, d = 260°$
3 a $13\,cm$          b $120\,cm^2$
5 $a = 32°, b = 64°, c = 26°, d = 58°$
7 3.875 ft or 3 ft, 10.5 in.
9 $3.71\,cm$

## EXERCISE 88       (page 241)

1 a

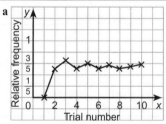

p(biased) $= \dfrac{3}{5}$

b Heidi's suspicion seems to be true. More trials would improve the experiment.

**3**  $p(\text{vowel}) = \dfrac{9}{20}$

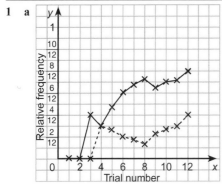

## EXERCISE 88★ (page 242)

**1 a**

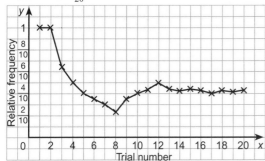

**b** Learning curve, so warm up before playing. Practise more from RHS.

**3 a**  $p(W) = \dfrac{12}{20} = \dfrac{3}{5}$;  $p(P) = \dfrac{8}{20} = \dfrac{2}{5}$

**b** No. of W = $\dfrac{3}{5} \times 100 = 60 \Rightarrow$ No. of P = 40

## INVESTIGATE (page 243)

$A = \dfrac{1}{2}$, $B = \dfrac{1}{4}$, $C = \dfrac{3}{4}$, $D$ = almost zero, $E$ = almost 1

## EXERCISE 89 (page 244)

**1 a**  $p(g) = \dfrac{4}{10} = \dfrac{2}{5}$   **b**  $p(a) = \dfrac{3}{10}$

**c**  $p(t) = 0$   **d**  $p(\overline{S}) = \dfrac{9}{10}$

**3 a**  $p(R) = \dfrac{1}{2}$   **b**  $p(K) = \dfrac{1}{13}$

**c**  $p(\text{mult of } 3) = \dfrac{3}{13}$   **d**  $p(AJQK) = \dfrac{4}{13}$

**5 a**  $\dfrac{1}{10}$   **b**  $\dfrac{1}{2}$   **c**  $\dfrac{3}{10}$   **d**  $\dfrac{2}{5}$

## EXERCISE 89★ (page 245)

**1 a (i)**  $\dfrac{5}{36}$   **(ii)**  $\dfrac{1}{12}$   **(iii)**  $\dfrac{1}{12}$   **(iv)**  $\dfrac{15}{36}$

**b** 7

**3**

| 2nd spin | 1st spin | | | | |
|---|---|---|---|---|---|
| | 1 | 2 | 3 | 4 | 5 |
| 1 | 1 | 2 | 3 | 4 | 5 |
| 2 | 2 | 4 | 6 | 8 | 10 |
| 3 | 3 | 6 | 9 | 12 | 15 |
| 4 | 4 | 8 | 12 | 16 | 20 |
| 5 | 5 | 10 | 15 | 20 | 25 |

**a**  $\dfrac{9}{25}$   **b**  $\dfrac{14}{25}$   **c**  $\dfrac{6}{25}$   **d**  $\dfrac{9}{25}$

**5**

| Spinner | Die | | | | | |
|---|---|---|---|---|---|---|
| | 1 | 2 | 3 | 4 | 5 | 6 |
| 2 | 2 | 2 | 3 | 4 | 5 | 6 |
| 4 | 4 | 4 | 4 | 4 | 5 | 6 |
| 6 | 6 | 6 | 6 | 6 | 6 | 6 |

**a**  $\dfrac{1}{2}$   **b**  $\dfrac{3}{18} = \dfrac{1}{6}$   **c**  $\dfrac{13}{18}$   **d**  $\dfrac{5}{18}$

**7**  $f = 5$

## INVESTIGATE (page 247)

Safest first (ratio of deaths per 5 yr period per 1 million adults): golf (0.009), tennis (0.022), running (0.045), badminton (0.051), gymnastics (0.071), cricket (0.1), football (0.109), rugby (0.167), hockey (0.222), swimming/diving (0.516), fishing (1.351), horse riding (1.59), boating/sailing (3), motor sports (5.909), mountaineering (8.5), air sports (51)

## REVISION EXERCISE 90 (page 247)

**1**  $\dfrac{2}{3}$; More trials for a better estimate

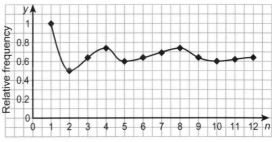

**3 a**  $\dfrac{7}{51}$   **b**  $\dfrac{1}{17}$   **c**  $\dfrac{1}{3}$   **d** 0

**5 a**  $\dfrac{1}{6}$   **b**  $\dfrac{1}{6}$   **c** 0   **d**  $\dfrac{11}{12}$

## REVISION EXERCISE 90★ (page 248)

**1 a** 2002, $\dfrac{7}{10}$; 2003, $\dfrac{13}{20}$; 2004, $\dfrac{17}{30}$

**b** Decrease in numbers from 2002 is suggested by the data

**3  a  (i)** $\frac{1}{11}$  **(ii)** $\frac{2}{11}$  **(iii)**  0

   **b**  $Z$ or $U$, $\frac{2}{11}$

**5**  £45

## Examination practice 4  (page 254)

**1  a  (i)**  0.6  **(ii)** $1 \times 10^5$
   **b  (i)**  0.177 886  **(ii)**  17 788.6
**3  a**  9.5 by 29.5  **b**  280.25 m²
**5  a**  $C = 15n + 4$  **b**  $n = \dfrac{C - 4}{15}$
**7**

| $t$ | 0 | 1 | 2 | 3 | 4 | 5 |
|-----|-----|------|-----|------|-----|------|
| $h$ | 1.5 | 0.75 | 0.5 | 0.75 | 1.5 | 2.75 |

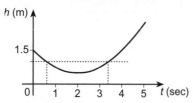

   **b**  1.5 m
   **c**  0.5 m at 2 seconds
   **d**  0.6 s to 3.4 s

**9  a**  £72.50  **b**  30 cm
**11  a**  $a = 2, b = 4$
**13  a**  $\frac{5}{18}$  **b**  $\frac{1}{12}$
**15  a**  150 cm  **b**  24 cm  **c**  $-6\frac{2}{3}$
   The last answer is not possible, as it is negative.

# Unit 5

## Number 5

### ACTIVITY 29 (page 258)
Edible chicken (to 2 s.f.)
a Chicken breast 1.8 g/p
b Boneless chicken breast 1.1 g/p
c Whole chicken 2.1 g/p
d Chicken leg 1.8 g/p

The whole chicken is obviously the best buy, because it has not been prepared for cooking. However, it is not that much cheaper than the leg or the breast. The boneless breast is nearly twice as expensive as the other three cuts.

### EXERCISE 91 (page 258)
1 X $4.50/litre, Y $4.20/litre. Y better value
3 A $6.50/m, B$6.25/m. B better value
5 110 000 m, 3600 s; 30.6 m/s
7 2.74 litres/day
9 38.6 eggs/min
11 28 cm
13 80 cm
15 1:100

### ACTIVITY 30 (page 259)
Man: 0.22 m; greyhound: 4.1 m; bullet: 180 m;
Earth: 90 m; thunderclap: 66 m; lightning: $6 \times 10^7$ m

### EXERCISE 91★ (page 259)
1 Marble $75/m$^2$, slate $72/m$^2$, limestone $72.22/m$^2$.
  Slate is cheapest.
3 9600 shilling
5 Eurozone $600, Malaysia $590, UK $580.
  UK cheapest.
7 0.694 mm/min    9 0.26 m$^3$/min
11 ≈ 500 000 km

### INVESTIGATE (page 261)

| 1 euro on 1.1.2000 | | DM | Fr | Lira | Ptas | |
|---|---|---|---|---|---|---|
| | £0.636 | 1.956 | 6.56 | 1936 | 166 | $1.03 |
| % change of euro | 10% down | 0% | 0% | 0% | 0% | 12% down |

There was no change in value between the fr, DM, lira, ptas and the euro because these currencies were pegged to the euro. A 10% drop in the value of the euro made exports from the UK to Euroland 10% more expensive, and imports from Euroland to the UK 10% cheaper.

### ACTIVITY 31 (page 261)
Aluminium: 2.7 g/cm$^3$; brass: 8.5 g/cm$^3$; gold: 19 g/cm$^3$;
platinum: 21 g/cm$^3$

| Substance | Mass (g) | Volume (cm$^3$) | Density (g/cm$^3$) |
|---|---|---|---|
| **Aluminium** | 135 | 50 | **2.7** |
| **Brass** | 459 | 54 | **8.5** |
| **Platinum** | **210** | 10 | 21 |
| Oak | 46.2 | 55 | **0.84** |
| Petrol | 46.8 | **65** | 0.72 |
| Air | 1200 | **$10^6$** | 0.0012 |
| Water | **any value** | **same value** | **1** |
| $D$ | $M$ | $V$ | $\dfrac{M}{V}$ |

$D = \dfrac{M}{V}$

### EXERCISE 92 (page 262)
1 156 g    3 208 g    5 84 joules
7 1.5 m

### EXERCISE 92★ (page 262)
1 1900 cm$^3$    3 3.2 m    5 13 g/cm$^3$
7 2520 joules

### EXERCISE 93 (page 264)
1 $2^4 = 16$    3 $2^5 = 32$    5 $2^2 = 4$
7 $5^3 = 125$    9 $3^6 = 729$    11 $2^{10} = 1024$
13 $(0.1)^3 = 0.001$    15 $2.1^2 = 4.41$    17 $4^4 = 256$
19 $20^4 = 160 000$

### EXERCISE 93★ (page 264)
1 $8^3 = 512$    3 $7^3 = 343$    5 $5^3 = 125$
7 $6^1 = 6$    9 $5^3 = 125$    11 $10^6 = 1 000 000$
13 $2^0 = 1$    15 $5^3 = 125$    17 9
19 10000
21 $8^{10} = 2^{30}$ and $4^{14} = 2^{28}$ ∴ $8^{10} > 4^{14}$

### EXERCISE 94 (page 265)
1 $\frac{1}{3}$    3 $\frac{15}{99}$    5 $\frac{621}{999} = \frac{23}{37}$

### EXERCISE 94★ (page 265)
1 $\frac{1}{90}$    3 $\frac{4321}{9999}$    5 $\frac{11}{15}$

### REVISION EXERCISE 95 (page 265)
1 $2^8 = 256$    3 $3^6 = 729$
5 Tin A 167 g/$, tin B 176g/$. Tin B gives better value.
7 $\frac{56}{99}$    9 25.6 km/h
11 7.22 m/s    13 79.8 grams

## REVISION EXERCISE 95★ (page 266)

**1** $20^3 = 8000$

**3** $2^{10} = 1024$

**5** $60.8$

**7** $\$30.9$, $60.8$ francs

**9** $\dfrac{52}{9000}$

**11** $100\,000$ tonnes; $40\,000$ m³

## Algebra 5

## EXERCISE 96 (page 270)

**1** $x^2 + 5x + 4$

**3** $x^2 - 4x - 21$

**5** $x^2 - 4x - 12$

**7** $x^2 - 8x + 15$

**9** $x^2 + 6x + 9$

**11** $x^2 - 8x + 16$

**13** $x^2 - 25$

**15** $x^2 - 6x - 16$

**17** $15x^2 - 7x - 2$

**19** $x^3 + 2x^2 - 5x - 10$

## EXERCISE 96★ (page 270)

**1** $x^2 + 4x - 21$

**3** $x^2 - 9$

**5** $x^2 + 24x + 144$

**7** $12x^2 - 25x + 12$

**9** $x^2 + x(b - a) - ab$

**11** $16x^2 - 40x + 25$

**13** $15x^3 + 21x^2 + 5x + 7$

**15** $8x + 8 = 8(x + 1)$

**17** $\dfrac{a^2}{4} - \dfrac{ab}{5} + \dfrac{b^2}{25}$

**19** $10x^5 + 11x^4 + 3x^3$

**21** $4$

**23** $x = -\dfrac{5}{3}$

**25** $a = 3$, $b = 1$

## ACTIVITY 33 (page 270)

$(x^2 + 2x + 3)(x + 1) = x^3 + 3x^2 + 5x + 3$

$(x + y + 3)(x - 2y) = x^2 + xy + 3x - 2y^2 - 6y$

$(x^2 + 2x - 3)(x^2 - 2x + 3) = x^4 - 4x^3 + 12x - 9$

## EXERCISE 97 (page 271)

**1 a** $x^2 + 3x + 2$   **b** $3x + 2$

  **c** $x = 3$

**3 a** $5x^2 + 25x + 30$   **b** $2x^2 + 22x + 42$

**5** $x = 6$

## EXERCISE 97★ (page 272)

**1 a** $\pi(x^2 + 12x + 36)$   **b** $x = 1$

**3** $x = 6$

**5 a** $2n$ is divisible by 2 and so is even; $2n + 1$ is then odd.

  **b** $(2n + 1)(2m + 1) = 4mn + 2n + 2m + 1$; $4mn$, $2n$ and $2m$ are even so this is odd.

  **c** $(2n - 1)(2n + 1) + 1 = 4n^2 = (2n)^2$

## INVESTIGATE (page 273)

Order of brackets is not important because $ab = ba$.

$-3 \times +3 = -9$; this is always the connection.

$x^2 + 9$ cannot be factorised.

## EXERCISE 98 (page 273)

**1** $x(x - 3)$

**3** $x(x + 2)$

**5** $x(x - 31)$

**7** $x(x + 42)$

**9** $(x - 4)(x + 4)$

**11** $(x - 7)(x + 7)$

## EXERCISE 98★ (page 273)

**1** $x(x - 312)$

**3** $x(x + 51)$

**5** $(x - 8)(x + 8)$

**7** $(x - 11)(x + 11)$

## EXERCISE 99 (page 274)

**1** $a = 1$

**3** $a = 4$

**5** $a = -1$

**7** $a = -2$

**9** $a = 2$

**11** $a = -1$

## EXERCISE 99★ (page 274)

**1** $a = 3$

**3** $a = 4$

**5** $a = -7$

**7** $a = -3$

**9** $a = -8$

**11** $a = \dfrac{1}{2}$

## EXERCISE 100 (page 275)

**1** $(x - 2)(x - 1)$

**3** $(x - 1)(x - 3)$

**5** $(x - 4)(x - 3)$

**7** $(x + 4)(x + 4)$

**9** $(x - 1)(x - 8)$

**11** $(x - 1)(x - 1)$

## EXERCISE 100★ (page 275)

**1** $(x + 7)(x + 3)$

**3** $(x - 2)(x - 6)$

**5** $(x - 8)(x - 8)$

**7** $(x - 6)(x - 12)$

**9** $(x + 9)(x + 5)$

**11** $(x + 12)(x + 12)$

## EXERCISE 101 (page 276)

**1** $(x + 3)(x - 2)$

**3** $(x + 2)(x - 5)$

**5** $(x + 2)(x - 6)$

**7** $(x + 1)(x - 10)$

**9** $(x + 7)(x - 2)$

**11** $(x + 8)(x - 1)$

## EXERCISE 101★ (page 276)

**1** $(x + 6)(x - 5)$

**3** $(x + 4)(x - 6)$

**5** $(x + 12)(x - 5)$

**7** $(x + 5)(x - 14)$

**9** $(x + 8)(x - 15)$

**11** $(x - 5)(x + 15)$

## EXERCISE 102 (page 276)

**1** $(x - 1)(x - 2)$

**3** $(x + 3)(x - 1)$

**5** $(x + 1)(x + 12)$

**7** $(x - 2)(x - 6)$

**9** $(x - 4)(x - 4)$

## EXERCISE 102★ (page 276)

**1** $(x + 10)(x - 2)$

**3** $(x + 2)(x - 9)$

**5** $(x + 9)(x + 4)$

**7** $(x - 4)(x - 8)$

**9** $(x + 12)(x - 4)$

**11** $(3 - x)(x + 1)$

## INVESTIGATE (page 277)

$a$ and $b$ (integers) are factors of 12: $(12, 1)$, $(6, 2)$, $(4, 3)$, $(3, 4)$, $(2, 6)$, $(1, 12)$, $(-12, -1)$, $(-6, -2)$, $(-4, -3)$, $(-3, -4)$, $(-2, -6)$ or $(-1, -12)$.

## EXERCISE 103 (page 277)

**1** $x = -1$ or $x = -2$

**3** $x = -4$ or $x = 1$

**5** $x = 7$ or $x = 2$

**7** $x = -8$

**9** $x = 0$ or $x = 10$